T0232077

European Studies in Philosophy of Science

Volume 8

This new series results from the synergy of EPSA - European Philosophy of Science Association - and PSE - Philosophy of Science in a European Perspective: ESF Networking Programme (2008–2013). It continues the aims of the Springer series "The Philosophy of Science in a European Perspective" and is meant to give a new impetus to European research in the philosophy of science. The main purpose of the series is to provide a publication platform to young researchers working in Europe, who will thus be encouraged to publish in English and make their work internationally known and available. In addition, the series will host the EPSA conference proceedings, selected papers coming from workshops, edited volumes on specific issues in the philosophy of science, monographs and outstanding Ph.D. dissertations. There will be a special emphasis on philosophy of science originating from Europe. In all cases there will be a commitment to high standards of quality. The Editors will be assisted by an Editorial Board of renowned scholars, who will advise on the selection of manuscripts to be considered for publication.

More information about this series at http://www.springer.com/series/13909

Amanda Guillán

Pragmatic Idealism and Scientific Prediction

A Philosophical System and Its Approach to Prediction in Science

 Springer

Amanda Guillán
Faculty of Humanities
University of A Coruña
Ferrol, Spain

ISSN 2365-4228 ISSN 2365-4236 (electronic)
European Studies in Philosophy of Science
ISBN 978-3-319-87461-6 ISBN 978-3-319-63043-4 (eBook)
DOI 10.1007/978-3-319-63043-4

Printed on acid-free paper

This Springer imprint is published by Springer Nature
The registered company is Springer International Publishing AG
The registered company address is: Gewerbestrasse 11, 6330 Cham, Switzerland

Preface

The issue of scientific prediction is undoubtedly a central topic for the philosophy and methodology of science. It is particularly important for scientific practice in the empirical sciences (natural, social, and artificial), where prediction has several crucial roles. It can be a test for the scientific character of hypotheses and theories (basic science), it precedes the prescription oriented toward the solution of concrete problems (applied science), and it serves as a support for decision-making in practical acting contexts (application of science).

Furthermore, when the *goals* of scientific research are considered, it is usual to highlight two of these: the explanation of past phenomena and events and the prediction of future phenomena and happenings. However, there has been very little attention to scientific prediction in comparison to the effort devoted to the study of scientific explanation, as is seen from the far higher number of publications on the latter. But it is also the case that prediction frequently appears as a key concept in the research into other issues of the philosophy and methodology of science, such as scientific progress, complexity, or the limits of science.

Nicholas Rescher's contribution to the philosophical reflection on the problem of prediction must be highlighted. These contributions are both theoretical (regarding a large number of problems) and practical, because he was one of the designers of the Delphi predictive procedure. Among his theoretical contributions, his paper "On Prediction and Explanation" (1958) is noteworthy. In this paper, he calls into question the logical symmetry between prediction and explanation, which was the dominant thesis at that moment. His book *Predicting the Future* also needs to be highlighted, in which a systematic philosophico-methodological analysis of scientific prediction is offered for the first time. This analysis is made from various relevant perspectives (especially, the epistemological, methodological, and ontological ones).

On the one hand, scientific prediction appears as a key concept within his philosophy and methodology of science. In effect, his pragmatic idealism emphasizes the importance of scientific prediction as a topic of analysis in the philosophy of science. On the other, several of his publications are devoted to the study of scientific prediction from different angles. He thus offers a rigorous conception of prediction, in which his analysis goes into a great depth of detail.

Besides the attention to the problem of scientific prediction in its different dimensions, Rescher has offered a *system of thought*; that is, he has provided his own
conception of philosophy, in general, and philosophy of science, in particular. His
proposal – pragmatic idealism – encompasses very influential contemporary elements, and very often his approach has original features.

These reflections highlight the relevance of developing a philosophico-
methodological analysis of Rescher's system of pragmatic idealism and its approach
to prediction in science. He is one of the authors who have made most contributions
to the study of this topic, which is crucial for science and which has been little considered in the philosophy of science (at least, in comparison with other problems,
such as scientific explanation). Here, the research is focused on the philosophy of
science, although some of Rescher's contributions to philosophy, in general, receive
attention, to the extent that they are connected with his philosophy of science (pragmatic idealism) and can be relevant for the study of scientific prediction.

Within this key philosophico-methodological topic of an influential contemporary
philosopher, this book has two main aims. Firstly, the study is oriented toward the
analysis of the philosophico-methodological characters of scientific prediction in
Rescher's conception. The research is focused on various thematic realms: semantic,
logical, epistemological, methodological, ontological, axiological, and ethical. Those
thematic realms of the analysis of scientific prediction are grounded in the different
components of science (language, structure, knowledge, processes, activity, ends, and
values), which can be oriented toward the future, and so are relevant for prediction.

Secondly, the investigation aims to offer a critical reconstruction of Rescher's philosophy of science, which can be characterized as a pragmatic idealism that is open
to some realist elements. This second line is developed in parallel with the first, since
in Rescher's approach the pragmatic idealism modulates the characters of prediction.
In effect, his philosophy of science is related to a *system*, so the thematic realms of
the analysis of scientific prediction are interrelated within a system of thought.

In order to develop these two axes of the research in a way that the relations
between them can be appreciated, the book is organized into three parts, according
to a thematic criterion. The thematic order is as follows: (I) General Coordinates,
Semantic Features, and Logical Components of Scientific Prediction, (II) Predictive
Knowledge and Predictive Processes in Rescher's Methodological Pragmatism, and
(III) From Reality to Values: Ontological Features, Axiological Elements, and
Ethical Aspects of Scientific Prediction.

I would like to express my gratitude to the various persons and institutions that, in
one way or another, have contributed to make this book possible. In this regard, I must
recognize the importance of Wenceslao J. Gonzalez, professor of logic and philosophy
of science at the University of A Coruña, who has supervised my Ph.D. research on
"Scientific Prediction in Nicholas Rescher's Conception: Philosophico-Methodological
Analysis" ("La Predicción Científica en la Concepción de Nicholas Rescher: Análisis
Filosófico-Metodológico"). This research made at the University of A Coruña is the
basis of this volume. Besides Gonzalez's remarkable help in developing this book, I
have also benefited from his research on scientific prediction, which has been crucial
in the development of this book, as is borne out in the bibliographical sections. In this

regard, I would stress the importance of his book *Philosophico-Methodological Analysis of Prediction and Its Role in Economics* (Springer, 2015).

I also want to thank Nicholas Rescher, who showed a great interest in this project from the beginning. I have had the opportunity to work with him in two research stays at the University of Pittsburgh in 2014 and 2015. I appreciate his support in making those stays possible, as well as his kindness and the time he devoted to this research. The conversations I had with him have been decisive for this book. I have had the opportunity to clarify aspects of his conception with him. Moreover, I must emphasize in this regard that he was always willing to solve any and all the questions I posed. I am likewise grateful for his very helpful comments on previous versions of the chapters of this book.

This research was developed within the framework of the *Programa FPU* of the Spanish Ministry of Education, Culture and Sport. I wish to express my recognition to this program of the ministry, which also supported my research stays at the Department of Philosophy of the University of Pittsburgh.

Ferrol, Spain   Amanda Guillán

Contents

Part I
General Coordinates, Semantic Features, and Logical Components of Scientific Prediction

Chapter 1
Scientific Prediction in a System of Pragmatic Idealism

Abstract Nicholas Rescher has developed "pragmatic idealism" as a philosophical system. In this original system, the problem of scientific prediction appears as an element that is part of a whole. Following the idea of a system as a backdrop, this chapter offers the philosophico-methodological coordinates for the analysis of scientific prediction in Rescher's thought. Several steps are followed: (1) There is a description of his academic and intellectual trajectory. (2) His system of pragmatic idealism is analyzed within the contemporary context, with emphasis on its originality. (3) The idealistic aspect of his philosophy is seen in regard to the role of concepts in the articulation of knowledge, and the pragmatic aspect is considered with the problem of scientific progress at stake. (4) The main philosophico-methodological characters of scientific prediction are addressed. This includes paying attention to the semantic, logical, epistemological, methodological, ontological, axiological and ethical features of prediction, which in Rescher are closely related.

Keywords Nicholas Rescher • Pragmatism • Idealism • Scientific prediction • Prescription

Undoubtedly, Nicholas Rescher is one of the most productive of contemporary philosophers. His long list of publications covers a wide variety of realms. He has addressed all the areas of the philosophy and methodology of science: semantics of science, logic of science, epistemology, methodology of science, ontology of science, axiology of scientific research, and ethics of science. Additionally, he has discussed other philosophical realms, such as logic, metaphysics, history of philosophy, and theory of knowledge.[1]

This broad thematic scope and variety of analyses have enabled Rescher to develop his own philosophical system. He is probably the contemporary philosopher who has addressed the most philosophical fields. In this regard, when he deals

---

[1] The extent of Rescher's work can be noticed in the bibliography, which is current to the year 2009, included in Jacquette (2009, pp. 633–643). An updated list of Rescher's publications can be found on his website: http://www.pitt.edu/~rescher/, (accessed on 26.10.2015).

© Springer International Publishing AG 2017
A. Guillán, *Pragmatic Idealism and Scientific Prediction*, European Studies in Philosophy of Science 8, DOI 10.1007/978-3-319-63043-4_1

with the problem of scientific prediction, prediction does not appear as something isolated, but as an element that is part of a whole (Gonzalez 2010, pp. 253–281). This means that, in order to analyze the concept of scientific prediction in his work, it is necessary to clarify the general coordinates of his philosophical proposal, which is configured as a system of "pragmatic idealism" (Rescher 1992a, 1993a, 1994).

This chapter seeks to offer the philosophico-methodological coordinates for the analysis of scientific prediction in Rescher's thought, which has the idea of *a system* as a backdrop. (i) A description of his academic and intellectual trajectory is offered. Through his philosophical work, he came to articulate a system of thought that is supported by two mainstays: the theory of knowledge of Immanuel Kant and the pragmatism of Charles Sanders Peirce. (ii) After presenting this *historical* context of his trajectory, his original system of pragmatic idealism is analyzed within the contemporary context. (iii) In order to go more deeply into his philosophical conception, the idealistic aspect of his philosophy can be seen with regard to the role of concepts in the articulation of knowledge, and the pragmatic aspect leads to the consideration of the problem of scientific progress. (iv) The main philosophico-methodological characters of scientific prediction are also considered. This involves paying attention to the semantic, logical, epistemological, methodological, onto-logical, axiological, and ethical features to prediction, which in Rescher are closely related.

1.1 Nicholas Rescher's Philosophy: General Coordinates

There are some background coordinates that modulate Rescher's thought, and from these it seems to me that his philosophical production, characterized by its ampli-tude and thematic diversity, has coherence and systematicity. In this regard, the reflections that he made in his intellectual autobiography on his academic and research trajectory can be emphasized.[2] His main effort, which is directed at clarify-ing the philosophico-methodological framework in which his contributions sit, is also important. This effort can be seen in his three volumes on *A System of Pragmatic Idealism* (1992a, 1993a, 1994).

In order to understand his proposal for scientific prediction properly, the first step is to describe his intellectual and academic trajectory. In this regard, the years when Rescher worked as a mathematician in the RAND Corporation are especially rele-vant, because it was there that he developed, with Olaf Helmer and Norman Dalkey, the predictive procedure called Delphi.[3] After this, an inquiry into his system of pragmatic idealism is also required, since it is the framework for Rescher's approach to scientific prediction.

---

[2] This autobiography has had several editions. See Rescher (1983a, 1986, 1996a, 2002, 2010).
[3] On the Delphi procedure of prediction, see Rescher (1998a, pp. 91–96), Rowe and Wright (2001, pp. 125–144), Ayyub (2001, pp. 99–105), and Bell (2003, pp. 261–272).

1.1.1 Academic Training and Career

Nicholas Rescher was born on July 15th 1928 in Hagen, a German city in the region of Westphalia. Faced with the troubles of the rise of Nazism, his family emigrated to the United States of America. Erwin Hans Rescher was the first to cross the Atlantic, and he was followed a year after by his wife, Meta Anna, and his son, Klaus Helmut Erwin Rescher (who was named Nicholas Rescher after his arrival in the United States) on July 8th 1938 when Rescher and his mother embarked on the *USS President Roosvelt* for North America, arriving on the morning of July 16th 1938.

From an early age Rescher was interested in mathematics and philosophy.[4] He took a degree in Mathematics at the *Queens College* in Flushing, New York, between 1946 and 1949; although he also attended some philosophy lessons. Herbert G. Bohnert (who was a student of Carnap) and Carl Gustav Hempel, one of the main representatives of the Berlin School (led by Hans Reichenbach) were among his teachers in those years. Once Rescher had his degree, he received offers from the departments of Mathematics and Philosophy of Harvard, Yale, and Princeton. Rescher choose Princeton, where he was a PhD student from 1949 and 1951.

He attended the courses on Logic of Alonzo Church at Princeton. Alonzo Church was of special relevance in Rescher's career because of his decisive contribution to increasing Rescher's interest in Logic. In 1950 he was awarded his Master's degree, and he began lecturing at the university. By this time, he was collaborating with Paul Oppenheim. In 1951 he obtained his PhD with a dissertation on "Leibniz's Cosmology: A Study of the Relations between Leibniz's Work in Physics and his Philosophy." In so doing, he became, at the age of 22, the youngest student to be awarded a PhD at the Department of Philosophy in Princeton.

Between 1952 and 1954 he served in the United States Marine Corps. This meant a break in his academic career. In 1954 he was offered a job in the Mathematics Division of RAND Corporation, directed by John D. Williams. Rescher accepted the offer and moved to Santa Monica, where he stayed until 1956. During his years at RAND Corporation, Rescher became interested in the problem of scientific prediction.

This is relevant insofar as RAND Corporation (*Research and Development*) is a good example of what has been labeled a *think tank* (an organized group to provide ideas): it is a research institution that offers ideas and advice on political, trade or military interests. RAND Corporation was initially set up to offer research support to the US Armed Forces, and this was its main role when Rescher joined the Corporation as a researcher.

When he started his work at the RAND Corporation, Rescher was interested in issues related to game theory. Later on, he collaborated with Fred Thompson in economic issues related to air war. After a second project, where he collaborated

---

[4] In his autobiography, Rescher links his interest in philosophy with the reading of the *History of Philosophy* by Will Durant. Cf. Rescher (2002, p. 50).

again with Fred Thompson and Frederick B. Moore, Rescher started to work on a project that he designed at the beginning of 1955. It was "a speculative assessment of how, given current intelligence assessments of then-extant Russian military capabilities, a preemptive nuclear 'counterforce' attack against U.S. retaliatory potential might be designed" (Rescher 2002, p. 90).

RAND Corporation would develop a great interest in prediction as an important part of its research support for the US Air Force. According to Rescher, "predicting enemy intentions has always been a key task of military intelligence, but in the modern technological world forecasting the development and deployment of weapons systems became no less crucial a mission. Accordingly the issue of identifying and validating prediction methods evolved as an area of RAND interest" (2002, p. 92).

Within the RAND Corporation, the set of investigations related to prediction were known under the code name "Delphi Project." The judgmental predictive procedure that Rescher developed together with Olaf Helmer and Norman Dalkey was named the "Delphi Method."[5] Helmer was the most influential person for Rescher during the years he worked at RAND Corporation and both shared a common interest in the theoretical aspects of scientific prediction, which led them to decide to meet once a week for work sessions in which they could go more deeply into aspects related with the Delphi procedure.

The Delphi procedure is a predictive procedure based on the interaction among a group of predictors, who are experts in the issue that the prediction is about. Predictors do not confront each other. Instead, they answer a series of questionnaires in an individual and anonymous way. These questionnaires are presented to them in several successive rounds. After each round is finished, the predictors can know the group results, so that they can review their own initial answers. The final goal is to achieve an "aggregate prediction" that is supported by all the experts (Rescher 1998a, p. 92).

The study of this predictive procedure was divided in two successive levels: on the one hand, Rescher collaborated with Helmer in the theoretical analysis of the Delphi procedure; and, on the other, Helmer collaborated with Dalkey—and, thereafter, with Bernice Brown and Theodore Gordon—in the study of concrete cases. The first of these levels basically has to do with the epistemology of prediction. One of the first papers that Rescher published on prediction (with the collaboration of Helmer) was on this issue: "On the Epistemology of the Inexact Sciences" (1959).

In 1957, after leaving the RAND Corporation, Rescher started his academic career at the University of Lehigh, in Bethlehem, Pennsylvania. There, he taught philosophy to undergraduate students until 1961. In this period, he met Adolf Grünbaum, a fellow lecturer in the department. Rescher's philosophical interests were at this moment very diverse, and among them his research on Arabic logic can

---

[5] Rescher takes into account two mayor methodological approaches: (i) judgmental procedures, which are based on the estimation of experts who use a kind of unformalized reasoning; and (ii) discursive or scientific methods, which are based on the correctness of inferential principles. See Rescher (1998a, p. 87). This idea appears in one of his first papers on prediction (1967).

be highlighted. Moreover, his philosophical publications in this period are about this topic.[6]

Following Grünbaum's suggestions, who had accepted a position at the University of Pittsburgh and recommended Rescher, he was named Professor of philosophy at the University of Pittsburgh in 1961. Since then, he has been in Pittsburgh, where he has been a Distinguished University Professor since 1970. He then started a prestigious and prolific career as professor and researcher, that led to his being named *Honoris Causa* by eight universities: the University of Loyola, Chicago (1970); *Universidad Nacional* of Córdoba, Argentina (1992); University of Lehigh (1993); University of Konstanz (1996); *Queens College* (1999); University of Hagen (2001); University of Helsinki (2006); and University of Cleveland (2007).

Besides these *Honoris Causa*, Rescher has also received many prestigious awards; among them, the Alexander von Humboldt Humanities Prize (1983); the Medal of Merit for Distinguished Scholarship, University of Helsinki (1990); the Chancellors Distinguished Research Award, University of Pittsburgh (1990); and the Medal of Merit of the Federal Republic of Germany (*Bundesdienstkreuz erster Klasse*), 2011. It should be highlighted that in 2010 the University of Pittsburgh established the biennial Nicholas Rescher Prize for Contributions to Systematic Philosophy.

Further recognition of his academic and research career is found in his being president of numerous associations: *Charles Sanders Peirce Society* (1983–1986); *G. W. Leibniz Society of America* (1983–1986); *American Philosophical Association, Eastern Division* (1989–1990); *American Catholic Philosophical Association* (2003–2004); and *American Metaphysical Society* (2004–2005). He is also an elected member of many Academies, among which can be highlighted the *Institut International de Philosophie,* the *Academie Internationale de Philosophie des Sciences*, the *European Academy of Arts and Sciences*, the *Royal Society of Canada*, and the *American Academy of Arts and Sciences*.

He has been the editor of several journals, such as *American Philosophical Quarterly* (1964–1994), *History of Philosophy Quarterly* (1983–1992), and *Public Affairs Quarterly* (1986–1991). In addition, he is co-editor of the *Pittsburgh Series in Philosophy and History of Science*, of the *University of California Press*, since 1980, and of *C.P.S. Publications in Philosophy of Science* of the *University of America Press*, since 1982. He is in the editorial committee of many scientific journals, among them *Epistemologia, Mind and Society, Journal of the Philosophy of Management, History of Philosophy and Logical Analysis, Idealistic Studies,* and *Philosophisches Jahrbuch*.

This long, fruitful academic trajectory reveals a key feature that Rescher himself acknowledges: "I was unwilling or unable to settle down to one particular specialty" (2002, p. 115). Nevertheless, he came to articulate his own system of thought, which has some well-defined coordinates, so his philosophical contributions form part of that system. In this regard, the contact with some important thinkers of neopositivism

---

[6] See, for instance, Rescher (1962, 1963a, b). A complete list of his publications in the sixties can be found on his website: *http://www.pitt.edu/~rescher/*, (access on 15.12.2014).

and logical empiricism is relevant. Rescher himself, in a paper entitled "The Berlin School of Logical Empiricism and its Legacy," says that he is part of the "younger generation" of the Berlin School insofar as he was a student of Hempel (2006c)· This is especially relevant in his studies on prediction, because Rescher considers that they come "in the wake of [Hans] Reichenbach's work" (Rescher 2006c, p. 298).

Nevertheless, his system of though is not in tune with empiricism. His quite different approach endorses a "pragmatic idealism." This system that Rescher proposes has his roots in his own conception about the task of a philosopher: "A good philosopher (…) must be many-sided because the impetus to philosophizing is ultimately a search for systematic principles underlying the jumbled profusion of phenomena" (Rescher 2002, p. 174). In this he acknowledges a Leibnizian influence: "The inspiration of Leibniz is clearly present in some of my books (e.g., *The Coherence Theory of Truth*) and is discernible in my general approach to the conduct of philosophical work" (Rescher 2010, p. 69).

However, Immanuel Kant and Charles Sanders Peirce are, undoubtedly, the philosophers who have influenced most of Rescher's proposals. *De facto*, there is in his work an explicit concern to elaborate a system of thought where idealism is compatible with pragmatism: "I have gradually acquired the vision of a system of philosophy geared to the idealistic tradition from Leibniz and Berkeley through Kant to Hegel and Peirce, with the German idealists on the left side, the English Hegelians on the right, and the American pragmatists to the front" (2002, p. 174). It is a philosophical system of pragmatic idealism built on the basis of two mainstays: Kant's theory of knowledge and Peirce's pragmatism.

1.1.2 A Kantian Pragmatism: The Primacy of Practice

By means of a large number of publications, Rescher has configured his own philosophy. It is a "'Kantian pragmatism' open to realist contributions" (Gonzalez 2010, p. 254). In effect, in the three volumes entitled *A System of Pragmatic Idealism* (1992a, 1993a, 1994), he sees human knowledge from an idealistic perspective, according to which our categories and concepts have a decisive role in characterizing reality. But Rescher's idealism admits realist notions, such as "fact" or "objectivity." Thus, Kantism is open to realist contributions.[7] This is possible because "realism is compatible with pragmatism in the style of that proposed by Charles S. Peirce" (Gonzalez 2010, p. 256).

Within this framework of a Kantian pragmatism, Rescher offers an approach to human rationality that moves away from maximization, which is the favorite conception of rationality in neoclassical approaches to economics (Rescher 1988. See

---

[7] Certainly, he rejects naïve realism, which claims that is easy to access reality (and also considers that it is easy to identify true statements). On the varieties of realism related to science, cf. Gonzalez (1993). See also Gonzalez (2006).

also Moutafakis 2007). In this regard, his proposal on rationality is a pragmatic one, insofar as the rational agent is that who "proceeds on the basis of the grounds that are available to him (which may well also be imperfect)" (Rescher 1988, p. 7). Thus, he takes into account the limitations of the knowing subject as well as the information limitations to which the subject can be exposed.

Due to these limitations, Rescher maintains that rationality demands an "optimization" with regard to the circumstances, which is not maximization in the strict sense (Rescher 1987b, pp. 55–84; especially, pp. 71–79). He raises the issue from the primacy of practice: "Being rational consists in the disposition to make good reasons constitute the motives for what one does. Since this is something we can achieve only within limits, one must regard perfect rationality as an idealization and acknowledge that we humans are 'rational animals' because of our *capacity* for reason, and certainly not because of our achievement of perfected rationality" (Rescher 1988, p. 10).

Based on rationality as limited, Rescher suggests a holistic conception of rationality, where the role of values is fundamental. Thus, there are—in his judgment— three kinds of rationality, depending on the object of rational deliberation: "Philosophical tradition since Kant sees three major contexts of choice, those of *belief*, of accepting or endorsing theses or claims, of *action*, of what overt acts to perform, and of *evaluation*, of what to value or disvalue. These [contexts] represent the spheres of cognitive, practical, and evaluative reason, respectively" (1988, pp. 2–3).[8]

The pragmatic and Kantian influence is noticeable. On the one hand, Rescher gives priority to practice in his approach to human rationality, in general, and to scientific rationality, in particular (Marsonet 1996). This leads him to insist on the role of economic rationality, which refers to the instrumental component of rationality. In effect, he thinks that both actions and beliefs should be evaluated in accordance to their effectiveness and efficiency to achieve ends. Thus, in order to achieve a goal, "a rational creature will prefer whatever method process or procedure will, other things equal, facilitate goal realization in the most effective, efficient, and economical way" (Rescher 2004, p. 44).

But, on the other hand, Rescher thinks that "a really thorough pragmatism must dig more deeply" (Rescher 2000, p. 168). Thus, he considers that the practice—the rational human activity—requires the selection of the best means in accordance with rational beliefs, appropriate values, and valid goals. Therefore, human rationality is not, in his judgment, just an instrumental rationality, which only selects the means according to given ends. Its realm is certainly wider, since there is a rationality of ends or *evaluative* rationality that leads to selecting the appropriate ends.

In this regard, Rescher defends the existence of a clear nexus between rationality, science, and human values. He sets this nexus in an explicit way on the basis of two

---

[8] This involves a holistic view of rationality according to which "cognitive, pragmatic, and evaluative rationality constitute a unified and indissoluble whole in which all three of these resources are inseparably co-present. Good reasons for believing, for evaluating, and for acting go together to make up a seamless and indivisible whole" (Rescher 1988, p. vii).

fundamental proposals: "1) that rationality includes not only correct reasoning, but also adequate evaluation; and 2) that *praxis*—the effective implementation of practice into action—is ultimately the criterion of evaluation" (Rescher 1999a, p. 48).

Certainly, there is in his work an explicit criticism of those approaches to rationality understood as a merely instrumental rationality (i.e., a rationality centered only on the process and that do not take into account the issue of the value of the result).[9] The thing is that Rescher thinks of science as an activity oriented towards *ends*. From this perspective, "values play a crucial role in science, and (...) this role is not something arbitrary or added, but it is inherent to the goal structure that defines science as a rational search" (Rescher 1999a, p. 95).

He maintains then that it is not good enough to evaluate diverse courses of action regarding given ends. Instead, the election of ends should be evaluated as well and this should be done from a pragmatic perspective. Thus, an "axiology of purposes" is required, which he sees as "a normative methodology for assessing the legitimacy and appropriateness of the purposes we espouse" (Rescher 2004, p. 45). According to this "axiology of purposes", the assessing of the ends is a pragmatic assessing that evaluates the appropriateness of the concrete ends with regard to human needs and interests (Rescher 2004, p. 44–47).

Together with the Kantian and pragmatic influences—which can be seen in his approach to rationality—there are also realist elements in Rescher's philosophical proposal. In this regard, *objectivity* of values is a key feature. On the one hand, he insists on science as *our* science, since it is indebted to the conceptual categories of human beings (Rescher 1992c, 1993d). And, on the other hand, it is a human activity of a teleological character that is modulated by values. Thus, both the ends sought by the research and the means oriented towards those ends should be selected in accordance with valid values. The validity of values has an objective basis, which lead to an ontological component: it is rooted in the human needs that are of a universal character.

This leads Rescher to maintain that there is a *plurality of values* that should modulate both the ends and the means of the scientific research. These values may be internal (cognitive, methodological, ...) or external (social, cultural, ecologic, economic, etc.). Among them, Rescher gives priority to the values that are internal to scientific activity (above all, the epistemological and methodological values). This highlights his Kantism, "insofar as the content of science is more important than the socio-historical milieu" (Gonzalez 2010, p. 269).

Despite this primacy of the internal component, Rescher's axiological approach, which connects to his proposal on scientific rationality, leads to a holism of values (Rescher 1993c). Thus, although we can make distinctions among values—there are internal values and external values—a complete separation of them is not possible. The reason for this criterion is a pragmatic one: in practice, the set of values is

---

[9] Rescher criticizes the purely instrumentalist approaches to rationality. This can be seen in Herbert Simon's proposal, insofar as he does not accept a rationality of ends but only a rationality of means (Rescher 1988, p. viii). An analysis of the notion of "rationality" in Simon and a comparison with Rescher's account is in Gonzalez (2003).

linked to human needs that are of a universal kind (Gonzalez 1999a, pp. 11–44; especially, p. 22). Nevertheless, as long as the values shape a system, it is possible to establish a hierarchy or scale of preferences. Consequently, the principal values, which modulate the ends and means of the research, are, for Rescher, the values internal to scientific activities.

1.2 Pragmatic Idealism in the Contemporary Context

Within these coordinates, when Rescher deals with scientific prediction, each one of the philosophico-methodological angles of analysis (semantic, logical, epistemological, methodological, ontological, axiological, and ethical) is placed in the framework of his own philosophical proposal, which can be characterized as a pragmatic idealism open to elements of realism. This means that the philosophico-methodological study of his concept of prediction must take into account the coordinates of a system of thought, which is supported by two major mainstays: idealism and pragmatism. But, at the same time, his view is open to realism with regard to relevant philosophical aspects.

1.2.1 A System Open to Realism Without Eclecticism

It happens that each one of these philosophical traditions (realism, idealism, and pragmatism) is characterized by having a heterogeneous character. In effect, there is a wide variety of realistic, idealistic, and pragmatic approaches within contemporary philosophy, in general, and contemporary philosophy and methodology of science, in particular. Thus, within the same philosophical tradition, each thinker can defend very diverse approaches regarding relevant points. This philosophical diversity leads to the existence of proposals that are, in principle, antagonistic (such as, for instance, realism and idealism). But, concerning some aspect, there might be some convergent points between them.

Regarding this issue of a possible convergence, it should be pointed out that Rescher's system of thought seeks such a combination of different conceptions. His view of pragmatic idealism open to realistic elements is not conceived as an eclectic proposal. Instead, he articulates his own philosophical system, where he chooses versions of idealism and pragmatism that are compatible with realistic elements. They also belong to a system, understood as a coherent philosophico-methodological conception about science.

Rescher avoids inconsistency when he claims that realism and idealism "need not be contradictory; indeed, both contain a substantial element of truth" (1992h, p. 304). In his judgment, in order to avoid the contradictory elements, we have "to opt for the middle ground and to combine a plausible version of realism with a plausible version of idealism. The issue is not one of the dichotomous choice of either realism or idealism but rather one of a compromising synthesis in the interests of a fruitful collaboration between these historically warring positions" (1992h, p. 324).

With his pragmatic view, Rescher accepts some realistic notions, such as "fact" or "objectivity." His position seeks to integrate them in a pragmatic idealistic proposal. Thus, his approach involves—as it happens in some pragmatic conceptions—the acceptance of an ontological variety of realism that acknowledges the existence of a reality that is independent to the mind of the knowing subject. Moreover, he thinks that reality has its own properties that are accessible to the subjects that want to know that reality, within some limits. For him, human capacity to know the reality is *limited*, so that "we cannot justifiably equate reality with what can, in principle, be known by us, nor equate reality with what can, in principle, be expressed by our language" (Rescher 1992d, p. 253).

Following a pragmatic vision, Rescher dismisses a naïve version of realism, which is incompatible with his epistemological proposal (which is fallibilistic). At the same time, he manifestly rejects scientific realism, a position which he characterizes as the philosophical doctrine that maintains that science provides us, in fact, *true knowledge* about the reality. Thus, in his judgment, scientific realism involves equating *reality as such* with *reality as we know it* through science. Consequently, "what decisively impedes the tenability of scientific realism is the fundamentally epistemological consideration that the world will doubtless eventuate as being very different from the way our best scientific theories currently represent it to be" (Rescher 1992f, p. 277).

From this perspective, Rescher reduces scientific realism to a version of naïve realism, which is incompatible with a fallibilistic approach to scientific knowledge. Thereby, he does not take into account other versions of scientific realism that are certainly more sophisticated.[10] Thus, he opts for a version of "metaphysical realism" that he describes as "the doctrine that the world exists in a way that is substantially independent of the thinking beings that inquire into it, and that its nature—its having whatever characteristics it does actually have—is also comparably thought independent" (1992e, p. 255).

This characterization is an approach that, in its development, goes beyond the "classical" metaphysical realism.[11] It is also different from the "metaphysical realism" defended by K. R. Popper (1974a). This is so because, in Rescher's thought, the acceptance of an ontological realism is connected with an epistemological conception of *conceptual idealism*, which gives primacy to Kantism regarding cognitive matters.

---

[10] In this regard the critical scientific realism of the Finnish School, which has been developed by authors such as R. Tuomela or I. Niiniluoto can be highlighted. Cf. Gonzalez (1993, pp. 47–50). In addition, Rescher does not analyze the recent debates on scientific realism, such as the "structural realism" proposed by John Worrall. Cf. Gonzalez (2006).

[11] On the characterization of metaphysical realism, see Gonzalez (1993, pp. 40–41 and 44–46). A criticism to metaphysical realism can be found in the "internal realism" proposed by H. Putnam. See Putnam (1990, pp. 30–42).

However, Rescher's epistemological approach involves elements of realism, such as the possibility of obtaining objective knowledge of the extramental reality. This is the case due to his acceptance that the ontological dimension of realism—the existence of a reality independent of the knowing subjects—is something inseparable from the epistemological dimension that involves the possibility of achieving to some extent adequate information about that mind independent reality. Furthermore, he thinks that the epistemological dimension presupposes the acceptance of the ontological component (Rescher 1992e, p. 256).

Consequently, Rescher considers that it is necessary to clarify which bases are needed to accept the existence of a mind independent reality. In this regard, he clearly acknowledges the Kantian influence, which is modulated by a pragmatic conception in the line of Peirce. Thus, Rescher thinks that "objectivity represents a postulation made on functional (rather than evidential) grounds: we endorse it in order to be in a position to learn by experience at all. As Kant clearly saw, objective experience is possible only if the existence of such a real, objective world is *presupposed* from the outset rather than being seen as a matter of ex post facto discovery about the nature of things" (1992e, p. 257).

Therefore, for Rescher, the independence of the extramental reality is something that we must accept *a priori*, on the basis of its practical utility, insofar as it makes possible from the beginning the intersubjective communication and the communal inquiry. This is because "only in subscribing to such a fundamental postulate of reality can we take the sort of view of experience, inquiry, and communication that we in fact have. Without it, the entire conceptual framework of our thinking about the world and our place within it would come crashing down" (Rescher 1992e, p. 266). One of his main concerns is present here: the rejection of skepticism, which calls into question the very possibility of achieving a true or verisimilar knowledge about reality (Rescher 1980a, 2003).

In order to reject a skeptic approach, Rescher maintains that appealing to the existence of a reality independent of the knowing subjects as a *necessary condition* for scientific practice is not good enough. What he asks for is a "retrojustification" on the basis of the results of the scientific research (Rescher 1992e, pp. 266–270). This "retrojustification" has a pragmatic dimension and a cognitive component. Thus, "on the *pragmatic* side we find that we obtain a world picture on whose basis we can operate effectively (pragmatic revalidation); on the *cognitive* side we find that we arrive at a picture of the world that provides an explanation of how it is that we are encouraged to get things (roughly) right—that we are in fact justified in using our phenomenal data as data of objective fact (explanatory revalidation). Accordingly, the success at issue is twofold—both in terms of understanding (cognition) and in terms of application (praxis). And it is this ultimate success that justifies and rationalizes, retrospectively, our evidential proceedings" (Rescher 1992e, p. 268).

On the basis of these considerations, the "metaphysical realism" of Rescher—understood as a version of realism that admits a mind independent reality and the accessibility of that extramental reality to the knowing subjects, within some limits—is supported by an epistemological idealism, which in turn is connected with a pragmatism of the primacy of practice. He emphasizes that "the sort of realism contemplated here is accordingly one that pivots on the fact that we *think* of reals in a certain sort of way, and that in fact the very conception of the real is something we employ because doing so merits our ends and purposes" (1992e, p. 270). Thus, for him, reality and concepts are eventually seen from the perspective of human practice.

1.2.2 An Idealism with Distinctive Features

Rescher offers a type of idealism with distinctive features, since he develops a view of idealism that is compatible with some realist elements in his philosophical conception. In this regard, he distinguishes two major types of idealism: ontological idealism and epistemic idealism. In turn, each can take different varieties of idealism. Within ontological idealism, Rescher distinguishes two varieties: (i) *causal idealism* and (ii) *supervenience idealism*. Both types of ontological idealism have in common that they consider that everything there is, apart from minds themselves, arises from the operations of minds, either causally or in a supervenient way (Rescher 1992h, p. 324). He rejects both kinds of ontological idealism, so his idealist proposal is within the coordinates of an epistemic idealism.

Within the framework of epistemic idealism, there are also several options: (a) *fact idealism*, which maintains that to be as a fact is to be a language-formable fact—that is, a truth; (b) *cognitive idealism*, which considers that to be as a truth is to be knowable; (c) *strong substantival idealism*, which is the option according to with to be as a thing or entity is to be actually discerned by some knower; (d) *weak substantival idealism*, which states that to be as a thing or entity is to be discernible; (e) *explanatory idealism*, which maintains that an adequate explanation of the material reality requires some recourse to mental characteristics or operations; and (f) *conceptual idealism*, which is the version of idealism that maintains that whatever is real is in principle knowable and that the knowledge of reality involves conceptualization (Rescher 1992h, 305).

Conceptual idealism is the version of idealism that modulates Rescher's system of pragmatic idealism. In order to deal with concepts, his proposal is based on two types of dependences between mind and matter (the extramental reality), which follow different directions. On the one hand, he maintains that "mind is *causally* dependent upon (i. e., causally requires) matter, in that mental process demands causally or productively the physical workings of matter"; and, on the other, "matter (conceived of in the standard manner of material substance subject to physical law) is *explicatively* dependent upon (i.e., conceptually requires) mind, in that the conception of material processes involves hermeneutically or semantically the mentalistic workings of mind" (Rescher 1992h, p. 318).

Therefore, Rescher thinks that our knowledge of the extramental reality is always mediated by our concepts and categories, so it is not possible to equate, in principle, *reality as such* and *reality as we know it*.[12] On this basis, he rejects scientific realism, which he sees as the thesis that science, in fact, describes now reality in an adequate way (Rescher 1992f). Because, in his judgment, "the world *that we describe* in science is one thing, the world *as we describe it* in science is another, and they would coincide only if our descriptions were totally correct—something that we are certainly not in a position to claim" (1992f, p. 279).

Although Rescher criticizes scientific realism—which, in his conception, is reduced to a version of naïve realism—he does not accept an instrumentalist approach to science (1992f, pp. 286–289). The reason is that his pragmatism is out of tune with instrumentalism, insofar as he admits realist elements in the worldview. Thus, he admits that the aims of science are in tune with realism, since science actively seeks an objective knowledge of reality. But those aims are only achievable within some limits, so real science—*our* science—must be distinguished from ideal or perfect science (Rescher 1992g).

Consequently, Rescher maintains that it should be accepted that "the cognitive enterprise is governed by ideals—in particular, those of knowledge/truth and of science/system. But in a community of *rational* agents, even ideals must pay their way by proving themselves to be efficient and effective in conducting to full realization of the goals and values in whose name they are instituted" (1992g, p. 300). This attention to efficacy and efficiency in goal realization led to the pragmatic dimension of his thought, which is clearly influenced by Peirce.

In effect, together with an ontological conception, which is mainly realist in the central tenets, and an epistemology, which gives clear primacy to a conceptual idealism, the pragmatic dimension eventually gains primacy in his methodological account. This feature is especially important in order to understand his contributions to the semantic, logical, axiological, and ethical realms. So, when Rescher develops his methodological pragmatism, he explicitly declares that it is a return to the Peircean roots of pragmatic tradition.[13] In this regard, the Peircean influence leads him to reject explicitly other versions of pragmatism, such as the subjective pragmatism of W. James, pragmatism as social and cultural construction of J. Dewey, and the relativistic proposals by F. Schiller and R. Rorty (Rescher 2000, ch. 1, pp. 1–56; especially, pp. 15–31 and 44–47; and ch. 2, pp. 57–80).

The differences between the version of pragmatism that Rescher subscribes and other proposals of pragmatism—subjective, of social and cultural construction, and relativistic—can be seen in the realist elements of his system of pragmatic idealism. They are differences between such conceptions regarding notions such as "truth," "fact," "objectivity," and "value." Thus, Rescher develops a realist account of those notions that is, in fact, compatible with a pragmatic approach to the rationality of the human beliefs, actions, choices, and evaluations.

To sum up, in my judgment, Rescher manages to coordinate in a coherent way nuanced philosophical positions from traditions such as idealism and pragmatism

---

[12] To go more deeply into Rescher's conceptual idealism, see Rescher (1973).

[13] On Rescher's methodological pragmatism see Chap. 5, Sect. 5.1. His view on scientific progress, which is of a pragmatic kind, is also developed in Sect. 1.3.2.

that are, in principle, very diverse. In addition, he accepts central tenets of realism insofar as they are in tune with pragmatism. He does this in a way that allows him to avoid a merely eclectic approach. Instead, he seeks to combine idealism and pragmatism with realist elements such that they could be mutually compatible. Thus, he configures his own system of thought, which is the framework of his conception regarding scientific prediction. In effect, the different realms of analysis of prediction (semantic, logical, epistemological, methodological, ontological, axiological, and ethical) are interrelated in Rescher's thought, which is oriented toward *a system* of pragmatic idealism that is open to realist elements.

1.3 A Systematic Conception of Science and Philosophy

The idea of a "system" has a clear presence in Rescher's work. In effect, he thinks that the aim of his task as a philosopher "is to become clear about the import and the credentials of various sorts of human knowledge—above all in the sciences, in everyday life, and in philosophical reflection itself. The adequate comprehension of the character of these various realms of inquiry—especially of the mutual interrelationships—is the formative purpose of the enterprise" (Rescher 2002, p. 174).

His philosophical system is supported by two fundamental mainstays: Kantism and pragmatism. On the one hand, Rescher thinks that human knowledge is modulated by concepts and ideas, so that "science is indebted to the conceptual categories of the human beings and it is different from [the kind of science] that would be made by other agents with other conceptual configuration" (Gonzalez 2010, p. 254). In this regard, the Kantian influence of his thought has primacy. And, on the other, when he focuses on the progress in science, the pragmatic dimension of his thought is highlighted. The nexus between scientific progress and technological innovation is then emphasized; and, furthermore, he highlights the link between scientific progress and scientific prediction.

1.3.1 The Role of Concepts in the Development of Knowledge

As a Kantian philosopher, Rescher insists on the relevance of the concepts to the articulation of knowledge. Categories and concepts in general allow us to articulate the reality. Therefore, science is a human product, which is above all of an intellectual kind and related with certain practices. The scientific view of the world is not absolute in cognitive terms, because science is "our" science. This means that it is the result of the interaction between the researcher and the environment (in principle, natural), according to our conceptual scheme (Rescher 1990, pp. 77–104).

Rescher's acceptance of elements of ontological realism is based in pragmatic reasons. They encompass our notions of truth, communication, fact, or research, insofar as they require *presupposing* the notion of *reality* (Rescher 2006b, pp. 386–

397; especially, pp. 388–393). At the same time, he defends a conceptual idealism: we know reality through our mental categories, our concepts to characterize real things. Knowledge of reality can only be reached through the resources that human beings have: "our only access to information about what the real is through the mediation of mind" (Rescher 1992h, p. 324).

Through this, Rescher establishes a distinction between reality *as such* and reality *as it presents itself to us* (Rescher 1990, p. 77). This involves that "the range of *fact* is always broader than that of *knowledge*" (1990, p. 77).[14] Certainly, he does not call into question the existence of a reality that is independent of the knowing subject (Rescher 1993d, p. 2). Thus, he accepts the notion of "objectivity", which he associates with impartiality. He does this for pragmatic reasons, because objectivity is—in his judgment—a "functionally useful instrumentality" to guide research (Rescher 2006b, p. 388).

In his thought, the acceptance of an objective reality has to do with its utility. Thus, he takes into account six reasons by which the notion of a *mind-independent reality* is required: "1) to preserve the distinction between true and false with respect to factual matters and to operate the idea of truth as agreement with reality; 2) to preserve the distinction between appearance and reality, between our picture of reality and reality itself; 3) to serve as a basis for intersubjective communication; 4) to furnish the basis for a shared project of communal inquiry; 5) to provide for the fallibilistic view of human knowledge; and 6) to sustain the causal mode of learning and inquiry and to serve as basis for the objectivity of experience" (Rescher 2006b, pp. 390–391).

Consequently, Rescher accepts the existence of a reality that is independent of the subjects that try to know that reality. But the knowledge of reality is always mediated by the categories and concepts of human beings, in such a way that it "represents information about an inquiry-relative *empirical* reality" (Rescher 1990, p. 80). Therefore, the acceptance of an objective reality and a fallibilistic view of knowledge go hand-in-hand, since the human being articulates reality through an imperfect conceptual scheme (Rescher 2006b, p. 390).

This approach involves a view of science as a human product, where agents prevail as producers of science and recipients of the things achieved. In this way, Rescher maintains that "the limits of our experience set limits to our science" (1999b, p. 216). From this perspective, the ideal of a perfect science can be ruled out; i.e., the possibility of a fully completed science is rejected (Rescher 1999b, p. 216). So in his approach the characterization of science as *our* science leads to a view of scientific knowledge as imperfect and incomplete (Rescher 1992b).

There is a clear connection between this issue of "perfect science" and the problem of the limits of scientific research. With regard to the limits of science, there are initially two different sides: the limits as "barriers" (*Schranken*)—what separates science from non-science—and the limits as "confines" (*Grenzen*), which deal with the final frontiers of the scientific research (Radnitzky 1978).[15] Usually, when

---

[14] This idea is also developed in Rescher (1987a).

[15] An analysis of the limits of science as a matter that involves these two dimensions (the "barriers" and the "confines") is in Gonzalez (2016).

Rescher analyzes the limits of science, his attention is focused on the second side of the problem: the possible "confines" or the ceiling for scientific activity. His effort leads him to insist on the fact that we cannot know now the science that we will have in the future (Rescher 1983b, 2012). In addition, insofar as he sees science as a system, it is implicit in his view that there are also "barriers," which separate science from other human activities.[16]

With regard to the confines, the perspective of a system highlights the distinction between the "internal" obstacles—those that are due to scientific activity itself—and the "external" obstacles, which are those that come from the environment. Usually, Rescher focuses on the internal obstacles to scientific activity, which he analyzes in accordance with the distinction between limits *in the weak sense*—the current difficulties to solve a problem—and limits *in the strong sense* (the unsolvable problems of science) [Rescher 2001, pp. 73–75]. In this regard, there are internal limits to science that are rooted in its constitutive elements, so we can find obstacles due to the language of science, its structure, its knowledge, its processes, its activity, and its values (among then, ethical values) [Gonzalez 2010, pp. 275–276].[17]

But there are also external limits to science that have to do with the relations between science and the environment (natural, social, or artificial). These limits have to do with the complexity that hinders scientific knowledge and that—in his proposal—has a strong impact on prediction (Rescher 1998a, chap. 8, pp. 133–156). Thus, when the future we try to predict is developmentary open, it can become unpredictable for us (or, at least, "not predictable"). This is what happens with future knowledge, which is not accessible to our current categories of knowledge, above all when the future is in the long run and we try to predict in a very detailed way (Rescher 2012).

Regarding the barriers—what separates science from pseudoscience or other legitimate ways of knowledge—Rescher points out that science is also a limited endeavor, since scientific knowledge is just one human good among others (1999a, pp. 103–105). He takes into account other ways of knowledge, so science is just one possible way of knowing. Nevertheless, scientific knowledge is—in his judgment—an especially important good due to its high instrumental value. Thus, from a pragmatic point of view, the pursuit of knowledge as a good "in no way hinders the cultivation of other legitimate goods; on the contrary, it aids and facilitates their pursuit, thereby acquiring an *instrumental* value in addition to its value as an absolute good in its own right" (Rescher 1999b, p. 243).

Thus, Rescher acknowledges that there can be legitimate knowledge outside science, such as philosophy and the humanistic disciplines (Rescher 1999a, pp. 106–

---

[16] The existence of some kind of "barriers" follows from his acceptance that science cannot cover the full field of human knowledge and specific human activities. See, in this regard, Rescher (1999a). An analysis of Rescher's proposal on the limits of scientific knowledge both as "confines" and "barriers" can be found in Guillán (2016).

[17] On the constitutive elements of science, see Gonzalez (2005, pp. 3–49; especially, pp. 10–11).

113). However, he does not usually take into account the historicity of knowledge. In particular, regardless of his objections to Peter Strawson related to the processes (Rescher 1996b, pp. 60–64), *historicity* is not adequately stressed in his view.[18] Above all, the lack of attention to the notion of "historicity" can be seen when he characterizes scientific change, which he connects to the idea of "progress." In this regard, his approach is in terms of "process," instead of being an approach focused on historicity (see Guillán 2016, pp. 144–146).

1.3.2 Scientific Progress and the Limits of Science

Scientific progress is one of the topics that receive most attention in the contemporary philosophy and methodology of science. Since 1978, Rescher has published several monographs where he analyzes scientific progress (1978a, b, 1989, 1996c, 2006a). Usually, he sees progress from a perspective focused on the economic dimension, so that he offers an approach to scientific progress where the analysis in terms of costs and benefits is especially relevant.[19]

Generally speaking, it is possible to claim that the term "'progress' is a normative or goal-relative—rather than a purely descriptive—term" (Niiniluoto 1980, p. 427). So when it is claimed that there has, in fact, been "scientific progress," it is assumed that the new things achieved are an *improvement* in comparison with the old things; that is, there has been a scientific change and that change has a positive character (at least in epistemological terms).[20] Thus, the ends are closer than they were before the change had taken place.

In this regard, Ilkka Niiniluoto maintains that "'progress' can be contrasted with such neutral terms as 'development', and a philosophical analysis of scientific progress is tantamount to a specification of the *aims* of science" (1980, p. 428). That is, if we want to be able to recognize the progress, the aims sought must also be recognizable. Rescher's notion of "scientific progress" goes in the same way, because—in his judgment—scientific progress is relative to the ends of science (Rescher 1999b, pp. 145–165). For him, these ends are basically four: description, explanation, prediction, and control over nature (1999a, p. 138).[21]

There are several aspects that characterize Rescher's notion of "scientific progress": (1) Scientific progress is potentially unlimited, since it is impossible for us to achieve a perfect science (that is, a complete science), so it is always possible to enlarge or improve the available knowledge (Rescher 1999b, pp. 5–18; and 1978a, pp. 38–53). (2) Regarding the limits that hinder scientific progress, the economy of research must be considered. There are economic limits that involve a deceleration

---

[18] On the role of historicity in scientific change, cf. Gonzalez (2011a).

[19] For a synthesis of the content of Rescher's monographs on scientific progress, see Wible (2008).

[20] On the notion of "scientific progress", see Gonzalez (1990, 1997, 2015, pp. 29–32) and Niiniluoto (2011).

[21] This highlights that his approach is principally oriented toward the sciences of nature.

of scientific progress as its costs increase (Rescher 1978a, pp. 79–94). 3. Scientific progress has basically a conceptual character, although the practical dimension of scientific research should be stressed to assess progress: above all, the improvements with regard the ability of prediction and control over nature (Rescher 1977. See also Guillán 2016).

1. On the first feature—scientific progress as potentially unlimited or endless— Rescher thinks that science is always open to future developments (Guillán 2016, pp. 139–141). In this regard, there is an important relevant difference between two different senses of limits: the limits *in the weak sense* and the limits *in the strong sense*. There are limits *in the weak sense* when we are not able now to answer questions, because we do not have the knowledge required to answer them. Meanwhile, there are limits *in the strong sense* when we think that there are questions that we will not answer in the future, even in the long run (Rescher 2001).

When Rescher compares the limits in the weak sense to the limits in the strong sense, he considers that the later are more problematic than the former. Limits in the strong sense are associated with *insolubilia* (the unsolvable problems of science). In this regard, when we accept that science is subject to limitations in the strong sense, it is accepted that there is now or there will be in the future significant questions that never will be answered. The solution to these questions is beyond the limits of science, as a human activity. This is also the case in the long run.

However, Rescher points out that "there is no reason to think, on the basis of general principles, that any issues within the domain of natural science lie beyond its capabilities" (1999b, p. 3). From this point of view, limits that affect science are always limits in the weak sense. In this case, this proposal—that science is subject to limits in the weak sense—is compatible with the claim that whatever be the question posed at a specific moment of scientific research, we should think that we will be able to answer it, at least in the future.

But, on the basis of the *unpredictability* of the future science, it cannot be claimed that all the questions posed by science will eventually be answered.[22] Thus, Rescher accepts, in principle, the possibility that there are limits in the strong sense, so there can be unsolvable problems in scientific research. However, he points out that these are two different theses: (i) that there could be unsolvable problems, and (ii) that those unsolvable problems can be identified. The later thesis—that we can identify now the questions that science will never be able to solve—is called "hyperlimitation" by Rescher (1999b, p. 113).

Certainly, this topic has to be seen in connection to the difficulties in predicting future knowledge. As Rescher states this problem, it is possible to claim that there

---

[22]There is an important distinction between "unpredictable" and "not predictable." "Unpredictability" involves the complete impossibility of predicting. It is mainly due to the presence of anarchic phenomena. Meanwhile, "non predictability" is related to the current impossibility of achieving a prediction, which is usually due to the instability of the phenomena. This distinction is in Gonzalez (2010, p. 289). See also Eagle (2005).

are difficulties in identifying the *insolubilia* (1998a, pp. 186–188, 1999b, pp. 111–127), insofar as there are also difficulties in predicting how future science will be (Rescher 1998a, pp. 177–183). Indeed, he points out that "the prospect of present knowledge about future discoveries is deeply problematic since the future of knowledge is fundamentally unpredictable. The details of the cognitive future are hidden in an impenetrable fog" (2001, p. 64).

This issue is connected with the Kant's *principle of question propagation*, which—in Rescher's judgment—has two main consequences for science: (a) the unpredictability of future science, and (b) the impossibility of achieving a perfect science. Thus, in the first place, on the basis of current knowledge, Kant's principle of question propagation involves the impossibility of predicting the questions we will ask in the future and, certainly, we do not know now the answers to these questions. For this reason, we cannot identify the *insolubilia* or assure their existence (that is, not as a current incapability that will be overcome in the future).

Obviously, the advancement of science can be seen as a wide cognitive process based on a "Kantian inspiration," because it can be seen as a process of questions and answers, where each new answer influences the question that can be posed. In this regard, Rescher points out different ways by which new knowledge can affect the questions that we consider: (I) it can give new answers to old questions; (II) it can generate new questions; and (III) it can show that old questions are improper or illegitimate (Rescher 1999b, p. 12).

Therefore, Rescher considers that we cannot predict now with accuracy the content of future knowledge. In consequence, we cannot predict what *questions* we will consider in the future. In addition, we do not know (and cannot know) if all the questions we have not answered yet are legitimate. Thus, "the task of specifying the limits of scientific capability in the production of knowledge is itself one that transcends the limits of our cognitive powers" (Rescher 2009, p. 18). Likewise, the task of identifying what the *insolubilia* problems of science are also transcends our capabilities, although we can assume that these problems exist.

It should be pointed out that Kant's principle of question propagation involves the impossibility of the completeness of science. Rescher relates this issue to the existence of limits in the weak sense. Thus, scientific knowledge is open to the future, and this means that perfect science is an unfeasible ideal; that is, he does not accept the possibility of a completed science.[23] This is because each scientific improvement will generate new questions that require answers, and so on. In this

---

[23] Rescher calls into question the possibility of achieving a perfect science from theoretical and practical viewpoints. His starting point is in the goals of science, which in his judgment are four: description, explanation, prediction, and control. Then, he identifies four conditions that a scientific discipline should meet in order to be considered complete: erotetic completeness, predictive completeness, pragmatic completeness, and temporal finality (the omega-condition). He considers that the first three are problematic from a theoretical point of view and unfeasible from a pragmatic viewpoint. Meanwhile, temporal finality is unfeasible if we take into account the internal dynamics of the connection between scientific progress and technological innovation. See Rescher (1999b, pp. 145–176).

way, science develops through a question-answering process, insofar as it overcomes the limits that hinder its advancement.

The distinction between limits *in the weak sense* and limits *in the strong sense*—where *hyperlimitation* can be included—has major implications for scientific progress. Rescher's proposal is clear: "the distinction between these various types of limits thus carries the important lesson—already drawn by Kant—that even the resolution of all *our* scientific problems would not necessarily mean that science as such is finite or completable" (2001, p. 74). Thereby, he maintains an approach to scientific progress as being unlimited, insofar as it is always open to the future (1978a).

On the one hand, Rescher sees science as subject to limits *in the weak sense*. They are *current* limits that can be found in each historical stage of scientific development and will also affect science in the future. But there are obstacles that can be overcome through scientific development itself. In his own words: "to maintain (…) the essential limitlessness of science on the side of terminating limits—the feasibility of unending scientific progress—is not to deny the prospect of problems whose solution lies beyond the physical and/or economic limits of man's investigative capacities. The existence of actually unanswerable questions in science—problems whose solution lies forever on the inaccessible side of a technologically imposed data-barrier—would not mean an eventual end to scientific progress" (Rescher 1979, p. 32).

Accordingly, Rescher thinks that, *de facto*, there will be always limits that hinder scientific progress; but, at the same time, he considers that scientific progress is always possible to the extent that those limits can be eventually overcome. Concurrently, he calls into question that there are, strictly speaking, limits *in the strong sense*. This is because, in the first place, when we consider that a question is in principle beyond the powers of science, it is difficult to maintain that this question is a legitimate scientific one (i.e., that it belongs to the scientific domain); and, in the second place, even if those limitations exist, it is impossible for us to identify them (Rescher 2001, p. 75). Accordingly, it is questionable whether there is now or will be in the future a "ceiling" to scientific research, even when we must acknowledge the present existence of limits to scientific progress.

2. When Rescher considers the limits in the weak sense (that is, those limitations that affect science in each concrete stage), the second feature of his characterization of scientific progress appears: the relevance of the economy of research. In effect, "Rescher's approach on scientific progress is frequently based on the language of cost and benefit, because—for him—cost-effectiveness is a salient aspect of rationality, where the benefits of knowledge can be theoretical (or purely cognitive) or practical (or applied)" (Gonzalez 2008b, p. 92).

In this regard, it can be pointed out that "on the one hand, among the *internal benefits* of sciences is the increasing capacity that a science has to provide explanation and prediction, which also contribute explicitly to the human worldview as well as the solution of many practical problems of everyday life. On the other hand, there are growing *external costs*, mainly in the natural sciences and in the sciences of the artificial, which are due to the enlarging complexity of the phenomena studied as well as the greater difficulty in learning and mastery" (Gonzalez 2008b, p. 92).

Therefore, with regard to scientific progress, the analysis of the external costs is a fundamental issue. Regarding this problem, Rescher pays attention to the relations between science (above all, natural sciences) and technology. He thinks that science and technology are like "two legs of the same body" since each of them needs the other and contributes to its development (Rescher 1999a, p. 100). Again, his pragmatic component is clear: they are interrelated endeavors.

Because Rescher's analysis is focused on the natural sciences, he maintains that technologies for observation, experimentation, and the subsequent data-processing have a key role in scientific progress. In this way, scientific progress and technological innovation are two notions that go together in his approach.[24] As J. F. Wible points out, in Rescher's proposal "scientific progress depends on scientific observation and it in turn depends both on the availability and types of resources required for scientific observation" (2008, p. 446).

In effect, Rescher thinks that there cannot be scientific progress without technology. "On the one hand, the transforming resources of technology use and exploit our scientific understanding of the world's processes. But, on the other hand, it turns out that science cannot progress without technology, because we can only obtain information about reality by interacting with it. We can only theorize about nature in an effective way to the extent that we can detect its processes (by 'observation') and manipulate its phenomena (by 'experimentation')" (Rescher 1999a, p. 100).[25]

However, this relation of inter-dependence between science and the available technology implies a major limitation to scientific progress, since there are increasing costs related to data acquisition and management technologies as science advances. On the other hand, technological innovation enlarges complexity, because each solution given to the problems related to information processing and the control of the processes generates new problems of complexity (Rescher 1999a, pp. 118–121).[26]

In Rescher's judgment, "Kant's principle of question-propagation" is also present in the technological realm, and it is related to complexity. In his words, "throughout the progress of science, technology, and human artifice generally, complexity is self-potentiating because it engenders complications on the side of problems that can only be addressed adequately through further complication on the side of process and procedure. The increase in technical sophistication confronts us with a dynamic feedback interaction between problems and solutions that ultimately

---

[24] "Although scientific progress is always possible in principle (…) the achievement of this permanent possibility demands a continuous improvement of the technological capability of data extraction and exploitation," Rescher (1999a, p. 126).

[25] Rescher calls this approach the "thesis of the technological dependency," which maintains that *"progress in the theoretical superstructure of natural science hinges crucially upon improvements in the technological basis of data-acquisition and processing,"* Rescher (1978a, p.142).

[26] On scientific creativity and technological innovation in the context of complexity, cf. Gonzalez (2013b).

transforms each successive solution into a generator of new problems" (Rescher 1999a, p. 120).

This problem implies that there is, or can be in principle, a deceleration in scientific progress due to economic elements. Thus, the costs related to scientific progress increase at the same time as the benefits decrease, since "each successive order-of-magnitude step involves a massive cost for lesser progress; each successive fixed-size investment of effort yields a substantially diminished return" (Rescher 1999b, p. 59). From this point of view, the major limits to scientific progress are practical limits that rest basically on the physical-economic limitations to data acquisition and processing (Rescher 1978a, p. 236).

3. The third basic feature that characterizes Rescher's proposal on scientific progress is double sided: on the one hand, he emphasizes that science progresses, fundamentally, through conceptual change; and, on the other, he considers that scientific prediction and control over nature are the best criteria at our disposal to assess scientific progress. Thus, he thinks of scientific progress as a *process* where changes occur and these changes are basically related to concepts.[27]

In effect, Rescher explicitly states that "scientific change (...) is not just a matter of marginal revisions of opinion within a fixed and stable framework of concepts; the crucial developments involve a change in the conceptual apparatus itself" (1999b, p. 39). This leads to a view of scientific progress as "a process of *conceptual* innovation that always places certain developments outside the cognitive horizons of earlier workers because the very concepts operative in their characterization become available only in the course of scientific discovery itself" (Rescher 2012, p. 151).

But, at the same time, Rescher thinks that to *assess scientific progress* on the basis of conceptual change is problematic. This is because, in his judgment, "when the 'external' element of *control over nature* is given its due prominence, the substantiation of imputations of scientific *progress* becomes a more manageable project than it could ever possibly be on a 'internal,' context-oriented basis" (1977, p. 188).

This approach is—in my judgment—problematic if we consider the framework of thought that Rescher offers. Clearly, he acknowledges that concepts can have an *objective content*. So it could be possible to assess, on the basis of the objectivity of the concepts, scientific progress in connection with conceptual changes, where historicity (of science, the agents and the researched reality itself) is compatible with objectivity.[28] Instead, Rescher claims: "the progress of science will be taken to center on its pragmatic aspect—the increasing success of applications in problem solving and control" (Rescher 1977, p. 185).

Niiniluoto, among other authors, has criticized this proposal of Rescher. In his judgment, Rescher's insistence on prediction and the ability of control as criteria to assess the verisimilitude of theories results in a biased view of scientific progress.

---

[27] Conceptual progress from the point of view of processes is not the same as *conceptual historicity* as the driving force of scientific change. Cf. Thagard (1992) and Gonzalez (2011b).

[28] On an analysis of conceptual change from the notion of "historicity," see Gonzalez (2008a and 2011a).

From this perspective, *pragmatic success* would be at most a criterion to evaluate *cognitive success*, but it is neither the only criterion nor the most important (Niiniluoto 1997, p. 402). So Niiniluoto considers that there are a wide variety of reliable criteria. These are "cognitive factors such as truth, information, explanatory power, predictive capacity, precision, and simplicity" (Niiniluoto 1997, p. 402).

Regarding Niiniluoto's criticism, I should point out that Rescher does not maintain, strictly speaking, that predictive success is, by itself, a criterion to assess the verisimilitude of scientific theories (see Chap. 5 in this book, Sect. 5.1.2). What Rescher actually thinks is that "only a complex, reciprocally interactive gearing of explanation, prediction, and control can in the final analysis provide a satisfactory standard of scientific adequacy" (Rescher 1998a, p. 165). Nevertheless, he thinks that its role as a criterion to assess scientific progress is fundamental. Thus, "predictive efficacy is the best available token for the explanatory adequacy of our theories" (1998a, p. 164).[29]

1.4 The Main Philosophico-methodological Elements of Prediction in Rescher's Conception

Certainly, the problem of scientific prediction is among the relevant topics of the philosophy and methodology of science in the 20th century and the beginning of 21th century. The relevance that the notion of "prediction" has to scientific activity has been highlighted from various philosophical approaches.[30] In effect, prediction has several relevant roles in science: (i) it is an important aim of scientific research; (ii) in basic science, prediction is usually used as a test of theories; (iii) it precedes the prescriptive task of applied sciences (Gonzalez 2010, p. 11); and (iv) it can be a starting point for decision making in the realm of the application of science (Gonzalez 2013b, pp. 11–40; especially, pp. 17–18).

These considerations are based on the fact that the different components of science (language, structure, knowledge, method, activity, ends, and values) can be oriented towards the future (Gonzalez 2005). Consequently, scientific prediction can be analyzed from a variety of realms that concern the aforementioned components. Thus, the study of scientific prediction has been addressed from the semantic,

---

[29] It seems odd to maintain that "predictive efficacy" is a key factor for "explanatory adequacy," above all if we take into account that Rescher supports the asymmetry between explanation and prediction.

[30] However, when it is compared with scientific explanation, the problem of prediction has undoubtedly received less attention in the contemporary philosophy and methodology of science: "Despite the fact that most philosophers acknowledge the general importance of prediction for science, the vast majority of the intellectual focus between the two goals rests on explanation. Prediction is rarely a topic in its own right, appearing mainly in discussions of confirmation, realism, and other topics. It has been this way for over 40 years," Douglas (2009, p, 445).

logical, epistemological, methodological, ontological, axiological, and ethical realms.[31]

Within Rescher's philosophical conception, the problem of scientific prediction is placed in a prominent position. Worhty of note is his monograph on scientific prediction, where he offers many elements for the analysis of prediction from the perspective of the diverse constituents of science: language, structure, knowledge, method, activity, ends, and values (Rescher 1998a). As he explicitly states, the book seeks "to provide a *theory of prediction*" (1998a, p. 1). These components considered are interrelated within his proposal—a pragmatic idealism—since he is interested in a *system*.

Thus, they are a set of interdependent elements in his conception, so "each factor can be distinguished from the others, but it cannot be properly separated: it is a part of a whole" (Gonzalez 2010, p. 259). In fact, albeit to a greater or lesser extent, Rescher deals with the semantic, logical, epistemological, methodological, ontological, axiological, and ethical components of scientific prediction. His goal is to offer a systematic conception of prediction; where the different levels are related, so giving rise to an interdependence network. *De facto*, he reaches the proposed aim: to provide a theory of prediction, which is analyzed here.

1.4.1   Semantic and Logical Features of Prediction

From a semantic viewpoint, two main issues should be considered: Rescher's proposal about language and the semantic characteristics that he gives to scientific prediction. When he analyses prediction from language, his starting point is a pragmatic conception, where the view of meaning as use has primacy. Again, the influence of Peirce is crucial here, because his version of pragmatism is open to realist notions (Rescher 1998b). Thus, he admits that language can be evaluated in terms of objectivity and truth (or, at least, truthlikeness). In my judgment, this is a wise choice in his approach, since he avoids reducing meaning to the mere use of language, insofar as he admits an objective basis in its content.

In this way, his approach makes it possible to characterize scientific prediction as a statement that can have an *objective content*. Moreover, Rescher accepts Israel Scheffler's idea of scientific prediction as a *statement* oriented towards the future (Scheffler 1957). Therefore, in contrast to authors like Milton Friedman (1953) or Stephen Toulmin (1961), he rejects the possibility of a "prediction about the past." On this basis, the predictive language has specific features, so the predictive statements can be distinguished from other kind of scientific statements, such as the descriptive or the explanatory ones.

---

[31] On the roles of prediction and the diversity philosophical analyses regarding its different realms, see Gonzalez (2010). These realms of philosophical analyses have a direct repercussion in sciences such as economics. See Gonzalez (2015).

But, in order to achieve a comprehensive account of prediction, a higher rigor regarding language is required. On the one hand, the barriers between scientific prediction and non-scientific prediction should be further elaborated (for example, taking into account the different level of precision and accuracy in the sense and reference of the terms used). On the other, the research into the types of prediction — such as "foresight," "prediction," "forecasting," and "planning" (Fernández Valbuena 1990) — should also be developed.

Logically, the attention goes to the structural aspects of the scientific theories. In this regard, the problem of the "well-structured" theories oriented toward prediction should be considered. In this sense, one question is whether a "well-structured" theory can have an inductive structure (for example, the hypothetical-inductive [Niiniluoto and Tuomela 1973]) or, on the contrary, the deductive structure (especially, the hypothetical-deductive) is the only valid structure. Regarding the problem of *induction*, Rescher offers a characterization as an estimative procedure, instead of considering induction as a form of inference (Rescher 1980b). From this perspective, the justification of induction as a predictive procedure appears in practical terms. It is ultimately rooted in the capacity of the procedures supported by induction to obtain successful predictions.

Rescher's acceptance of induction for prediction leads to the problem of the limits of deductivism. This issue is especially important for him, since his methodological pragmatism cannot be supported by only deductive bases. This feature is an important contribution, because it seems to me that the elements offered by Rescher—and Salmon's criticism in this regard—emphasize the insufficiency of deductivism to solve the problems posed by scientific prediction (Rescher 1998a, Salmon 1981). Therefore, a logical-methodological approach to scientific prediction of an exclusively deductive character—like that suggested by Popper (1974b, 1979)—does not allow us to grasp all the elements that are at stake when prediction is analyzed from a logical perspective.

Again from a logical perspective, some major problems are (i) the structural relations between scientific explanation and scientific prediction, and (ii) the role of induction and deduction for scientific prediction. Regarding the first problem, it has been widely discussed whether prediction and explanation are symmetrical or asymmetrical processes. In this respect, Rescher is right when he maintains that there is a *logical asymmetry* between explanation and prediction, so a structural equivalence between both processes can be rejected. Moreover, his criticism to the symmetry thesis — originally developed by Hempel and Oppenheim (1948) — goes further than the logical realm, so he also gives reasons that show asymmetry in the semantic, epistemological, methodological, and ontological levels (Rescher 1958, 1970, pp. 32–34, 1998a, pp. 165–166).

Furthermore, the perspective of *temporality*—the temporal anisotropy between explanation and prediction—may give more elements in favor of the asymmetry thesis than those suggested by Rescher. On the basis of this perspective, there is an adequate framework in order to analyze another two questions of logical character that are not exhaustively considered by him: (a) the possible equivalence between "retrodiction" and scientific explanation; and (b) the problem of the logical equivalence between prediction and "retrodiction," which is usually connected with the acceptance of a genuine "prediction of past." These are questions that have been discussed by Grünbaum (1962) and Salmon (1993), among others.

1.4.2   Prediction in the Epistemological and Methodological Realms

Epistemologically, the research into scientific prediction goes to the kind of *cognitive content* offered by prediction and the related problems. Regarding this issue, scientific prediction is related to the theory of rationality in Rescher's conception. In this respect, his approach is broader than the proposals of other philosophers, such as Herbert A. Simon (1982a, b, 1997). The main difference between them is the evaluative rationality, which Rescher expressly assumes. With evaluative rationality the preferability of the ends is taken into account, instead of only considering what is preferred or just accepting the ends as given (Rescher 1988).

The attention to the rationality of ends is connected with the axiology of research, since the selection of ends must be made according to values (internal and external). Thus, the epistemological and axiological realms are closely related in Rescher's philosophy of science. When this approach is applied to scientific prediction, the most important aspect is—in his proposal—the cognitive content of the prediction, which should have a series of epistemological values. But he is certainly aware of the difficulties in predicting the possible future in a reliable way. Thus, his approach is in no way naïve, but he maintains a fallibilistic view of scientific knowledge, especially when this knowledge is about the future, which, in principle, has many possibilities.

Within a fallibilistic epistemological framework, Rescher defends the possibility of obtaining true (or, at least, truthlike) predictions. For this reason, he emphatically rejects the "non-reasoned predictions:" these do not make science (Rescher 1998a, p. 38).[32] However, he excessively highlights the pragmatic dimension, since he maintains that the main difference between reasoned and non-reasoned predictions is rooted in the fact that the second have no practical utility (Rescher 1998a, p. 55). In this respect, he disregards, to some extent, the theoretical dimension of prediction. This means that the demarcation between scientific predictions and non-scientific predictions is not always clear in his approach.

Besides the importance of the rational bases to achieve successful predictions, Rescher highlights the epistemological obstacles to prediction. In effect, the analysis of the limits to predictability (above all, epistemological and ontological) is one of his main contributions to the problem of scientific prediction (Rescher 1998a, pp. 133–156). However, two limitations should be mentioned. (1) Rescher mainly focuses on the natural sciences within a context that is directed, above all, to basic research, so he does not take into account all the empirical sciences, or the different

---

[32] In this way, Rescher rules out the thesis of D. H. Mellor, according to which "predictions don't need reasons," (Mellor 1975, p. 221).

types of scientific activity (basic, applied, and of application). (2) His perspective on the epistemological limits is mainly structural, so he does not address the problems related to the historicity of knowledge.

Methodologically, Rescher sees scientific prediction in the framework of a methodological pragmatism that highlights the importance of prediction for scientific practice, above all since predictive success can be used as an indicator to evaluate scientific progress (Rescher 1977). Thus, his methodological approach involves an *instrumental* approach to prediction,[33] which emphasizes the value of predictive success as a criterion to assess the comparative theoretical adequacy of theories.

Besides the conception of methodological pragmatism as the framework for the methodological analysis of scientific prediction, Rescher's proposal is concerned with clarifying the common bases of the different predictive procedures. In my judgment, one of his most valuable contributions is in the "preconditions for rational prediction," which are the *previous* and *necessary* conditions for the processes oriented toward predicting: data availability, pattern discernability, and pattern stability (Rescher 1998a, p. 86). I think that this is an adequate synthesis of the necessary preconditions for predictive success, and one which takes in the different relevant factors from a methodological perspective.

After the analysis of the preconditions for rational prediction, Rescher's methodological approach focuses on the different predictive procedures and methods and their scientific importance. In this regard, he offers a very detailed framework of the different predictive processes, which involves attention to their specific features. He classifies the predictive processes into two main groups: (i) *estimative procedures* and (ii) *discursive methods* (1998a, ch. 6, pp. 85–112; especially, pp. 86–88). The former are developed on the basis of the personal estimations of the experts, such as in the case of the Delphi procedure. Nevertheless, the formalized methods follow a series of well-articulated rules or inferential principles. In turn, these can be divided into two types: (a) *elementary* discursive processes, such as trend projection or the use of analogies; and b) *scientific* discursive processes, such as inference from laws and predictive models (1998a, 97–102).

Rescher's framework for the predictive processes is important. It allows us to highlight the reliability and characteristics of the different predictive *procedures* and *methods* (Gonzalez 2015, ch. 10). Certainly, the level of rigor and sophistication in the processes has repercussions in the reliability of the predictions. Thus, predictions obtained through estimative procedures (where some cognitive biases can intervene) seem to be, in principle, less reliable than predictions that are the result of the use of genuinely scientific methods.

---

[33] "Prediction, in sum, is our instrument for resolving our meaningful questions about the future, or at least of *endeavoring* to solve them in a rationally cogent manner," Rescher (1998a, p. 39).

1.4.3   Ontological Features of Prediction and the Realm of Values

Ontologically, the problem of prediction connects with the reality of the phenomena. This feature implies that the study of scientific prediction must take into account the characteristics of the reality the prediction is about, which can be a natural, social, or artificial reality. For Rescher, the characteristics of the phenomena that we want to predict have epistemological and methodological repercussions on prediction. In this regard, he considers that natural phenomena are generally more stable than social phenomena.[34] Consequently, the unreliability of prediction is a more frequent problem in social sciences than in natural sciences.

But Rescher does not go deeply into the specific characteristics of the social reality and does not consider the artificial realm. In this regard, he does not pay attention to a crucial element for prediction in these realms: *historicity*. Thus, there is a component of variability that adds complexity to prediction in social sciences and the sciences of the artificial, and that is crucial for understanding some of the obstacles to scientific prediction in those realms, such as human creativity. It seems clear to me that *complexity* is a crucial aspect for the study of the ontological characters of scientific prediction and it is especially important in order to clarify the ontological limits of prediction (such as creativity).

Certainly, complexity (both epistemological and ontological) has relevance for the analysis of scientific prediction and its characteristics. So, to the extent that prediction has to do with a complex reality (natural, social, or artificial), two problems appear: (1) complexity has repercussions on the very possibility of predicting, and (2) it has also repercussions on the kind of prediction that is achievable (in relation to its reliability, accuracy, precision, etc.). This is clearer when complexity is seen as a twofold notion: it has a structural dimension and a dynamic component (Gonzalez 2013b).

It seems to me that Rescher's approach to complexity is more thorough than other authors' conceptions (for example, Simon's proposal) [Simon 1996, 2001]. However, his proposal is too focused on the structural dimension of complexity and he frequently disregards the dynamic component. In my judgment, a broader approach to complexity should also take into account complex dynamics, which are connected with the notion of historicity. Indeed, as Wenceslao J. Gonzalez has emphasized, historicity allows us to characterize the change in complex dynamics (both "internal" and "external") as well as to recognize its repercussions on scientific prediction, in general, and on prediction in social sciences and the sciences of the artificial, in particular (Gonzalez 2012a, b, c, 2013c).

Axiologically, the analysis of scientific prediction in a system of pragmatic idealism should take into account, first, Rescher's proposal regarding the axiology of research, in order to see, secondly, how he modulates the axiological features of prediction. Rescher's axiological proposal follows a structural approach

---

[34] Cf. Rescher, N., *Personal Communication*, 29.7.2014.

preferentially, so the internal values have primacy, above all the epistemological and methodological ones. But, although Rescher clearly gives primacy to the structural dimension, his proposal has original features: he offers an approach of holism of values (Rescher 1993b). This allows him to configure a *broad* axiology of research, where the internal dimension and the external component of analysis are interrelated.

Insofar as Rescher's conception is offered as a pragmatic idealism, his view on values is idealistic to the extent that it is a *system*, and its pragmatic component leads him to think in terms of activity and *primacy of practice*. Thus, prediction is related with this system of values and, furthermore, it appears as a necessity of the scientist in order to address different problems. From this perspective, his axiology of scientific research is compatible with a dynamic approach. However, he does not develop that dynamic dimension of the relation between prediction and values in an articulate way precisely because he does not highlight the *historicity* of the scientific activity (Rescher 1998a, pp. 113–115).

Consequently, his approach does not emphasize the external values that accompany science, in general, and scientific prediction, in particular. For this reason, attention to the dynamic component (internal and external) is, in my judgment, required. In my view, this leads to the need to broaden his proposal. This should be done through the analysis of prediction and the connected values in a conception that is open to the dynamic perspective, both internal and external (Gonzalez 2013a). To do this, the differences between three types of activities must be taken into account: basic science, applied science, and the applications of science.[35] When the attention goes to the internal dynamics, it can be seen how prediction and the connected values have a crucial role in the evaluation of scientific activity (basic, applied, or of application), where there is an articulation of aims and processes, which can lead to some results, and a relation with a context, which is changeable.

From an ethical perspective, it should be noticed that, although Rescher's philosophy of science pays special attention to the ethics of scientific research, his theory of prediction does not develop the ethical perspective. In my judgment, his ethics of science is coherent with his pragmatic idealism that is open to elements of realism. Again, the acceptance of elements of realism allows him to configure an assumable approach, where ethical values have objective bases (see, for example, Rescher 1999a). However, Rescher gives excessive primacy to the internal factors of scientific activity and gives less weight to the external factors. This feature means that his proposal regarding the ethics of science cannot grasp the entire field when the ethical aspects of scientific prediction are considered.

A broader approach to the ethics of science should, in my judgment, take into account two dimensions of analysis: (a) the endogenous ethics, which is oriented toward *scientific activity* by itself, so it can be seen that there are values in the aims, processes, and results of the scientific endeavor; and (b) the exogenous ethics, which analyses *science as activity*, in order to highlight that there are ethical values that

---

[35] On the distinction between basic science, applied science, and the application of science, see Niiniluoto (1993) and Gonzalez (2013b, 2015, pp. 32–40).

connect science with the context (Gonzalez 1999b). In order to develop such a broader approach, the study of the relations between science and ethical values should take into account three different realms: basic science, applied science, and the application of science. The relation between scientific prediction and ethical problems (which has specific features in each one of these human activities) is modulated by the different internal configuration of each of the three activities, as well as by their own features regarding the external dynamics.

To sum up, when Rescher addresses the problem of scientific prediction, his proposals are modulated by his own system of thought. Thus, the aspects of scientific prediction are interrelated, since they are integrated within a pragmatic-idealistic approach. This allows us to make a critical reconstruction of Rescher's philosophical system of pragmatic idealism from his approach to scientific prediction, where several thematic realms are considered: semantic, logical, epistemological, methodological, ontological, axiological, and ethical.[36] On the one hand, Rescher has developed his own conception regarding each one of these thematic realms; and, on the other, they are relevant for the analysis of scientific prediction, since the different components of science can be oriented towards the future.

References

Ayyub, B.M. 2001. *Elicitation of expert opinions for uncertainty and risks.* Boca Raton: CRC Press.

Bell, W. 2003. *Foundations of futures studies. History, purposes, and knowledge. Human science for a new era,* vol. 1. Piscataway: Transaction Publishers (5th rep., 2009; 1st ed., 1997).

Douglas, H.E. 2009. Reintroducing prediction to explanation. *Philosophy of Science* 76 (4): 444–463.

Eagle, A. 2005. Randomness is unpredictability. *British Journal for the Philosophy of Science* 56: 749–790.

Fernández Valbuena, S. 1990. Predicción y Economía. In *Aspectos metodológicos de la investigación científica,* ed. W.J. Gonzalez, 2nd ed., 385–405. Madrid-Murcia: Ediciones Universidad Autónoma de Madrid and Publicaciones Universidad de Murcia.

Friedman, M. 1953. The methodology of positive economics. In *Essays in positive economics,* M. Friedman, 3–43. Chicago: The University of Chicago Press (6th reprint, 1969).

Gonzalez, W.J. 1990. Progreso científico, autonomía de la Ciencia y realismo. *Arbor* 135 (532): 91–109.

———. 1993. El realismo y sus variedades: El debate actual sobre las bases filosóficas de la Ciencia. In *Conocimiento, Ciencia y Realidad,* ed. A. Carreras, 11–58. Zaragoza: Seminario Interdisciplinar de la Universidad de Zaragoza-Ediciones Mira.

———. 1997. Progreso científico e innovación tecnológica: La 'Tecnociencia' y el problema de las relaciones entre Filosofía de la Ciencia y Filosofía de la Tecnología. *Arbor* 157 (620): 261–283.

---

[36] These thematic realms of the analysis of scientific prediction are grounded in the different components of science (language, structure, knowledge, processes, activity, ends, and values). See Gonzalez (2005).

————. 1999a. Racionalidad científica y actividad humana: Ciencia y valores en la Filosofía de Nicholas Rescher. In *Razón y valores en la Era científico-tecnológica*, N. Rescher, 11–44. Barcelona: Paidós.

————. 1999b. Ciencia y valores éticos: De la posibilidad de la Ética de la Ciencia al problema de la valoración ética de la Ciencia Básica. *Arbor* 162 (638): 139–171.

————. 2003. Racionalidad y Economía: De la racionalidad de la Economía como Ciencia a la racionalidad de los agentes económicos. In *Racionalidad, historicidad y predicción en Herbert A. Simon*, ed. W.J. Gonzalez, 65–96. A Coruña: Netbiblo.

————. 2005. The philosophical approach to science, technology and society. In *Science, technology and society: A philosophical perspective*, ed. W.J. Gonzalez, 3–49. A Coruña: Netbiblo.

————. 2006. Novelty and continuity in philosophy and methodology of science. In *Contemporary perspectives in philosophy and methodology of science*, ed. W.J. Gonzalez and J. Alcolea, 1–27. A Coruña: Netbiblo.

————. 2008a. El enfoque cognitivo en la Ciencia y el problema de la historicidad: Caracterización desde los conceptos. *Letras* 114 (79): 51–80.

————. 2008b. Economic values in the configuration of science. In *Epistemology and the social, Poznan studies in the philosophy of the sciences and the humanities*, ed. E. Agazzi, J. Echeverría, and A. Gómez, 85–112. Amsterdam: Rodopi.

————. 2010. *La predicción científica. Concepciones filosófico-metodológicas desde H. Reichenbach a N. Rescher*. Barcelona: Montesinos.

————. 2011a. Conceptual changes and scientific diversity: The role of historicity. In *Conceptual revolutions: From cognitive science to medicine*, ed. W.J. Gonzalez, 39–62. A Coruña: Netbiblo.

————., ed. 2011b. *Conceptual revolutions: From cognitive science to medicine*. A Coruña: Netbiblo.

————. 2012a. Las Ciencias de Diseño en cuanto Ciencias de la Complejidad: Análisis de la Economía, Documentación y Comunicación. In *Las Ciencias de la Complejidad: Vertiente dinámica de las Ciencias de Diseño y sobriedad de factores*, ed. W.J. Gonzalez, 7–30. A Coruña: Netbiblo.

————. 2012b. La vertiente dinámica de las Ciencias de la Complejidad. Repercusión de la historicidad para la predicción científica en las Ciencias Diseño. In *Las Ciencias de la Complejidad: Vertiente dinámica de las Ciencias de Diseño y sobriedad de factores*, ed. W.J. Gonzalez, 73–106. A Coruña: Netbiblo.

————. 2012c. Complejidad estructural en Ciencias de Diseño y su incidencia en la predicción científica: El papel de la sobriedad de factores (*parsimonious factors*). In *Las Ciencias de la Complejidad: Vertiente dinámica de las Ciencias de Diseño y sobriedad de factores*, ed. W.J. Gonzalez, 143–167. A Coruña: Netbiblo.

————. 2013a. Value ladenness and the value-free ideal in scientific research. In *Handbook of the philosophical foundations of business ethics*, ed. Ch. Lütge, 1503–1521. Dordrecht: Springer.

————. 2013b. The roles of scientific creativity and technological innovation in the context of complexity of science. In *Creativity, innovation, and complexity in science*, ed. W.J. Gonzalez, 11–40. A Coruña: Netbiblo.

————. 2013c. The sciences of design as sciences of complexity: The dynamic trait. In *New challenges to philosophy of science*, ed. H. Andersen, D. Dieks, W.J. Gonzalez, Th. Uebel, and G. Wheeler, 299–311. Dordrecht: Springer.

————. 2015. *Philosophico-methodological analysis of prediction and its role in economics*. Dordrecht: Springer.

————. 2016. Rethinking the limits of science: From the difficulties for the frontiers to the concern on the confines. In *The limits of science: An analysis from "barriers" to "confines"*, Poznan studies in the philosophy of the sciences and the humanities, ed. W.J. Gonzalez, 3–30. Leiden: Brill.

Grünbaum, A. 1962. Temporally-asymmetric principles, parity between explanation and prediction, and mechanism versus teleology. *Philosophy of Science* 29 (2): 146–170.

Guillán, A. 2016. The limits of future knowledge: An analysis of Nicholas Rescher's epistemological approach. In *The limits of science: An analysis from "barriers" to "confines"*, Poznan Studies in the Philosophy of the Sciences and the Humanities, ed. W.J. Gonzalez, 134–149. Leiden: Brill.

Hempel, C., and P. Oppenheim. 1948. Studies in the logic of explanation. *Philosophy of Science* 15: 135–175.

Jacquette, D., ed. 2009. *Reason, method, and value: A reader on the philosophy of Nicholas Rescher*. Frankfurt: Ontos Verlag.

Marsonet, M. 1996. *The primacy of practical reason. An essay on Nicholas Rescher's philosophy*. Lanham: University Press of America.

Mellor, D.H. 1975. The possibility of prediction. *Proceedings of the British Academy* 65: 207–223.

Moutafakis, N.J. 2007. *Rescher on rationality, values, and social responsibility. A philosophical portrait*. Heusenstamm: Ontos Verlag.

Niiniluoto, I. 1980. Scientific progress. *Synthese* 45: 427–462.

Niiniluoto, I., and R. Tuomela. 1973. *Theoretical concepts and hypothetico-inductive inference*. Dordrecht: Reidel.

Niiniluoto, I. 1993. The aim and structure of applied research. *Erkenntnis* 38 (1): 1–21.

———. 1997. Límites de la Tecnología. *Arbor* 157 (620): 391–410.

———. 2011. Scientific progress. In *The Stanford Encyclopedia of Philosophy*, Summer 2011 Edition, ed. E.N. Zalta. http://plato.stanford.edu/archives/sum2011/entries/scientific-progress/. Accessed 24 Nov 2011.

Popper, K.R. 1974a. Intellectual autobiography. In *The philosophy of Karl Popper*, ed. P.A. Schilpp, 3–181. La Salle: Open Court.

———. 1974b. Replies to my critics. In *The philosophy of Karl Popper*, ed. P.A. Schilpp, 961–1197. La Salle: Open Court.

Putnam, H. 1990. A defense of internal realism. In *Realism with a human face*, ed. H. Putnam, 30–42. Cambridge: Harvard University Press.

Radnitzky, G. 1978. The boundaries of science and technology. In *The Search for absolute values in a changing world. Proceedings of the VIth International Conference on the Unity of Sciences*, vol. II, 1007–1036. New York: International Cultural Foundation Press.

Rescher, N. 1958. On prediction and explanation. *British Journal for the Philosophy of Science* 8 (32): 281–290.

———. 1962. Some technical terms of Arabic logic. *Journal of the American Oriental Society* 82: 203–204.

———. 1963a. Al-Farabi on logical tradition. *The Journal of the History of Ideas* 24: 127–132.

———. 1963b. Avicenna on the logic of "conditional" propositions. *Notre Dame Journal of Formal Logic* 4: 48–58.

———. 1967. The future as an object of research. *RAND Corporation Research Paper*, P-3593.

———. 1970. *Scientific explanation*. New York: The Free Press.

———. 1973. *Conceptual idealism*. Oxford: Blackwell.

———. 1977. *Methodological pragmatism. A systems-theoretic approach to the theory of knowledge*. Oxford: Blackwell.

———. 1978a. *Scientific progress. A philosophical essay on the economics of the natural science*. Oxford: Blackwell.

———. 1978b. *Peirce's philosophy of science*. Notre Dame: University of Notre Dame Press.

———. 1979. Some issues regarding the completeness of science and the limits of scientific knowledge. In *The structure and development of science*, ed. G. Radnitzky and G. Andersson, 19–40. Dordrecht: Reidel.

———. 1980a. *Skepticism*. Oxford: Blackwell.

———. 1980b. *Induction*. Oxford: Blackwell.

———. 1983a. *Mid-journey: An unfinished autobiography*. Lanham: University Press of America.

———. 1983b. The unpredictability of future science. In *Physics, philosophy and psychoanalysis*, ed. R.S. Cohen et al., 153–168. Dordrecht: Reidel.

———. 1986. *Ongoing journey: An autobiographical essay.* Lanham: University Press of America.

———. 1987a. *Scientific realism.* Dordrecht: Reidel.

———. 1987b. Maximization, optimization, and rationality. On reasons why rationality is not necessarily a matter of maximization. In *Ethical idealism. An inquiry into the nature and function of ideals,* ed. N. Rescher, 55–84. Berkeley/Los Angeles: University of California Press.

———. 1988. *Rationality. A philosophical inquiry into the nature and the rationale of reason.* Oxford: Clarendon Press.

———. 1989. *Cognitive economy. The economic dimension of the theory of knowledge.* Pittsburgh: University of Pittsburgh Press.

———. 1990. *A useful inheritance. Evolutionary aspects of the theory of knowledge.* Savage: Rowman and Littlefield.

———. 1992a. *A system of pragmatic idealism. Vol. I: Human knowledge in idealistic perspective.* Princeton: Princeton University Press.

———. 1992b. The imperfectibility of science. In *A system of pragmatic idealism. Vol. I: Human knowledge in idealistic perspective,* N. Rescher, 77–95. Princeton: Princeton University Press.

———. 1992c. Our science as *our* science. In *A system of pragmatic idealism. Vol. I: Human knowledge in idealistic perspective,* N. Rescher, 110–125. Princeton: Princeton University Press.

———. 1992d. Cognitive limits. In *A system of pragmatic idealism. Vol. I: Human knowledge in idealistic perspective,* N. Rescher, 243–254. Princeton: Princeton University Press.

———. 1992e. Metaphysical realism. In *A system of pragmatic idealism. Vol. I: Human knowledge in idealistic perspective,* N. Rescher, 255–274. Princeton: Princeton University Press.

———. 1992f. Scientific realism. In *A system of pragmatic idealism. Vol. I: Human knowledge in idealistic perspective,* N. Rescher, 275–295. Princeton: Princeton University Press.

———. 1992g. Science and idealization. In *A system of pragmatic idealism. Vol. I: Human knowledge in idealistic perspective,* N. Rescher, 296–303. Princeton: Princeton University Press.

———. 1992h. Realism and idealism. In *A system of pragmatic idealism. Vol. I: Human knowledge in idealistic perspective,* N. Rescher, 304–327. Princeton: Princeton University Press.

———. 1993a. *A system of pragmatic idealism. Vol. II: The validity of values: Human values in pragmatic perspective.* Princeton: Princeton University Press.

———. 1993b. How wide is the gap between facts and values. In *A system of pragmatic idealism. Vol II: The validity of values,* N. Rescher, 65–92. Princeton: Princeton University Press.

———. 1993c. Values in the face of natural science. In *A system of pragmatic idealism. Vol II: The validity of values,* N. Rescher, 93–110. Princeton: Princeton University Press.

———. 1993d. Nuestra Ciencia en tanto que *nuestra. Daimon, Revista de Filosofía* 6: 1–9.

———. 1994. *A System of pragmatic idealism. Vol. III: Metaphilosophical inquires.* Princeton: Princeton University Press.

———. 1996a. *Instructive journey: An autobiographical essay.* Lanham: University Press of America.

———. 1996b. *Process metaphysics. An introduction to process philosophy.* Albany: State University of New York Press.

———. 1996c. *Priceless knowledge? Natural science in economic perspective.* New York: Rowman and Littlefield.

———. 1998a. *Predicting the future. An introduction to the theory of forecasting.* New York: State University of New York Press.

———. 1998b. *Communicative pragmatism and other philosophical essays on language.* Lanham: Rowman and Littlefield.

———. 1999a. *Razón y valores en la Era científico-tecnológica.* Barcelona: Paidós.

———. 1999b. *The limits of science,* revised edition. Pittsburgh: University of Pittsburgh Press.

———. 2000. *Realistic pragmatism.* Albany: State University of New York Press.

———. 2001. On learned ignorance and the limits of knowledge. In *Cognitive pragmatism. The theory of knowledge in cognitive perspective,* N. Rescher: 63–80. Pittsburgh: University of Pittsburgh Press.

————. 2002. *Enlightening journey. The autobiography of an American scholar.* Lanham: Lexington Books.

————. 2003. Skepticism and Its deficits. In *Epistemology. An introduction to the theory of knowledge*, N. Rescher: 37–70. Albany: State University of New York Press.

————. 2004. Pragmatism and practical rationality. *Contemporary Pragmatism* 1 (1): 43–60.

————. 2006a. *Epistemetrics.* New York: Cambridge University Press.

————. 2006b. Pragmatic idealism and metaphysical realism. In *A companion to pragmatism*, ed. J.R. Shook and J. Margolis, 386–397. Oxford: Blackwell.

————. 2006c. The Berlin School of logical empiricism and its legacy. *Erkenntnis* 64: 281–304.

————. 2009. *Unknowability. An inquiry into the limits of knowledge.* Lanham: Lexington Books.

————. 2010. *Autobiography.* Heusenstamm: Ontos Verlag.

————. 2012. The problem of future knowledge. *Mind and Society* 11 (2): 149–163.

Rescher, N., and O. Helmer. 1959. On the epistemology of the inexact sciences. *Management Sciences* 6: 25–52.

Rowe, G., and G. Wright. 2001. Expert opinions in forecasting: The role of the Delphi technique. In *Principles of forecasting: A handbook for researchers and practitioners*, ed. J.S. Armstrong, 125–144. Boston: Kluwer.

Salmon, W. C. 1981. Rational prediction. *British Journal for the Philosophy of Science* 32: 115–125. (Reprinted in 1988: *The Limitations of deductivism*, ed. A. Grünbaum and W.C. Salmon, 47–60. Berkeley: University of California Press.)

Salmon, W.C. 1993. On the alleged temporal anisotropy of explanation. A letter to Professor Adolf Grünbaum from his friend and colleague. In *Philosophical problems of the internal and external worlds. Essays on the philosophy of Adolf Grünbaum*, ed. J. Earman, A. Janis, G. Massey, and N. Rescher, 229–248. Pittsburgh: University of Pittsburgh Press.

Scheffler, I. 1957. Explanation, prediction, and abstraction. *British Journal for the Philosophy of Science* 7: 293–309.

Simon, H.A. 1982a. *Models of bounded rationality. Vol. 1: Economic analysis and public policy.* Cambridge, MA: The MIT Press.

————. 1982b. *Models of bounded rationality. Vol. 2: Behavioral economics and business organization.* Cambridge, MA: The MIT Press.

————. 1996. *The sciences of the artificial*, 3rd ed. Cambridge, MA: The MIT Press (1st ed., 1969; 2nd ed., 1981).

————. 1997. *Models of bounded rationality. Vol. 3: Empirically grounded economic reason.* Cambridge, MA: The MIT Press.

————. 2001. Science seeks parsimony, not simplicity: Searching for pattern in phenomena. In *Simplicity, inference and modeling. Keeping it sophisticatedly simple*, ed. A. Zellner, H.A. Keuzenkamp, and M. McAleer, 32–72. Cambridge: Cambridge University Press.

Thagard, P. 1992. *Conceptual revolutions.* Princeton: Princeton University Press.

Toulmin, S. 1961. *Foresight and understanding.* Bloomington: Indiana University Press.

Wible, J.R. 2008. How is scientific knowledge economically possible?: Nicholas Rescher's contributions to an economic understanding of science. In *Rescher studies. A collection of essays on the philosophical work of Nicholas Rescher*, ed. R. Almeder, 445–476. Heusenstamm: Ontos Verlag.

Chapter 2
Analysis of Scientific Prediction from Language

Abstract Within the system of pragmatic idealism, the viewpoint of language is used for the analysis of prediction, in general, and scientific prediction, in particular. This chapter initially offers the general coordinates of a pragmatic idealist approach to language. Thereafter, it seeks to shed light on the semantic features of scientific prediction. First, Rescher's characterization of prediction is analyzed. This leads to us to consider the distinction between "scientific" and "non-scientific" prediction, as well as the differences between "prediction" and "retrodiction." Secondly, his contribution to the analysis of scientific prediction as a statement is made explicit. In this regard, the distinction between "qualitative prediction" and "quantitative prediction" is relevant, but we also need the distinction between "prediction," "foresight," "forecasting," and "planning." Finally, the limits of language are seen with regard to prediction, where the distinction between "non predictability" and "unpredictability" is highlighted.

Keywords Pragmatic idealism • Semantics • Scientific prediction • Retrodiction • Qualitative prediction • Quantitative prediction • Non predictability • Unpredictability

When Rescher analyzes prediction from language, his starting point is a pragmatic conception that gives primacy to the use of language regarding meaning (Rescher 1998b). This trait affects prediction, in general, and scientific prediction, in particular. In the first place, he is interested in the process of communication and, therefore, in prediction as a communicative content. He stresses the features that make the exchange and understanding of informative messages possible. Only in the second place are there other elements related to meaning, such as reference.[1] In his view, scientific prediction is considered within a pragmatic approach to meaning. It is the result of an activity that seeks to obtain justified answers to meaningful questions about future occurrences (Rescher 1998a, pp. 37–39).

---

[1] On the differences between the semantic and pragmatic approaches to reference, see Gonzalez (1986).

© Springer International Publishing AG 2017

A. Guillán, *Pragmatic Idealism and Scientific Prediction*, European Studies in Philosophy of Science 8, DOI 10.1007/978-3-319-63043-4_2

Following this framework of a pragmatic viewpoint, this chapter seeks to offer an analysis of Rescher's proposal on the features of scientific prediction related to language. In order to do this, several steps are followed: (1) the general coordinates of his approach to scientific language—where pragmatics has primacy over semantics—are considered. (2) The features that he assigns to scientific prediction are addressed within the option that considers meaning as use. (3) His characterization of prediction is analyzed, which leads us to consider the distinction between "scientific" and "non-scientific" prediction, as well as the differences between "prediction" and "retrodiction." (4) Rescher's contribution to the analysis of scientific prediction as a statement is also analyzed. In the first place, the distinction between qualitative prediction and quantitative prediction is discussed. This is followed by a consideration of the distinction between "prediction," "foresight," "forecasting," and "planning". The distinction is established on the basis of the degree of control of the relevant variables. (5) Finally, the limits of language are seen with regard to prediction, where the distinction between "non predictability" and "unpredictability" is highlighted.

2.1 Characteristic Features of Meaning in Rescher

Concerning meaning—both human and scientific—Rescher adopts a pragmatic perspective. So he suggests a view of meaning as use and an approach to language as communication. *Communicative Pragmatism* (1998b)—the title of his main work devoted to language—is indicative of his proposal on meaning. His view of language is in line with his approach to knowledge. Thus, as a pragmatist philosopher, he highlights the realm of *human activity*; and, as a Kantian author, he insists that *human knowledge* is modulated by our mental categories and concepts. On the one hand, the emphasis on the use prevails in his view of meaning; and, on the other, the role of the ideas—which he considers as decisive in the characterization of reality—is highlighted regarding knowledge.

2.1.1 Communication as Activity and Meaning with Cognitive Content

This emphasis on communication as human activity involves an instrumental account of language, since language appears as an "instrument" for human communication, instead of being mainly a way to represent reality.[2] Thus, for Rescher,

---

[2] These two different usages of language appear in various authors, among them Michael Dummett. Within his large intellectual production on philosophy of language, it could be highlighted Dummett (1981). G. Frege and L. Wittgenstein, including his latter conception, reject a psychologist approach to language. Thus, concepts cannot be reduced to pure mental acts or mere psycho-

communication is the process that allows us to *share* what we know about the world through the use of language: "our knowledge regarding the world's things is always developed within a linguistic-systematic system of reference" (Rescher 1999a, p. 54).

However, this pragmatic account of Rescher's idealism admits some elements of realism.[3] In his approach to meaning, the notion of "objectivity" has an important role. In effect, this notion of objectivity—which influences the realist view of "fact"—appears as a necessary condition for communication: "Human cognition as we understand it would be impossible without communal inquiry into and interpersonal communication about an objective order of reality. And without a presupposition of ontological objectivity the very idea of investigating a shared world would become inoperable" (Rescher 1998e, p. 94). This is important because M. Dummett maintains that objectivity is the key to semantic realism.

In his paper "Pragmatic Idealism and Metaphysical Realism," Rescher insists on this idea of ontological objectivity as a support for human communication. He maintains in his paper that the existence of an objective reality that is independent of the knowing subject is "a postulate whose justification pivots—in the first instance—on its functional utility in enabling us to operate as we do with respect to inquiry and deliberation" (2006, p. 386).

The opening of his thought on realist contributions can be seen in his approach to meaning, where the notion of objectivity is crucial as content of communication with an ontological basis. He considers that the acceptance of an objective reality is "*presupposed* from the outset rather than being seen as a matter of ex post facto discovery about the nature of things" (1992b, p. 257). He also maintains that it is a necessary condition for communication.

For Rescher, commitment to objectivity is an instrument that makes it possible to carry through any cognitive venture, since meaning itself involves a cognitive content. In this way, the concept of "objective reality" is justified on *functional basis*, instead of inferential basis: "We require this postulate to operate our conceptual scheme, and its validation accordingly lies in its utility. We could not form our existing conceptions of truth, fact, inquiry, and communication without presupposing the independent reality of an external world" (Rescher 1998c, p. 36).

Therefore, the pragmatic approach to meaning does not lead Rescher to see communication in relativistic terms. He points out that language has cognitive content that might be objective. For him, our access to the extramental reality is modulated by our categories and concepts. Although human knowledge is always fallible, it is possible to achieve objective knowledge. In this way, true statements are those that describe reality as it is (Rescher 1992a, pp. 243–244). As Wenceslao J. Gonzalez points out (2010, p. 256), Rescher accepts P. F. Strawson idea that "facts are what

---

logical thoughts. Cf. Dummett (1985). Rescher shares with them the viewpoint of language conceived from a non-psychologist perspective. This explains why he is pragmatic on language and, at the same time, he accepts elements of the semantic view.

[3] On realism and meaning two papers of M. Dummett can be highlighted. Cf. Dummett (1963, 1982). However, Dummett's personal philosophy endorses a semantic anti-realism.

statements (when true) state." (Strawson 1950, p. 136). Thus, he distinguishes the notions of "truth" and "fact." Because "truth" is, for Rescher, a linguistic notion: "it is the representation of a fact through its statement in some real language. Any correct statement in some real language formulates a truth" (1999a, p. 54). Meanwhile, a "fact" is a real circumstance that exists in an objective way and, therefore, it exceeds the limits of language and knowledge.

Nonetheless, we can distinguish three aspects: (a) truth in language as such (truthfulness); (b) truth in knowledge, which is expressed through the adequacy of statements and transmits objectivity regarding reality; and (c) truth in the real things, which is made explicit in terms of authenticity. The first is, strictly, the "semantic" truth, in the sense of the language as the expression of an actual content. The second is a cognitive (or epistemic) truth, since it transmits an agreement between the statement and the described fact. The third is an ontological truth, where the real thing itself is what is true. Usually Rescher deals with the cognitive and ontological analysis of truth.

Even when Rescher admits truth in science, the process of information acquisition involves cooperation and communication. Thus, it is "a process of *conceptual* innovation that always places certain facts completely outside the cognitive range of the researches in any concrete period" (Rescher 1999a, p. 45). For this reason, he insists that reality as such cannot be equated with the things we know and can express through language. Reality exceeds the descriptive resources of language and those resources are in debt to our cognitive mechanisms for conceptualization. Reality is potentially emergent to language, but it should be first integrated in our conceptual scheme.[4] In this regard, it can be considered that ontological truth is broader than epistemic truth.

But it happens that meaning, knowledge, and reality are actually interrelated. In fact, Rescher considers that human beings develop their knowledge of the world within language. In this way, language reflects a conceptual system, at the same time that it is limited by that very system. It is an imperfect resource, but language can carry an *objective content*. Thus, in his account, meaning is not reduced to a mere intersubjective use of language, because he admits objective bases in it.

---

[4] In this regard, Rescher writes that "blood circulated in the human body well before Harvey; substances containing uranium were radioactive before Becquerel. The emergence at issue relates to our cognitive mechanisms of conceptualization, not to the objects of our consideration in and of themselves. Real-world objects must be conceived of as antecedent to any cognitive interaction—as being there right along, 'pregiven' as Edmund Husserl put it. Any cognitive changes or innovations are to be conceptualized as something that occurs on our side of the cognitive transaction, not on the side of the objects with which we deal" (1992a, p. 247).

2.1.2 The Primacy of the Pragmatic Dimension

Rescher gives primacy to the pragmatic dimension, since he thinks that language is mainly an instrument that makes communication possible. On this basis, he is fundamentally interested in two matters: (i) what are the conditions that provide an effective communication, and (ii) what are the conditions that allow us to carry through effective communication in an optimal way (Rescher 1998c, pp. 3–4). Within this framework, he seeks to clarify the normative issues that regulate communication; that is, he investigates the general principles that make the communicative practice possible (and also efficient).

Although his account of meaning involves a cognitive content and he also admits objectivity with ontological basis, Rescher's approach to meaning is not, properly speaking, a semantic approach. He acknowledges the distinction between semantics and pragmatics. This distinction leads him to admit that there are differences between the *use conditions* and the *truth conditions* of a statement.[5] Use conditions encompass a series of operational criteria that allow us to express properly a statement in a specific language. Truth conditions are those objective circumstances that make it possible to claim that a statement is true. Therefore, use conditions are oriented towards the users of a language, while truth conditions are oriented towards the reality of what is expressed.

According to Rescher, "for while truth conditions deal with the objective facts, use conditions deal with the linguistic properties" (1998d, p. 62). In this regard, he sees use conditions as more important to communication than truth conditions; although he admits that both of them should be taken into account in any approach to language. For him, "meaning is a comprehensive concept that embraces both semantical and pragmatic issues. (…) Any exclusivistic doctrine along the lines of meaning is use, or meaning is a matter of truth conditions, is one-sided, dogmatic, *and* inappropriate in its claim to exclusiveness" (1998d, p. 67).

Rescher insists that both truth conditions and use conditions are required. Because "the fact that both are inextricably interrelated in matters of meaning—that meaning analysis has a formal (semantic) and an informal (pragmatic) dimension that are inseparably interrelated—means that there is a symbiotic interconnection here that permits neither side to claim unconditional priority over the other" (1998d, p. 74). Therefore, he acknowledges the relevance of both types of conditions in the analysis of meaning. Nevertheless, his account is mostly focused on the *use conditions*.[6] Because, even when he expressly claims that both are equally important, he considers that an approach to meaning from use conditions has advantages over an analysis focused on truth conditions (Rescher 2008).

---

[5] On the differences between use conditions and truth conditions, as well as on the primacy that Rescher gives to the former over the later, see Rescher (1998d).

[6] On the primacy of the pragmatic approach in his philosophical proposal on language, see Rescher (1998c).

When Rescher notes that truth conditions of a statement are about "objective facts" (1998d, p. 62), it can be claimed that he is in tune with semantic realism. A conception of truth as "agreement" with reality underlies this issue.[7] Thus, the truth of a statement depends on the agreement of its content with the objective facts. To preserve this notion of truth implies, therefore, presupposing the existence of a reality that is independent of the knowing subject. In the same way, intersubjective communication and research as a community task are only possible if we all can access the same objective reality.

However, if we accept that truth conditions are about "objective facts," this implies, for Rescher, that the concept of "truth" is not applicable in certain contexts and, thus, it remains in the background with regard to meaning (Rescher 1998d, p. 62). This happens when the meaning of a question, an order or a counterfactual conditional is analyzed, insofar as they are linguistic forms that do not refer to an objective reality (1998d, pp. 72–74).[8] Here, the issues of "correctness" or "appropriateness"—that are oriented towards the practice of using the language—supersede the notion of truth in the analysis of meaning (1998d, p. 72).

On the basis of the advantages that Rescher sees in pragmatics over semantics, he offers a conception of meaning as use and an account of language as an instrument that facilitates communication. He is especially interested in the communicative use of language, so he focuses on the principles that regulate communication. Thus, his pragmatic perspective has an economic inspiration. He understands communication as a process that follows an economic rationality in terms of costs and benefits. A sender and a receiver intervene in this process, which should be ruled by economic values such as effectiveness and efficiency.[9]

But language is a means to transmit some content, so language is an instrument that makes the transmission of information possible. This concerns both ordinary language ("a general-purpose instrument") and scientific language ("a specialized [instrument].") [Rescher 1998c, p. 9]. The use of language makes the transmission of information possible, through the communication performed by a sender and a receiver. Economic principles are important in this process because "effective communication is throughout a matter of maintaining proper cost-benefit coordination" (Rescher 1998c, p. 7).

There is a close connection in Rescher between his account of language and his approach to rationality. This nexus is rooted in his view of language as being linked to communication and his account of knowledge as a human need. He sees rationality as "a means to adaptive efficiency, enabling us—sometimes at least—to adjust

---

[7] He maintains that one of the reasons to accept the assumption that there is an objective and mind-independent reality is "to preserve the distinction between true and false with respect to factual matters and to operate the idea of truth as agreement with reality," Rescher (2006, p. 390).

[8] When he analyses prediction from language, his approach is also pragmatic, since predictive statements do not usually refer to an objective existing reality, but they has to do with the possible future.

[9] On the economic features of communication, cf. Rescher (1989; especially, chapter 2, pp. 47–68).

our environment to our needs and wants rather than the reverse" (Rescher 1988, p. 2).

On the one hand, language makes human communication possible; and, on the other hand, that human beings meet their need of obtaining information is something that depends on effective communication. Thus, he considers that "given our need for information to orient us in the world (on both pure and practical grounds), the value of creating a community of communicators is enormous. We are rationally well advised to extend ourselves to keep the channels of communications to our fellows open, and it is well worth expending much for the realization of this end" (1989, p. 53).

Thus, Rescher sees communication as a rational process that is oriented towards an aim. From the point of view of the sender, the aim is to transmit information to the receiver. For the receiver, the aim is to obtain information from the sender (Rescher 1998c, p. 15). In the case of science, communication is especially important, since the production of scientific knowledge is a community process (1998c, p. 14). Both the sender and the receiver are interested in exchanging information in an effective and efficient way, since they both obtain benefits from this process.

In economic terms, sharing information is the rational option: "It is far easier, cheaper, and more convenient for people to get information by sharing than by themselves having to undertake the often laborious inquiries and researches needed to develop it *de novo*" (Rescher 1989, pp. 47–48). For him, communication is a human activity whose aims, processes, and results should be evaluated in economic terms (i.e., criteria based on economic values).

In his judgment, communication is not costless, since it involves costs in terms of time and effort. To carry out the practice of using language in an effective and efficient way is a question that depends of the acceptance of a series of assumptions, which allow us to minimize the costs that are inherent to the communication process. These assumptions are independent of the special features of the discourse. They are inserted in the general context of communication: "They are forthcoming not from the specific content of the message at issue but from the contextually indicated presuppositions we make on our own responsibility" (Rescher 1998c, p. 6). Therefore, there are general principles in communication of normative character such as credibility, reliance, clarity, and contextualization (1998c, pp. 7–8).[10]

Nevertheless, since language is an "imperfect resource" (1998c, p. 8), it is not always possible to express oneself in a clear and explicit manner. Furthermore, a statement can be susceptible to several interpretations, at least in some cases. Hence, a proper interpretation of a statement involves knowing the communication context. For Rescher, "it is fair to say that *in interpretation context is not just important, it is everything*" (1998c, p. 9). Thus, he notices that a text transmits an informative message in two different ways: (1) the substance of what it says; that is, the information it conveys *directly* through its explicit meaning; and (2) the message it conveys *obliquely* by saying what it says in a particular way (1998c, p. 9).

---

[10] On reliance and cooperation as principles that make it possible to minimize the costs of research and communication see Rescher (1989, pp. 33–46).

When a statement is taken with independence of the context, it can admit different interpretations. Besides the aim sought by communication, two aspects intervene in the processes of communication: on the one hand, the explicit content of what is stated; and, on the other, the context in which that content is sent. In this way, the message obtained is the result of the interpretation of the content in relation to the context of discourse. Consequently, the receiver must carry out a process of interpretation in order to select one of the many alternative constructions that a statement or a set of statements can admit. For this reason, the suitable transmission of information depends on the correct interpretation of the content in relation to the context.

Besides the context, Rescher highlights the role of reliance and credibility in communication (1989, pp. 33–46). The sender must strive to have credibility and the receiver must trust in the sender. A high cost comes from a systematic critical position on the statements of other people. To proceed always (not only in the case that we have good reasons) under the assumption that we cannot trust the sender has a high cost; because a completely skeptical attitude would deprive us of any possibility of obtaining information.

In accordance with his pragmatism with economic components, Rescher thinks that rational behavior leads us to obtain information in an efficient and effective way. To achieve this goal, credibility is highly important: "We adopt an epistemic policy of credence in the first instance because it is the most promising avenue toward our goals, and then persist in it because we subsequently find, not that it is unfailingly successful, but that it is highly cost-effective" (Rescher 1998c, p. 16).

Considered this issue from an economic viewpoint, an effective activity of communication requires conventions such as: (a) the sender expresses what he or she understands is the truth (truthfulness in language); and (b) the sender expresses himself or herself in an accurate and non-misleading way (which involves truth as agreement) [1998c, p. 8].[11] These conventions are justified on economic grounds, since they are "practices that represent the most efficient and economical way to accomplish our communicative work" (1998c, p. 8). The issue is to obtain the highest benefit from the information transmitted, so minimizing the inherent costs to the process of information acquisition and transmission. It is possible to think that there is profitability in stressing the truth in language.

Consequently, the pragmatic dimension has primacy in Rescher's account of language. He deals with language as an "instrument" for communication and with those conditions that make it possible to exchange messages with informative content in the communicative practice. The most important thing is, then, to achieve effectiveness and efficiency in the process. Although he acknowledges that language is an imperfect resource, he also admits that it can carry an *objective content*. In this way, even when he gives primacy to the pragmatic dimension, meaning is not reduced to a mere intersubjective use of language because he admits objective bases in its content.

---

[11] This is especially important in the case of scientific prediction, because it is not possible to determine *now* if what the predictive statement says is true or not.

2.2 Scientific Prediction in a Theory of Meaning

From the point of view of language, scientific prediction is about a possible future and "its sense—the content expressed—and the referent towards which it is oriented belong to the realm of what is expected" (Gonzalez 2010, p. 284).[12] When Rescher considers the referent of a prediction, he thinks that it does not agree with an available reality—something that has already happened or that is happening now—but with a *possible future*. For this reason, in his approach, truth conditions of a predictive statement are in the background. Even more, when he analyses scientific prediction from the viewpoint of language, his account is principally "pragmatic." It is not, strictly speaking, a "semantic" account.

In addition, Rescher places prediction in an active context (Gonzalez 2010, p. 260). Predicting is, in his judgment, an activity whose aim is to achieve meaningful claims with regard to future occurrences. In order to predict, we have to endeavor "to provide warranted answers to detailed substantive questions about the world's future developments" (Rescher 1998a, pp. 37–38). So, faced with a question about a future occurrence, prediction seeks to offer an answer on the basis of the available knowledge. Additionally, he gives prediction an instrumental component: "prediction, in sum, is our instrument for resolving our meaningful questions about the future, or at least of *endeavoring* to resolve them in a rationally cogent manner" (1998a, p. 39).

2.2.1 Context of Use

As happens in the case of language in the general level, the language of prediction cannot be analyzed without taking into account the context of use. As an intellectual activity, prediction is carried out in a communicative context—in every kind of language—and a research context (in scientific language). In this way, to a large extent, the *value* that Rescher confers to prediction is due to its practical utility, because he considers that, in the realm of daily life, obtaining information about the future events is a human need.

We have meaningful questions about future developments, and we need answers to those questions. This is not, for Rescher, just a matter of curiosity, but a matter of survival. Every human action needs to some extent information about the future, and practical reasoning is basic in this regard. He considers that "to act, to plan, to survive, we must anticipate the future, and the past is the only guide to it that we have" (1998a, p. 65).[13] But this is not entirely accurate, because there is human

---

[12] What is expected can be in an ontological, epistemological, or heuristic sense.

[13] From this perspective, an inductive inference is required, which allows us to obtain statements about the future on the basis of past experience. See Chap. 3, Sects. 3.2 and 3.3.

creativity (and history shows that the future can be different from what was thought of in a historical moment).

He also values prediction according to its *utility* in the scientific realm. He thinks that scientific prediction can be used mainly in two directions "as a test of the acceptability of theories and as a guide to discovery" (1998a, p. 160). In the first case, the referent of the prediction is usually something that does not happen yet, so prediction deals with an ontological novelty (for example, in the prediction of an eclipse or the climate change). Meanwhile, in the second case, the novelty is epistemological, so the prediction allows us to discover a reality that has not been observed yet.

From his pragmatic viewpoint, to predict is an activity oriented towards an aim— "to provide warranted answers to detailed substantive questions about the world's future developments" (Rescher 1998a, pp. 37–38) and that aim is basically justified with regard to its utility. For this reason, Rescher considers that predictive knowledge is itself valuable. However, he highlights that "the fact that virtually all *action* is in some way future oriented endows our predictive knowledge with special practical potency" (Rescher 1998a, p. 12).

Prediction is an aim of science, but it is one aim *among others*, since the structure of ends of scientific activity is also oriented towards description, explanation, and control over nature (Rescher 1999a, p. 138). But prediction is an especially important aim, because it can be used as a guide for prescription in applied science; and its role as a test in basic science allows us to evaluate the comparative theoretical adequacy of scientific theories (Rescher 1998a, p. 161). In this way, the meaning of the prediction is seen from the perspective of the use. Therefore, it is possible to claim that scientific prediction is one of the realms where—in his judgment—an analysis of meaning focused on use conditions has advantages over an analysis centered on truth conditions.

2.2.2   Statement About Novel Facts

As a statement about the future or claim about novel facts, a predictive statement can be true—if what prediction claims happens in the future—but we cannot say that a prediction is true before the predicted phenomenon or development does happen. For this reason, Rescher maintains that "correctness" is more important than "truth" to prediction (1998a, p. 70). It is said that a prediction is "correct" when, on the basis of the available information, it is possible to claim that it adapts well to what we know about how the future facts could be. But we have to wait for those facts happen in order to assess if the prediction is actually true.

For Rescher, what makes that a statement about the future has, in fact, predictive character is not something linked to its sense and reference, but something that has to do with the nexus between language and action. So he accepts that there can be meaning without referent. In that case, scientific prediction can be meaningful even when we cannot confirm now if its sense is related with a real referent. Then, the reaction of receiver of the prediction is more important that prediction itself (its sense), insofar as the receiver attributes to the statement the condition of predictive assertion and he has to decide on its correctness (1998a, pp. 38–39).

Correctness and credibility must go together—in Rescher's judgment—in a successful prediction. To be *credible*, a prediction must have a plausible grounding. Instead of *correctness*, "it is credibility that is the cardinal predictive virtue" (1998a, p. 122), because it can be determined at the present time. Credibility is based on evidence and probability that support the predictive statement, which are the rational support for the prediction. In this way, practical utility of a predictive statement rests on its credibility, since only those predictions that are credible will be used as test for theories and as a guide for action.

Although it is commonly easier to achieve a successful prediction when it is not very informative (i.e., when it is general or without many details), science seeks informative definiteness. Scientific language seeks accuracy and precision, but—in Rescher's judgment—this involves taking risks. This is because, in principle, the more informative a prediction is—that is, the more precise, detailed, etc.—the less secure it is. Security is determined on the basis of its probability or its degree of acceptability (Rescher 1998c, pp. 19–24).

Generic predictive statements are commonly the most accepted ones, since they are, in principle, more credible with regard to their eventual correctness. However, credibility cannot be obtained by diminishing informativeness, which is an "indispensable criterion for a good prediction" (1998a, p. 120). This means that predictions should seek to be both epistemically secure and informative. To achieve an optimal equilibrium between informativeness and credibility is, therefore, one of the main aims of prediction; but it is also one of the main difficulties it must tackle.

There are, according to Rescher, two ways to establish the credibility of a prediction: the evidential and the authoritative (1998a, p. 123). These two options are related with the procedures and methods that are used to predict. He divides the predictive processes into two groups: judgmental or estimative procedures—where prediction is rooted in the personal estimation of the experts—and the formal or discursive methods that follow processes that are explicitly detailed (1998a, pp. 85–112).[14] On this basis, a prediction is credible if one of these possibilities is available: (i) it is considered that it has an evidential basis that supports the statement, or (ii) it is thought that the predictor is a reliable source.[15]

Every predictive statement can be seen as content or as a result. It is revisable, but it should be evaluated in terms of objectivity and truth (or, at least, correctness). This is possible if the scientific prediction is, in effect, a rational prediction, so we can ascribe to it values such as objectivity and truth. Therefore, even when he adopts a clearly pragmatic approach, it is important to highlight that Rescher's account does not reduce prediction to the mere use of language. He accepts, in effect, the

---

[14] On this distinction, see Chap. 6.

[15] From this perspective, it seems that Rescher considers that the methodological dimension is more relevant than the language when the demarcation between scientific and non-scientific prediction is at stake.

objectivity of the knowledge about the future, which could be true. This is because scientific prediction is a statement that we obtain as a result of a rational process supported by evidence (either theoretical or empirical). It deals with "novel facts" in some relevant sense (ontological, epistemological or heuristic).

2.3   The Language of "Prediction"

When "prediction," in general, is considered, it is understood that prediction encompasses a series of features. Thus, to offer a characterization of the concept of "prediction," Rescher suggests four main features: (a) it is future-oriented; (b) it is correct or incorrect; (c) it is meaningful; and d) it is informative (1998a, pp. 54–55). In my judgment, it is possible to enlarge this characterization if we stress that prediction has to do with *something expected*. A predictive statement implies that something is expected in the future. This linkage with something expected entails *novelty*, so prediction is connected with the notion of "novel facts," since prediction is about not observed or now unobservable things.

Besides the general characteristics of "prediction," the demarcation between "scientific" and "non-scientific prediction" should be further developed. In this regard, the role of language is considered in order to distinguish "prediction," in general, from "scientific prediction," in particular. Moreover, a characterization of prediction should take into account the concept of "retrodiction" as well. Regarding this issue, although Rescher's main interest is not the scientific language, his view of scientific prediction as a statement that is oriented towards the future should be emphasized. Thus, from a pragmatic perspective—the use of language—he rejects that we can have a genuine "retrodiction" or "prediction of past:" prediction involves the cognitive anticipation of a possible future and retrodiction is, in principle, oriented towards the past.[16]

2.3.1   The Concept of "Prediction"

When Rescher suggests the concept of "prediction," he places it within a framework with regard to language where the pragmatic dimension has primacy. As a statement,[17] scientific prediction is a content oriented towards the future, and it can be correct or incorrect; because it involves a meaning with an informative content (Rescher 1998a, pp. 54–55). As a statement oriented towards the future, prediction

---

[16] However, according to A. Grünbaum (1962), Hempel defends the possibility that "retrodiction" and "prediction" can be equivalent. See Chap. 3, Sect. 3.5.2.

[17] This involves a difference between prediction and scientific explanation, which can be understood as an argument. Cf. Gonzalez (2010, p. 260).

should be supported by rational bases. It is the result of an inference made from the data available regarding the facts of the past and the present (1998a, p. 86). It is not possible, in Rescher's judgment, to predict without reasons, because to predict— either scientifically or on the basis of the everyday experience—is, *eo ipso*, a rational activity.

It happens that, even when scientific prediction is oriented towards a potential future—the first feature pointed out—its content can be objective, since it is the result of a rational process. In my judgment, the acknowledgement that scientific prediction is supported by reasons—theoretical or empirical bases that justify an anticipation of the possible future—is especially important. Because both the use of prediction as a test for the validity of the theories and its use as a guide for prescription in applied sciences can only be justified if prediction has, in effect, rational bases. In addition, these rational bases could be corrigible and this implies that it is possible to obtain more or better information about the future in order to predict.

Besides the orientation towards the future, another feature that Rescher attributes to scientific prediction is that it can be correct or incorrect. These are conditions of the use of language, so truth conditions remain in the background. In his judgment, the meaningful character of a predictive statement has to do with the *possibility* that it turns out to be true, instead of being related with its actual truth. In that case, what makes the prediction meaningful is the possibility to prove, in the future, that it is true or false. Prediction establishes that something will happen instead of something else, and it makes this on rational bases that demarcate the predictive statement from the simple prophecy. In this way, successful prediction "is a matter of conjoining correctness and credibility" (Rescher 1998a, p. 56).

As usual, Rescher gives priority to the epistemological dimension. The credibility that is attributed to a prediction rests on its rational basis. This rational basis leads us to think that the statement is *correct*: rationality is the cement for the correctness of the prediction. In Rescher's judgment, "predictions are not (or should not be) categorized as being true / false but rather as correct / incorrect" (1998a, p. 70). This is because he is considering a notion of truth as correspondence. Thus, insofar as prediction is about future occurrences, the truth of a statement about the future cannot be established in the present. In effect, it only can be judged once the fact predicted by the statement has happened. So, for a prediction to be true depends on what the *future facts* will be, whereas its correctness depends on what we know about how those facts could be.

Therefore, the predictive statement should be made on the basis of reasons that make it credible before the fact or development predicted happens. In this way, it is also possible to attribute correctness to it. In Rescher's account, *credibility* has more weight than truth when prediction is analyzed from the point of view of language. A prophecy can be true; but it cannot serve as a guide for human action, because it does not have an inferential basis that allows us to think that it is credible and correct. As a consequence of this, the meaningful character of a prediction—the third feature pointed out above—rests (in Rescher's judgment) on use conditions instead of truth conditions. Because meaningfulness—in his approach—derives from the activity of communication and, in that case, is contextual.

In addition to the features that have been pointed out, scientific prediction must be informative. This means that it should meet several requirements, such as definiteness, exactness, detail, precision, etc. (Rescher 1998a, p. 62). It is difficult to obtain a very informative prediction, because of the problem of achieving an optimal equilibrium between predictive security and informativeness. The relation between these requirements responds to the following principle: *"the more informative a forecast is, the less secure it is, and conversely, the less informative, the more secure it is"* (1998a, p. 62). Thus, the achievement of one of these requirements in a high degree generally involves considerably diminishing the other. However, it is possible to achieve both requirements to a moderate degree. Rescher situates there the optimal point of equilibrium, which is the point where prediction is most effective as a guide for human action, in general, and scientific action, in particular.

In my judgment, there is another feature that might be added to the characterization offered by Rescher: prediction deals with something *expected*, so it is related with novelty. As a statement that, on a rational basis, is oriented towards the future, scientific prediction belongs to the realm of the things expected (Gonzalez 2010, 284). It is about not observed (or now unobservable) phenomena and it is therefore linked to the notion of "novel facts,"[18] which imply novelty.

Prediction is different from mere expectation, since it asserts something more than a reasonable possibility. Thus, prediction says that something will happen (given some conditions), and it usually does this on the basis of the regularities detected in the past and present facts. In this way, it has a cognitive content: it is linked to an objective basis so it cannot be reduced to the mere use of the language. Prediction not only anticipates a future fact, but it also asserts that we might expect it will happen.

2.3.2   The Distinction Between "Scientific" and "Non-scientific" Predictions

The distinction between "scientific" and "non-scientific predictions" is a particularly relevant issue for the characterization of the concept of prediction. However, Rescher does not develop exhaustively the barriers between scientific prediction and non-scientific prediction. In fact, he is more interested in the distinction between reasoned and unreasoned predictions. In this regard, he considers that being reasoned is a necessary condition for a prediction to be scientific; but it is not a sufficient condition, since not every reasoned prediction is a scientific prediction.

---

[18] Prediction involves some kind of novelty. In fact, it is possible to claim that it is a "research on novel facts," Gonzalez (2010, p. 11).

  The philosopher who gave most importance to the notion of "novel facts" was Imre Lakatos. See Lakatos (1978). On the notion of "novel facts" in Lakatos' conception, see Gonzalez (2001, 2010, pp. 179–184, 2014).

Regarding the distinction between "reasoned" and "unreasoned" predictions, Rescher points out that a prediction is different from a mere "precognition" or "clairvoyance" (1998a, pp. 53–56). Although he does not stress the problem of demarcation between scientific and non-scientific prediction, he does admit differences with regard to language. Thus, there are "*unreasoned* predictions" that are different from scientific predictions (Rescher and Helmer 1959, p. 32). Unreasoned predictions lack rational basis, so it is not possible to determine if they are credible or not. These unreasoned predictions are also called "prophecies" by Rescher, and he sees them as mere conjectures that do not have practical utility. In effect, from a scientific perspective, "predictions whose merits can be recognized only after the fact with the wisdom of retrospective hindsight are effectively useless" (Rescher 1998a, p. 55).

In contrast with a meaningful prediction—and, therefore, a prediction with cognitive content—a prophecy is "useless." In effect, it is not credible and, consequently, we cannot assign it the value of correctness. Rational basis is what gives credibility to a prediction, which is supported by the knowledge about past and present facts. In this way, Rescher demands a "realistic foresight" (1998a, p. 40). Consequently, he criticizes D. H. Mellor's proposal, according to which "predictions don't need reasons" (1975, p. 221).

Certainly, without reasons that justify an inference oriented towards the future, prediction lacks credibility: "Outside the context of grammatical examples and imaginative fictions, neither statements nor predictions have any serious interest for us in the absence of reasons for seeing them as credible" (Rescher 1998a, p. 256, n. 81). This feature has direct repercussions in order to consider a prediction as a genuine "scientific prediction."

The problem of the unrealistic assumptions has been widely discussed in economics. This question arises with the publication in 1953 of Milton Friedman's work "The Methodology of Positive Economics." In this text, Friedman proposes a methodological instrumentalism, in the sense of subordinating scientific methods to the aim of predicting. He claims that "the only relevant test of the validity of a hypothesis is comparison of its predictions with experience" (1953, pp. 8–9). Thus, in his judgment, an economic model cannot be assessed on the basis of realistic assumptions, but through its predictive ability, which is understood as correctness in the results.

Rescher disagrees with this account of methodological instrumentalism, and he considers that the defense of predictive models with unrealistic assumptions is infeasible. In fact, he explicitly criticizes Friedman's theses in this regard (Rescher 1998a, pp. 109 and 194–196). Rescher maintains that "'models' that do not actually *model*—that is, do not isomorphically reflect the real world's arrangements in their own makeup—will for this very reason fail to parallel the real world's *modus operandi* and accordingly prove predictively failure prone" (1998a, p. 109). This leads him to maintain that a criterion of demarcation between scientific prediction and non-scientific prediction is the realism of the assumptions. Thus, a model that does not adequately reflect the reality it seeks to predict is not a scientific model.

But the realism of the assumptions is not—in Rescher's proposal—a sufficient criterion to demarcate scientific prediction from non-scientific prediction. However, it is a sufficient criterion to distinguish between rational and non-rational prediction. Thus, in his judgment, not every rational prediction is a scientific prediction. He admits two kinds of *rational predictions*: those that are based on everyday experience and scientific predictions (Rescher 1998a, p. 57). Scientific predictions are the result of using scientific methods and knowledge; and, in this sense, they are superior to non-scientific predictions. He thinks that scientific predictions are superior as *science*, but not necessarily as *prediction*, because "the fact that all genuine prediction is oriented toward the open and (as yet) observationally inaccessible future means that our predictions are in principle always fallible" (1998a, p. 57).

In this regard, it seems to me that language can have a role when the problem is to distinguish between scientific prediction and non-scientific prediction. It is an issue that Rescher does not develop, because—in his view—the differences between scientific and non-scientific predictions are basically of a methodological character.[19] I think that, in order to distinguish a scientific prediction from a non-scientific one, the rigor of the language used should also be considered (for example, the accuracy and precision in the sense and reference of the terms used).

Moreover, when the scientific character of a prediction is only assessed from a methodological viewpoint, this can lead to instrumentalist approaches, such as Friedman's (which are proposals that Rescher tries to avoid). But, although Rescher avoids adopting a position of instrumentalist predictivism, it seems advisable to go deeper in the distinction between scientific prediction and non-scientific prediction, where language can have an important role. This is an issue that Rescher does not, in my judgment, develop satisfactorily.

2.3.3 Prediction and Retrodiction

Rescher insists on prediction as a statement oriented towards the future. This temporal feature leads him to reject, *de facto*, that there can be a genuine "retrodiction."[20] In this sense, he does not accept that it is possible to predict with regard to past events. This thesis of the "prediction of past" has been maintained by Friedman, among other authors. In his well-known text on the methodology of positive economics (1953, p. 9), he stresses that prediction must not necessarily deal with future phenomena, but can be about past events. The possibility of predicting with regard to the past was also defended by Stephen Toulmin in his book *Foresight and Understanding* (1961, pp. 26–27).

Friedman considers that "the 'predictions' by which the validity of a hypothesis is tested need not be about phenomena that have not yet occurred, that is, need not

---

[19] Cf. Rescher, N., *Personal Communication*, 15.7.2014.

[20] The differences between "prediction" and "retrodiction" are addressed from a logical perspective in Chap. 3,, Sect. 3.5.2.

be a forecast of future events; they may be about phenomena that have occurred but observations on which have not yet been made or are not known to the person making the prediction" (1953, p. 9). Toulmin not only admits the "prediction of past," but also a "prediction of present." He thinks of prediction as an "assertion about the occurrence of a particular sort of event—whether in the past, present, or future" (1961, p. 31).

However, Rescher does not subscribe a possible "prediction of past." He also rejects a "prediction of present," because prediction is about future events or developments. In his judgment, the acceptance of a "prediction of past," which has been maintained by authors such as Friedman and Toulmin, is the result of a failure in the distinction between *an event as such* and *people's stance towards an event*. So we can predict future reactions to past events, but never something that has already happened (Rescher 1998a, p. 254, n. 66).

In my judgment, Rescher criticism is right. Nevertheless, the notion of "novel facts" should be emphasized when this problem is considered, because prediction deals with not observed or now unobservable things. In this way, prediction connects to the notion of "novel facts." Thus, it seems to me that this notion and its relation to scientific prediction is a basic issue, both to clarify the concept of "prediction" and to call into question the possibility of a genuine "retrodiction." The notion of "novel facts" can be seen, at least, in three different senses: (i) ontological, which deals with a future event, temporal in the strict sense; (ii) epistemological, which has to do with a phenomenon that exists from an ontological point of view, but that is unknown; and (iii) heuristic, whose novelty rests on being a fact that is novel for the theory.[21]

When prediction involves a novelty in the ontological sense, the anticipation of the possible future is clear; for example, a prediction about the winner of some election or an event in a time that is posterior to the present moment. When the novelty is in the epistemological sense, the inference made from the available data leads to conclude that it is expected that something exists or that a concrete entity will be discovered (for example, the prediction about the existence of the neutrino).[22] So together with the strictly temporal factor, prediction is also related to something expected. In this way, prediction is oriented towards the future and claims that it may be expected that something happens.

Usually, Rescher highlights the temporal factor. He distinguishes between a statement about the past and a prediction—a statement about the future—on the basis of claiming that only a prediction can be falsified by the future development of the events or phenomena. In effect, "only statements that reach beyond the facts of the past-&-present—statements that could, in principle, be falsified by yet unrealized developments—can qualify as genuinely predictive" (Rescher 1998a, p. 46). Thus, with regard to the language, scientific prediction is a statement about the

---

[21] On the different senses of the notion of "novel facts," cf. Gonzalez (2001, pp. 505–508).

[22] This feature is connected with the role of prediction as a guide for discovery, which is one of the roles of prediction that Rescher points out. Cf. Rescher (1998a, p. 160).

future, so it is not possible to have neither a prediction of past (a "retrodiction") nor a prediction of present.

2.4 Characterization of the Predictive Statements

When scientific prediction is considered as a statement oriented towards the future, there are, initially, two different directions: quantitative prediction and qualitative prediction. This important distinction regarding the predictive statements is considered here in order to analyze Rescher's contribution to the reflection on scientific prediction as a statement. But it is also advisable to distinguish among different kinds of scientific predictions. Thus, it is possible to differentiate several predictive notions: foresight, prediction, forecasting, and planning, according to the degree of control of the variables (Gonzalez 2015, pp. 68–72).

2.4.1 Quantitative Prediction and Qualitative Prediction

Rescher does not take into account expressly the distinction between quantitative prediction and qualitative prediction, but it is implicit when he addresses scientific prediction from a methodological viewpoint. However, the distinction between "qualitative predictions" and "quantitative predictions" is an especially important issue when prediction is seen from the point of view of language, since it conditions diverse approaches to scientific research. In addition, it is an issue that has clear epistemological and methodological repercussions. Therefore, the concepts of "qualitative prediction" and "quantitative prediction" should be clarified.

Basically, *qualitative predictions* have the following features. (i) They do not follow expressly defined rules, but we achieve them through an intuitive procedure, which seeks to grasp tendencies, rhythms or patterns in phenomena to anticipate their behavior in the future. (ii) Since they are not obtained through a formal process, the subjects who make the prediction are fundamental. In this way, the resulting prediction is based to a large extent on the expertise of the predictors. (iii) Usually, all the available information used for the prediction is not detailed (Gonzalez 2015, p. 58).

Insofar as predictions are qualitative, interpretation has more weight; so different predictions might disagree. For this reason, it is possible that different experts make different predictions, even when they have the same information.[23] Rescher addresses this problem in terms of "predictive scatter," which is related to uncertainty and adds difficulty to prediction. Thus, when we deal with a limited body of information, competing theories can arise that will lead to contradictory predictions.

---

[23] This has been analyzed by B. G. Malkiel in the case of financial and stock markets. Cf. Malkiel (1973).

Consequently, "the prospect of conflicting predictions has to be accepted as a pervasively recurrent phenomenon" (Rescher 1998a, pp. 135–136). This can happen in the natural sciences and, to a larger extent, in the social sciences.

In contrast to qualitative predictions, *quantitative predictions* have the following features: (a) they are supported by models that can include some kind of law and, in some cases, they have a clear mathematical expression; (b) the role of the agent who makes the prediction is mostly in the background, since the important thing is the model itself; and (c) the variables used for the prediction are well specified, because the model must offer the information that is relevant for its validity (Gonzalez 2015, pp. 60–63). Quantitative prediction has advantages over qualitative prediction, since its evaluation has fewer difficulties. Firstly, its quantitative character makes it possible to test its accuracy in the future in a more detailed way. Secondly, if prediction is supported by models that might involve laws—instead of being made on the basis of the knowledge of the experts—there will be a higher level of objectivity. Thirdly, it is possible to clearly assess to what extent the relevant variables are taken into account.

Rescher's framework to address this distinction between qualitative and quantitative prediction is methodological, instead of being a semantic framework. This is because when he addresses the different types of scientific prediction, his attention goes to the *process* that has been followed to obtain the prediction. Thus, he divides the processes of prediction in two groups: the judgmental procedures and the formalized or inferential methods (Rescher 1998a, p. 88). It is possible to maintain that, above all, judgmental procedures lead to qualitative predictions. Meanwhile, predictions obtained as a result of the use of formalized processes can be either qualitative or quantitative. But Rescher normally uses the "scientific" term when the methods have an important mathematical component.

Usually, the features of the qualitative predictions can be seen in those predictions that, from a methodological viewpoint, Rescher calls "judgmental."[24] In this kind of prediction, the credibility and correction of the prediction depend directly on the confidence in the experts, because the predictor's expertise is basic in a judgmental prediction. Thus, it is possible to claim that judgmental prediction is supported by "their intuitive awareness of detectable patterns in the phenomena" (Rescher 1998a, p. 89). In this case, there is no formal process, and the information used or the inference made is not generally shown in an explicit way.

For Rescher, judgmental procedures should be valued to the extent that their usage "extends our predictive range by dispensing with the need for detailed theories and/or models to provide the theoretical underpinning of prediction" (1998a, p. 110). However, these predictions are little valued from a scientific point of view.[25]

---

[24] The main features of this predictive procedure, which is developed on the basis of predictor's expertise, already appear in one of the first papers on prediction by Rescher (1967).

[25] For example, in economics qualitative prediction has been seen as *complementary* to quantitative prediction: "En general, podemos decir que el único procedimiento de predicción que se presta a ser analizado y evaluado de acuerdo con unos criterios científicos es el de los modelos econométricos. Sin embargo, estos procedimientos coexisten con una pléyade de instrumentos subjetivos,

In fact, his methodological conception places them out of science, because in his approach only the predictions that are the result of a formal process (on the basis of models open to laws and mathematical regularities) can be called "scientific."

With a certain redundancy, it is possible to say that, in Rescher's judgment, those methods of prediction that "proceeds on the basis of scientific principles" are "scientific methods," and they usually have a mathematical component (Rescher 1998a, p. 106). In this case, he considers as scientific those predictions that are the result of processes that are mainly based on laws and models. He thinks that law-based predictions have a high value: "our most sophisticated predictive method is that of *inference from formalized laws* (generally in mathematical form), which govern the functioning of a system" (1998a, p. 106).

Certainly, Rescher shows a certain preference for the methods that he calls "scientific." Consequently, he is inclined to see quantitative predictions as more valuable than qualitative predictions. To a large extent, this is because he stresses accuracy and precision as the values that should characterize scientific prediction. However, this preference is more implicit than explicit in Rescher, and it is not as noticeable as in other authors. Kuhn, for example, emphasized to a greater extent the importance of the quantitative predictions in comparison with qualitative predictions.[26] Thus, he maintained that "probably the most deeply held values concern predictions: they should be accurate; quantitative predictions are preferable to qualitative ones; whatever the margin of permissible error, it should be consistently satisfied in a given field; and so on" (Kuhn 1970, p. 185).

In this regard, Rescher's position is more qualified than Kuhn's viewpoint. In my judgment, the key is that Rescher is in some sense more pluralistic. He notices that predictions that are the result of "scientific" processes—that can be associated with quantitative predictions due to the usage of models that have a mathematical expression—are the predictions that provide more "rational comfort" (1998a, p. 110). Moreover, he considers that "it is fortunate that the use of experts is no tour only predictive resource" (1998a, p. 97). Concurrently, he insists that every prediction is, in principle, fallible. So the most important thing is to acknowledge the limits that affect prediction and try to overcome them.

From this perspective of predictive pluralism, qualitative predictions are valuable, since they may extend our predictive range (Rescher 1998a, p. 110). In this way, qualitative predictions are something more than a simple complement to quantitative prediction, because they allow us to anticipate phenomena that are not predictable on the basis of formalized methods of prediction. However, Rescher's approach has several ambiguities: (1) he does not delimit in a clear way the differences and relations between quantitative and qualitative predictions; (2) he admits that there are, *de facto*, predictive procedures and predictive methods, and thinks that the later are more reliable; and (3) he does not characterize what are the thematic

---

paneles de opinión, encuestas, valoraciones de expertos, etc., que muchas veces complementan o son complementados por los resultados de las predicciones basadas en modelos," Fernández Valbuena (1990, p. 386).

[26] On Kuhn's approach to prediction, see Gonzalez (2010, chap. 4, pp. 127–159).

realms that can obtain some benefit form qualitative predictions, even when it seems that social sciences and the sciences of the artificial are the greater beneficiaries of the existence of qualitative predictions.

2.4.2 Foresight, Prediction, Forecasting, and Planning

From the point of view of content, although every prediction we can obtain always appears as fallible, there are differences between them. They are differences that depend on the degree of control of the variables. It seems clear that there are different types of predictions, depending on the phenomenon studied, the problem discussed, and the methodology that is used. All these issues are closely related to the degree of control of the variables that are important for the prediction. Consequently, it is possible to propose specific terms to refer to each type of prediction.

In a generic way, Rescher admits these differences with regard to variations in predictive reliability (see Rescher 1958). This feature leads him to distinguish, on the one hand, between quantitative and qualitative prediction; and, on the other, he notes the distinction between the generic and the specific prediction (Rescher 1998a, p. 198). But, to be more rigorous, it is possible to establish more distinctions. In this regard, a quadruple distinction has been proposed in economics between "foresight," "prediction," "forecasting," and "planning" (Fernández Valbuena 1990). In my judgment, this distinction makes the philosophical discussion on prediction more rigorous from the point of view of language.

To be sure, Rescher does not characterize the possible types of scientific prediction, and he uses the terms "prediction," "foresight," and "forecast" as synonymous. However, we can see in his approach a concern about assigning a specific term to each type of prediction (Rescher 1998a, pp. 53–56). But this is an issue that he does not develop, because, in his judgment, "the actual albeit regrettable fact is that English does not afford us this terminological luxury" (1998a, p. 55). It seems to me that this claim is questionable.

In effect, several distinctions are accepted in economics. Thus, it is possible to differentiate between "foresight," "prediction," and "forecasting," according to the degree of control of the variables that they achieve. *Foresight* is the most secure kind of prediction: "it is a presentation about the state of a variable within a period of time, when the variable is directly or indirectly under our control" (Fernández Valbuena 1990, p. 388). Thus, a foresight provides knowledge oriented towards the future about a variable that can be controlled (for example, VAT). For this reason, it is the most secure type of prediction, since—in principle—there will not be changes in the variables that affect the success of the foresight.

Sensu stricto, prediction is a specific type of statement about the future. Thus, unlike foresight, it does not involve the complete control of the variables that are relevant to the statement. In this case, there are factors of the variable that are not under the control of the predictor, but are subjected to variations, which can be due either to their endogenous behavior or to exogenous factors (for example, inflation or unemployment). For this reason, the reliability of the prediction—in this strict sense—is lower than in the case of a foresight.

Unlike foresight and prediction, *forecasting* has a margin of error associated with it. In this way, instead of providing a concrete number; the forecast establishes a margin where the forecasted phenomenon is expected to be placed (for example, a forecast about the unemployment rate). If we attend to Rescher's proposal, we have that, according to the relation between predictive security and informativeness, it is possible to think that, in principle, he seems to consider the forecast as more secure than the prediction, because the former is less precise and detailed than the latter.

Nevertheless, the difference between a prediction and a forecast can lie in the model used. In the case of economics, "that a particular presentation could be a prediction or a forecast depends on—almost always—the procedure used to make it. Thus, a determinist model (where there are no random variables) makes predictions, whereas a stochastic model (which includes random variables) makes forecastings" (Fernández Valbuena 1990, p. 389).[27] However, there is no unanimity in the terms within this discipline, where habitually "prediction" and "forecast" are used as if they were interchangeable notions (Gonzalez 2010, p. 262n).

Finally, it should be stressed that "foresight," "prediction," and "forecast" are different from "planning." *Planning* is made on the basis of the different statements about the future and it seeks to provide patterns for action in the realm of applied science and technology (Gonzalez 2010, pp. 261–263). So it encompasses a teleological approach, since it is oriented towards problem solving. In this way, within applied science it is possible to see prediction as the previous step to prescription, and this eventually serves the tasks of planning.

Rescher does distinguish between "prediction" and "planning." He associates planning to an intentional realm of the direction of action. It can be "positive," when the aim is make something happen; or "negative," when it tries to avoid something happening (Rescher 1998a, pp. 235–236). He assumes the importance of planning in economics: "*policy guidance* is one of the main aims of the macroeconomic enterprise" (1998a, p. 198). In his judgment, although economics does not achieve the desirable level of predictive success, it can indeed have success in policy guidance. Because he considers that "effective operation [of policy guidance] does indeed *not* demand categorical predictions, since even merely probabilistic considerations can provide serviceable and perfectly cogent guidance to action" (1998a, p. 198).

So Rescher distinguishes "prediction" and "planning." But he uses without distinction the terms "prediction," "foresight," and "forecast." However, sometimes he uses the term "forecast" to refer to a "specific sort of prediction which foretells the occurrence or nonoccurrence of a particular concrete eventuation at a particular definite time" (1998a, p. 42). Therefore, it might be a type of prediction that can be tested in a clearer way than in the case of less specific predictions. I think that this distinction is not good enough in philosophico-methodological terms.

---

[27] Quoted in Gonzalez (2015, p. 69).

To be more rigorous with regard to the language, a distinction between the diverse types of predictions is required. Regarding this point, Rescher's account is—in my judgment—revisable according to a more sophisticated analysis of language. It is possible to distinguish a qualitative prediction from a quantitative prediction, as well as to differentiate a generic prediction from a specific prediction. But it is also possible to establish differences between types of prediction such as "foresight," "prediction," "forecasting," and "planning." It should be emphasized that several distinctions are already accepted with regard to scientific explanation (Salmon 1990 and Gonzalez 2002) and achieving a typological variety in the realm of scientific prediction seems advisable.

2.5 Limits of Language and Prediction: "Not Predictable" and "Unpredictable"

Within his pragmatic orientation (which sees language from a pragmatic perspective), Rescher is interested in the limits of science. In fact, he is one of the authors who have paid more attention to the problem of the limits of knowledge. In his approach, science is a human product where agents prevail, because categories—and, in general, concepts—allow us to articulate the reality (categories and concepts are expressed through language). To the extent that the scientific vision of the world is a human product, it is not cognitively absolute, because science is "our" science. It is the result that arises from the interaction between the researcher and her context. For this reason, Rescher maintains that "the limits of our experience set limits to our science" (1999b, p. 216).

Above all, Rescher insists on the limits derived from agents' capabilities (mainly cognitive), although he also points out the obstacles that arise from the complexity of the phenomena that are researched.[28] Epistemological and ontological limits affect scientific knowledge, in general, and knowledge about the future, in particular. When he addresses the limits that affect prediction, he pays special attention to the natural science, although he also addresses the problem in relation to the social sciences, with attention to phenomena of economics and sociology (Rescher 1998a, ch. 11, pp. 101–208; especially, pp. 193–202).

According to Rescher, there are two main types of limits to prediction: epistemological and ontological. Epistemological limits are those limits that affect prediction insofar as it is made by agents with limited cognitive abilities.[29] However, the first limit is in the language: it is difficult to know something we cannot state. There are also ontological limits to predictability insofar as it deals with future phenomena, which have not happened yet and, therefore, they are still open.

---

[28] Rescher also addresses the ethical limitations to scientific activity and technological endeavor (1999a, pp. 151–203). His approach to the ethical limits of science and their incidence to scientific prediction are analyzed in Chap. 9, Sect. 9.3.

[29] On the epistemological limits to prediction, see Chap. 4, Sect. 4.4.

Nevertheless, it should be acknowledged that the limits of science are not only epistemological and ontological limits. This can be clearly seen when considering the distinction between the limits due to the agents and the limits of scientific activity itself (Gonzalez 2010, pp. 275–276). The limits due to the agents are related with the capabilities of scientists as agents with bounded rationality, who are faced with a varied context (cultural, social, etc.) in the knowledge of reality. Meanwhile, the limits of scientific activity itself are those limits involved in the scientific endeavor: they are rooted in the constitutive elements of science. These obstacles can be seen in the diverse realms of science: semantic, logical, epistemological, methodological, ontological, axiological, and ethical (Gonzalez 2010, pp. 277–281).

In other words, we can find obstacles due to the language of science, its structure, knowledge, processes, activity, and values (among them, ethical values). In this way, there are not only obstacles in the agents, but also in the activity itself developed by science. In Rescher's approach, to the extent that science is *our* science, the barriers between the limits due to the agents and the limits due to scientific activity itself tend to fade (Gonzalez 2010, p. 277). Thus, he insists on the limits due to the limited capabilities of the agents and, besides the epistemological and ontological realms, he pays much less attention to the obstacles present in the diverse realms of science.

Within this framework, it is important to analyze prediction from language: if words do not involve sense and reference, the advancement of science is really difficult. Thus, the issue of the semantic limits to prediction can be addressed. Strictly speaking, semantic obstacles to prediction are related to "the difficulties to identify new phenomena—their sense and reference" (Gonzalez 2010, p. 275). This gets complicated insofar as Rescher's approach to language is not properly a semantic approach, but a pragmatic one. His interest is not in the content of meaning, but rather in the nexus between two forms of impossibility of prediction: "non predictability" and "unpredictability."[30]

"Unpredictability" involves the total impossibility of prediction for human beings. It is mainly due to the presence of phenomena characterized by anarchy or

---

[30] In fact, Rescher uses the notions of "unpredictability" and "impredictability." In his judgment, "an important difference is neatly marked in English usage by the difference between *unpredictable* and *impredictable*, the former being geared to volatility, the latter to intractability. In London weather conditions are unpredictable in March: one minute it can be clear and sunny and ten minutes later there may be clouds and rain. Here instability is at work. On the other hand, the future of the American poetry is impredictable: we simply have no grip on any laws or regularities that provide for rational prediction. Both cases alike frustrate the project of prediction," Rescher (1998a, p. 137).

It seems to me that the terms used by Rescher can lead to confusion. For this reason, the notions chosen in this monograph are different, in order to distinguish something that we cannot predict now from something that will never be predicted. "Non predictability" is used here instead of what Rescher calls "unpredictability." Thus, a phenomenon is not predictable when there is a current impossibility of predicting it (either through a generic prediction or through a specific prediction). Meanwhile, "unpredictability" is used instead of "impredictability," when there is a complete impossibility of predicting a phenomenon or event (either in the short, middle, or long run). See Gonzalez (2015, p. 56).

lack of laws (Rescher 1998a, pp. 136–138). Meanwhile, "non predictability" refers to the current impossibility of stating a prediction, usually due to the instability of the phenomena that we want to predict (Rescher 1998a, pp. 79–82). The former is focused on the intractability of phenomena and the latter is oriented towards the volatility of the events.

When a phenomenon lacks regularity, there is no possibility of predicting its future behavior. Its "intractability" makes any attempt of prediction impossible. Anarchy prevails, which Rescher distinguishes from chaos. Chaos corresponds to extreme instability, and not to the complete lack of order.[31] Meanwhile, anarchy involves unpredictability. It is not possible to predict about an anarchic phenomenon or system, since there is no relation between its behavior in the past and its development in the future. In this case, the impossibility of prediction is inherent to the phenomenon at issue, and not to the current inability to predict it.

Meanwhile, *non predictability* is related to volatility, which has to do with the behavior of the processes over time. It is linked to the temporal projection of the prediction. Thus, commonly, processes are more stable in the short run than in the long run. But the volatility of phenomena is always a clear obstacle to prediction. In effect, stable processes are more predictable than those characterized by being volatile or unstable. In turn, when a phenomenon is stable, this makes it easier to clarify what elements give this continuity and contribute to a better projection into the future.[32]

But, even when Rescher admits this distinction between what is completely "unpredictable" and what is merely "not predictable" according to the available information and knowledge, he is not always rigorous in the usage of both notions (Rescher 1998a, pp. 138–140 and 146–148). In my judgment, this is because he usually thinks of this issue in terms of "non predictability." Thus, according to him, there are no reasons to think that science cannot answer any question that arises in its domain (if not now, at least in the future) [Rescher 1999b, p. 3]. In this way, it is problematic to claim that something is "unpredictable" in the strict sense.[33]

This can be seen with regard to the nexus he establishes between two aspects: on the one hand, the notions "non predictability" and "unpredictability;" and, on the other, the limits of science *in the weak sense* and *in the strong sense* (Gonzalez 2010, pp. 274–275). Thus, science is clearly subject to limits in the weak sense. It seems clear that we have questions that we cannot answer now, because nowadays

---

[31] On chaos as an obstacle to scientific prediction, see Werndl (2009).

[32] This can be seen in the case of economics, where stable elements are sought in order to overcome the obstacles to predictability, such as human rationality in decision-making. Cf. Gonzalez (2003).

[33] Moreover, the possibility of predicting the future depends on both epistemological and ontological issues. Regarding what we can predict or not in science, Rescher thinks that the best source of information we have comes from science itself: it is not an external issue. Moreover, only science itself can inform us about the achievable degree of precision for scientific prediction. This depends on circumstances such as the scope—short, medium or long run—, the available technology, etc. Additionally, it also depends on the question we want to ask. In principle, the more concrete the question is, the more complicated it will be to answer it accurately. Rescher, *Personal communication*, 17.6.2014.

we do not have knowledge enough to solve them. Meanwhile, when the limits are in the strong sense, we can establish now that there are questions that we will not be able to answer in the future, even in the long run.

With regard to the limits in the strong sense, Rescher considers that there are difficulties in identifying the "insolubilia" problems of science (1998a, pp. 186–188, 1999b, pp. 111–127), because there are also—to a greater or lesser extent—difficulties to predict how future science will be (Rescher 1998a, pp. 177–183).[34] If we cannot identify the unsolvable problems of science, because we do not know now what we will know in the future, then there are also difficulties to identify the unpredictable phenomena. That is, there are clear difficulties to establish what predictive problems are intrinsically unsolvable by human science, and not merely not-predictable in accordance with current limitations. However, we can claim that the complete predictability of phenomena is an unachievable goal.

In this case, it is assumed that science is subject to predictive incompleteness (Rescher 2010, pp. 149–163), not merely in descriptive terms but also in prescriptive ones. This means that we do not just reflect a factual situation—description at the current moment—but also that prescription is ruled out—the *ought to be* of science in the future—with regard to the possibility of achieving predictive completeness. It seems clear that, in order to achieve predictive completeness, there are not only problems of language (limits to identify all the affected elements of reality), but also epistemological and methodological problems.

Consequently, in order to be complete with regard to prediction, science has to achieve the goal of accurately predicting all things that, in principle, science itself considers predictable (Rescher 1999b, p. 146). But, both with regard to scientific knowledge, in general, and predictive knowledge, in particular, the fallibilistic position should be assumed. Therefore, he considers that all knowledge we accept is revisable, because it can turn to be false.[35]

Together with prediction there is metaprediction. In this case, he thinks that scientific activity itself is an unpredictable endeavor (Rescher 1983). "Kant's principle of question propagation" is also valid in the case of prediction: new predictive answers lead to new predictive questions that, in turn, require an answer. This leads Rescher to state that natural sciences are subject to predictive incompleteness (1998a, pp. 183–186); an approach which can be extended to the social sciences and the sciences of the artificial.

However, Rescher maintains that natural sciences are "our best predictive tool" (1998a, p. 187). But they are an imperfect tool, since they are unpredictable with

---

[34] In addition, Rescher considers that only science can inform us about its own limits, and this is always with regard to each particular moment. Cf. Rescher, N., *Personal communication*, 17.6.2014.

Further details on the problem of "insolubilia" have been developed in Sect. 1.3.2. of this book. See also Guillán (2016).

[35] The characterization of his own epistemology in terms of fallibilism can be seen in Rescher (1999b, ch 3, pp. 29–42).

regard to aspects of their very possible future (at least, with the desired level of accuracy and precision) [Rescher 2012]. This fact does not lead him to a skeptic position regarding the possibility of prediction, but to a realist approach. Thus, for him, "the inescapable imperfection of this instrument means that the predictive project too is imperfectable and that our aspirations in this direction must be kept within realistic bounds" (1998a, p. 188).

Therefore, Rescher accepts predictive and metapredictive limitations of science; that is, he admits that scientific theories imply limits to prediction and that we cannot predict future developments of the current sciences. Thus, he assumes that the complete predictability is a goal that cannot be achieved by means of "our" science. In his proposal, he insists on the limits that hinders the predictive task of science. He addresses this issue mainly in epistemological and ontological terms, although there are obstacles in the different realms of science, starting with language. This is because we cannot identify always all the elements at stake in a prediction (relevant variables, etc.).

According to the present analysis, it can be suggested that, besides the epistemological and ontological realms, the problem of the limits has to do also with the semantic, logical, methodological, axiological, and ethical fields. Certainly, they involve two types of impossibility of prediction: "non predictability" and "unpredictability," although it is not always easy to provide examples of this conceptual distinction.

On the one hand, there are "not predictable" phenomena due to the existence of current limitations; and, on the other, there are "unpredictable" developments, because there are things that we cannot predict neither now nor in the future. Although Rescher conceptually assumes this distinction, it is not really important in his approach. In addition, it is not clear enough insofar as one thing is not being able to offer now an accurate and precise prediction and another is to be "not predictable." Furthermore, in his proposal, we cannot establish the final limits of scientific knowledge. Consequently, claiming that there are phenomena that will *never* be predicted is also problematic.

References

Dummett, M. 1963. Realism (I), conference at the *Oxford University Philosophical Society*, March 8th 1963. Reprinted in Dummett, M. 1978. *Truth and other enigmas*, 145–164. London: Duckworth.

———. 1981. *Frege: Philosophy of language*. London: Duckworth (2nd ed. 1981).

———. 1982. Realism (II). *Synthese* 52 (1): 55–112.

———. 1985. Frege y Wittgenstein. *Anales de Filosofía* 3: 27–37.

Fernández Valbuena, S. 1990. Predicción y Economía. In *Aspectos metodológicos de la investigación científica*, ed. W.J. Gonzalez, 2nd ed., 385–405. Madrid-Murcia: Ediciones Universidad Autónoma de Madrid and Publicaciones Universidad de Murcia.

Friedman, M. 1953. The methodology of positive economics. In *Essays in positive economics*, ed. M. Friedman, 3–43. Chicago: The University of Chicago Press (6th reprint, 1969).

Gonzalez, W.J. 1986. *La Teoría de la Referencia. Strawson y la Filosofía Analítica*. Salamanca-Murcia: Ediciones Universidad de Salamanca and Publicaciones de la Universidad de Murcia.

———. 2001. Lakatos's approach on prediction and novel facts. *Theoria* 16 (3): 499–518.

———., ed. 2002. *Diversidad de la explicación científica*. Barcelona: Ariel.

———. 2003. Racionalidad y Economía: De la racionalidad de la Economía como Ciencia a la racionalidad de los agentes económicos. In *Racionalidad, historicidad y predicción en Herbert A. Simon*, ed. W.J. Gonzalez, 65–96. A Coruña: Netbiblo.

———. 2010. *La predicción científica. Concepciones filosófico-metodológicas desde H. Reichenbach a N. Rescher*. Barcelona: Montesinos.

———. 2014. The evolution of Lakatos's repercussion on the methodology of economics. *HOPOS: The Journal of the International Society for the History of Philosophy of Science* 4 (1): 1–25.

———. 2015. *Philosophico-methodological analysis of prediction and its role in economics*. Dordrecht: Springer.

Grünbaum, A. 1962. Temporally-asymmetric principles, parity between explanation and prediction, and mechanism versus teleology. *Philosophy of Science* 29 (2): 146–170.

Guillán, A. 2016. The limits of future knowledge: An analysis of Nicholas Rescher's epistemological approach. In *The limits of science: An analysis from "barriers" to "confines"*, Poznan Studies in the Philosophy of the Sciences and the Humanities, ed. W.J. Gonzalez, 134–149. Leiden: Brill.

Kuhn, Th. S. 1970. Postscript—1969. In *The structure of scientific revolutions*, Th.S. Kuhn, 174–210. Chicago: The University of Chicago Press (2nd ed., 1970).

Lakatos, I. 1978. *The methodology of scientific research programmes. Philosophical Papers*, vol. 1. Cambridge: Cambridge University Press.

Malkiel, B.G. 1973. *A random walk down Wall Street*. New York: W. W. Norton.

Mellor, D.H. 1975. The possibility of prediction. *Proceedings of the British Academy* 65: 207–223.

Rescher, N. 1958. On prediction and explanation. *British Journal for the Philosophy of Science* 8 (32): 281–290.

———. 1967. The future as an object of research. *RAND Corporation Research Paper*, P-3593.

———. 1983. The unpredictability of future science. In *Physics, philosophy and psychoanalysis*, ed. R.S. Cohen et al., 153–168. Dordrecht: Reidel.

———. 1988. *Rationality. A philosophical inquiry into the nature and the rationale of reason*. Oxford: Clarendon Press.

———. 1989. *Cognitive economy. The economic dimension of the theory of knowledge*. Pittsburgh: University of Pittsburgh Press.

———. 1992a. Cognitive limits. In *A system of pragmatic idealism. Vol. I: Human knowledge in idealistic perspective*, N. Rescher, 243–254. Princeton: Princeton University Press.

———. 1992b. Metaphysical realism. In *A system of pragmatic idealism. Vol. I: Human knowledge in idealistic perspective*, N. Rescher, 255–274. Princeton: Princeton University Press.

———. 1998a. *Predicting the future. An introduction to the theory of forecasting*. New York: State University of New York Press.

———. 1998b. *Communicative pragmatism and other philosophical essays on language*. Lanham: Rowman and Littlefield.

———. 1998c. Communicative pragmatism. In *Communicative pragmatism and other philosophical essays on language*, N. Rescher, 1–48. Lanham: Rowman and Littlefield.

———. 1998d. Truth conditions versus use conditions (a study on the utility of pragmatics). In *Communicative pragmatism and other philosophical essays on language*, N. Rescher, 61–75. Lanham: Rowman and Littlefield.

———. 1998e. Objectivity and communication. How ordinary discourse is committed to objectivity. In *Communicative pragmatism and other philosophical essays on language*, N. Rescher, 85–97. Lanham: Rowman and Littlefield.

———. 1999a. *Razón y valores en la Era científico-tecnológica*. Barcelona: Paidós.

———. 1999b. *The limits of science*, revised edition. Pittsburgh: University of Pittsburgh Press.

———. 2006. Pragmatic idealism and metaphysical realism. In *A companion to pragmatism*, ed. J.R. Shook and J. Margolis, 386–397. Oxford: Blackwell.

———. 2008. Linguistic pragmatism. In *Epistemic pragmatism and other studies in the theory of knowledge*, N. Rescher, 13–21. Heusenstamm: Ontos Verlag.

———. 2012. The problem of future knowledge. *Mind and Society* 11 (2): 149–163.

Rescher, N., and O. Helmer. 1959. On the epistemology of the inexact sciences. *Management Sciences* 6: 25–52.

Salmon, W.C. 1990. *Four decades of scientific explanation*. Minneapoilis: University of Minnesota Press.

Strawson, P. F. 1950. Truth (II). *Proceedings of the Aristotelian Society*, v. sup. 24: 129–156.

Toulmin, S. 1961. *Foresight and understanding*. Bloomington: Indiana University Press.

Werndl, Ch. 2009. What are the new implications of chaos for unpredictability? *British Journal for the Philosophy of Science* 60: 195–220.

Chapter 3
Logical Features of Scientific Prediction

Abstract Logically, scientific prediction is related to a series of problems that have to do with the internal articulation of the scientific theories. Among those problems, the relations between explanation and prediction have received particular attention. This chapter therefore analyzes the logical features of scientific prediction: (1) The possible limits of deductivism for scientific prediction are considered. (2) Rescher's conception on induction and its use for scientific prediction is taken into account. (3) It pays attention to the controversy among those who were in favor of the symmetry thesis of explanation and prediction and those who were not. In this regard, Rescher's original approach to the relations between explanation and prediction is considered. (4) The factor of temporality is studied, which leads to our addressing two problems: (a) is there a mere temporal anisotropy between explanation and prediction?; and (b) are retrodiction and prediction equal from a philosophical perspective?

Keywords Logic of science • Deduction • Induction • Scientific prediction • Scientific explanation • Symmetry thesis • Retrodiction

Logically, scientific prediction is related to a series of problems that have to do with the internal articulation of the scientific theories, whether they are conceived as isolated theories (in the Popperian way, for example) or as series of interrelated theories (for example, in the Lakatosian way or in another way that articulates theoretical frameworks and historicity). In this regard, when prediction is seen from the perspective of the logic of science, the debate on the "well-structured" theories oriented toward prediction need to be considered. So, one question is whether the only valid structure is the deductive one, especially, the hypothetical-deductive structure or, on the contrary, there could be an inductive structure (for example, the hypothetical-inductive) [Niiniluoto and Tuomela 1973]. In this regard, first, the possible limits of deductivism for scientific prediction are considered; and, secondly, Rescher's conception on induction and its use for scientific prediction is also analyzed.

Furthermore, it should be highlighted that, from a logical perspective, "the existence of well-structured scientific theories does not imply, in principle, that they should be predictive" (Gonzalez 2015a, p. 15). *De facto*, the nature of scientific theories can be diverse with regard to its configuration: (a) explicative; (b) explicative and predictive; and (c) predictive (Gonzalez 2015a, p. 15). Thus, the internal structure of scientific theories can be oriented towards explanation, prediction or both. For this reason, an important issue is that related to the possible structural similarities or differences between explanation and prediction, which includes its symmetry or its asymmetry.

This topic of the logical features of explanation and prediction became relevant for the status of scientific prediction.[1] In this regard, this chapter analyzes (i) the controversy among those who were in favor of the symmetry thesis of explanation and prediction and those who were not; (ii) Rescher's approach regarding the relations between explanation and prediction; and (iii) the factor of temporality and the relevance it has for the logic of prediction. On the factor of temporality two problems are considered: (a) is there a mere temporal anisotropy between explanation and prediction?; and (b) are retrodiction and prediction equal from a philosophical perspective?

3.1 Possible Limits of Deductivism to Scientific Prediction

Karl R. Popper maintained a logical-methodological approach to scientific prediction of an expressly deductivist character.[2] In effect, the Viennese philosopher thought that he had found a solution to the problem of induction,[3] which involved its complete rejection as a logical-methodological procedure.[4] This has clear consequences for prediction, since, in his judgment, the Logic of prediction is deductive Logic (Popper 1974, p. 1030). From this perspective, scientific prediction cannot be made on the basis of an inductive inference.

It is an approach that "assumes that, among the array of generalizations that are compatible with the available observational evidence, there is a *rational basis in favor of one unrefuted generalization* (conjecture, hypothesis, etc.) instead of others for use in a predictive argument" (Gonzalez 2015a, p. 86). Therefore, there are rational bases for the preference of a generalization instead of another one, where "corroboration" also has a role. In this regard, Popperian notion of *corroboration*

---

[1] See in this regard Barnes (2008) and Douglas (2009). On the approaches to scientific explanation, see Gonzalez (2002).

[2] On Popper's approach to scientific prediction, see Gonzalez (2004a, 2010, pp. 55–89). On the influence of falsificationism on prediction in economics, see Gonzalez (2015a, pp. 79–101). See also Martínez Solano (2005).

[3] "I think that I have solved a major philosophical problem: the problem of induction," Popper (1972, p. 1).

[4] Nevertheless, Popper finally admits that we need a whit of induction to achieve general statements. Cf. Popper (1974, p. 1193).

involves that corroboration only allows us to judge past performance of a theory, but it is not possible to use it to predict future performance (Popper 1976, p. 103).

Wesley Salmon made a valuable critique of this Popperian proposal. In his paper on rational prediction (1981), Salmon highlighted the problems that may have a logical approach to prediction that is developed on purely deductive basis, such as the Popperian approach (Gonzalez 2015a, pp. 85–87). In this regard, he goes more deeply in something that Hans Reichenbach had already done (Reichenbach 1978, p. 372): Salmon seeks to analyze the limits of deductivism to the problem of scientific prediction.

Within an approach that combines a defense of Logic as the basis of induction and probability as a key element for the philosophy of science, Reichenbach maintains that Popper fails when he discards the notion of probability: "Contrary to his claims: 1. The process of falsifying a theory contains the concept of probability. 2. The procedure for constructing a new theory contains the concept of probability" (1978 [1935], p. 373).[5] This is an element that supports the need for induction, since according to Reichenbach the inductive inference allows us to obtain probability statements that, in turn, are the basis of the statements about the future.[6]

Salmon also criticizes the insufficiency of deductivism to give an adequate basis for rational prediction. However, he does this from a perspective that is different from that adopted by Reichenbach, where it is required to take into account the purpose of the prediction. For Salmon, there are at least three reasons that can lead us to predict: (i) We make predictions because we want to satisfy our curiosity about the future happenings; (ii) in order to test a theory; and (iii) when we want to make an optimal decision and knowledge about the future occurrences is required (Salmon 1981, pp. 115–116).

Therefore, together with the future knowledge provided by prediction and its role as a test of scientific theories, Simon acknowledges the practical dimension, where prediction is linked with decision-making. Rescher also admits those roles of prediction (Rescher 1998).[7] However, these options do not exhaust the complete framework of the uses of prediction. Because in science, besides basic science, which seeks to explain and to predict, there is applied science, which relates prediction and prescription, and there is also the application of science, where predictions are required too.[8]

Salmon's proposal has consequences for the choice between generalizations that have not been refuted. Because in order to prefer one generalization instead of another, the purpose of the generalization needs to be taken into account. This implies that, on the basis of the different reasons to predict, the distinction between

---

[5] However, Popper developed his own approach to probabilty (Popper 1959). See also Gillies (2000, pp. 113–168).

[6] "Probability propositions express relative frequencies of repeated events, that is, frequencies counted as a percentage of the total. They are derived from frequencies observed in the past and include the assumption that the same frequencies will hold approximately for the future. They are constructed by means of an inductive inference," Reichenbach (1968 [1951], p. 236). On Reichenbach's approach to induction see Salmon (1991).

[7] He has also related his approach to prediction with decision-making. See Rescher (1995).

[8] See Gonzalez (2013a, pp. 17–18). When Salmon criticizes Popper, he only thinks of basic sciences and the ordinary uses of prediction (curiosity and decision-making in daily life).

two kinds of preferences should be considered: the *theoretical preference* and the *practical preference* (Salmon 1981, p. 118). Thus, with regard to the practical preference, Salmon considers that Popper was wrong when he maintained that corroboration cannot be used to predict future performance.

In Popper's judgment, corroboration can *motivate* the preference of a theory, although it has no predictive import; but, for Salmon, the problem is how it is possible to *justify* this preference (1981, p. 121). According to Salmon, a predictive argument requires us to choose a premise, a conjecture with predictive content. A rational prediction also requires a rational choice of that premise. However, no predictive appraisal follows from the observational evidence and the statements of corroboration of a conjecture (Salmon 1981, p. 119).

Through his alternative to Popper's approach, Salmon combines the positive character of experience with the pragmatic component of decision-making and the reliability of predictions. Thus, we need to solve "how it could be rational to judge theories *for purposes of prediction* in terms of a criterion which is emphatically claimed to be lacking in predictive import" (Salmon 1981, p. 122). Even more, in order to have an adequate conception of *rational prediction*, this problem of the *justification* of the choice between general statements must be solved.

For Salmon, the solution to this issue requires to distinguish between *predictive content* and *predictive import*. Because "statements whose consequences refer to future occurrences may be said to have predictive content; rules, imperatives, and directives are totally lacking in predictive content because they do not entail any statements at all. Nevertheless, an imperative—such as 'No smoking, please'—may have considerable predictive import, for it may effectively achieve the goal of preventing the occurrence of smoking in a particular room in the immediate future" (Salmon 1981, p. 123).

With this distinction between *predictive content* and *predictive import*, Salmon highlights how corroboration statements can have *predictive import*, even when they are not about future happenings.[9] On this issue, Rescher position is clear: "*of course* past performance is a predictive indicator. (What could possibly serve better?) What past performance does *not* enable one to do is to predict with failproof accuracy" (1998, p. 260, n. 112). Thus, Rescher insists in the legitimacy of the use of induction, although the fallibility of the inductive prediction must be acknowledged.

Criticism of Popper's deductivist conception is especially important for Rescher, since his methodological pragmatism cannot be supported by exclusively deductive bases.[10] There is room for induction in Rescher's methodological pragmatism. Thus,

---

[9] Rescher (1998, p. 161) makes a distinction similar to Salmon's. It consists of distinguishing between "predictive import" and "predictive inference". Thus, on the one hand, there are theories which make *predictive inferences*; that is, they are theories which can provide statements about the future. And, on the other hand, there are theories which deal with the past and, consequently, they do not make predictions; but they can have *predictive import*, so they can provide the content required to make predictive inferences (for example, the theory of evolution). The similitudes between them have been noticed by Wenceslao J. Gonzalez (2010, p. 266).

[10] Methodological pragmatism is developed by Rescher in some of his publications. Among them, the following could be highlighted: Rescher (1977, 2000, 2012a).

a thesis can be justified methodologically. The procedure follows two successive steps (Rescher 1977, p. 67). Firstly, a thesis is methodologically justified through the application of a cognitive method for the validation or substantiation of factual claims (Rescher 1977, p. 15). Secondly, the adoption of this method is justified on the basis of practical criteria, which are preeminently two: "success in prediction and efficacy in control" (1977, p. 67).

This procedure can be applied to the inductive practices. Instead of appealing to Logic, Rescher thinks that induction can be *methodologically* justified; that is, it can be supported by assessing if the inductive practice of obtaining generalizations on the basis of the available evidence works (Rescher 2012b, pp. 135–136). To do this, experience must have a positive role in the substantiation of generalizations, so— for him—falsification does not suffice in this regard.

The Popperian logical-methodological pairing of falsification-refutation is considered by Rescher as something marginal. Thus, "of course, an evidentially falsified or disconfirmed generalization must be ruled out, but this step does not take us far. The absence of evidential invalidation is no touchstone of truth—as any statistician knows from curve-fitting problems; *mutually* incompatible generalizations may yet be compatible with all the evidence" (Rescher 2012b, p. 136).

Because Rescher explicitly considers that falsification—and, therefore, refutation—does not suffice, experience must have a *positive role* for the justification of a generalization. But it happens that we cannot establish that a generalization is true on the basis of the available evidence: "Where generalizations are concerned, finite evidence cannot go as far as to yield assured truth, and so the 'it works' at issue cannot mean success at providing correct (true) empirical generalizations" (Rescher 2012b, p. 136). In this regard, his concern is the same as Salmon shows in his paper on rational prediction: how can we choose *in a rational way* among several generalizations that are incompatible amongst each other, but are compatible with the available evidence?

On this issue, Rescher again adopts a pragmatic criterion. Thus, for him, it is not an issue of "establishing an empirical generalization, but [of] validating the rationality of a practice (an epistemic practice, to be sure, by which generalizations are supported)" [Rescher 2012b, p. 139]. To do this, the purpose that is sought by the practice must be taken into account. This is because to establish the validity of an epistemic practice is a matter that directly depends on the efficacy and efficiency in the achievement of the aim sought, since epistemic practices are means that seek to achieve a certain aim (Rescher 1977).

From this perspective, the methodological pragmatism suggested by Rescher requires using the past performance of the methods (among them, inductive procedures) as an indicator of their predictive success in the future. But, "to be sure, its experiential support through evidence-in-hand regarding its record of success does not *prove* that a method will succeed in future applications. But the probative weight of experience cannot be altogether discounted: established success must be allowed to carry *some* weight" (Rescher 2012b, p. 141).

Therefore, Rescher notes that the impossibility of *proving* the truth of a generalization should not lead to our rejecting the positive role of experience nor the validity of inductive reasoning. This is because, on the basis of pragmatic considerations, he thinks that the use of past performance as a predictive indicator to establish generalizations is a valid procedure. Thus, through a different way, he achieves the same background conclusion that Salmon reached: deduction alone does not suffice for scientific prediction.

3.2   Role of Inductive Logic with Regard to Scientific Prediction

Habitually, the logical aspects of science are linked with deduction and induction, without rejecting a recent interest on abduction (Nepomuceno 2014, and Sintonen 2004). Certainly, logic has to do with deduction, and Popper insisted—as have other rationalist philosophers—on the nexus between prediction and deduction. Meanwhile, other empiricist-inspired authors, such as Reichenbach or Salmon, put effort into associating prediction with induction. If it is accepted that deduction is demonstrative whereas induction is not, then there are clear repercussions in terms of the validation of predictions.

Rescher's approach to induction has its own characters, since he offers an account which gives priority to the epistemological realm and, later, to the methodological level. This is because he is interested in a specific view on induction as a kind of *ampliative reasoning*. Thus, even when he acknowledges the logical basis of induction—and he also establishes several sorts of inductive inference—his attention is usually focused on the cognitive content provided by induction. He does this from a pragmatic approach, where the use of induction as a procedure to solve answers is highlighted.

Rescher addresses the two facets of the problem of induction: (a) its characterization, which has incidence for the subsequent use of induction in *our* science; and (b) its justification, since he researches the bases that support induction as a process. In this regard, there are two different dimensions, although they are intertwined. On the one hand, in his approach, induction can be characterized as a mode of reasoning and, consequently, as an issue that has to do with Logic. And, on the other, Rescher's justification of induction is pragmatic. Thus, the validity of an inductive argument is established on the basis of epistemological considerations (that is, with regard to the content of knowledge that it provides) and its methodological repercussions (Rescher 2012b).

When the focus is on the use of induction by science, then we have that Rescher acknowledges the role of induction with regard to scientific prediction. This is because induction can lead to statements about the future, based on the available information about the development followed by past and present phenomena. This activity, which gives relevance to induction for scientific prediction, can be seen at two successive levels, which have to do with the context of discovery and the context of justification, respectively: (i) induction as a procedure to achieve predictive statements; and (ii) the role of induction in the justification of prediction.

3.2.1   The Problem of Induction as Scientific Procedure

Undoubtedly, the problem of induction is one of the most debated and controversial issues in philosophy and methodology of science, in general, and in the sphere of logic of science, in particular. According to Hempel's view of this topic, the so-called "problem of induction" can be divided, in principle, into two sub-problems (Hilpinen 2000, p. 91). One has to do with the *characterization* of induction (what is induction and which kind of rules does inductive reasoning follow?). The other sub-problem, which depends on the solution to the previous one, deals with the *justification* of induction (that is, the possible validation of the inductive reasoning) [Hempel 1981, pp. 389–390].

When Rescher thinks of the *problem of induction*, he considers preferably the second sub-problem. Thus, in his book *Induction*, his focus of attention is on the problem of how to provide a pragmatic justification of induction. However, he also addresses the problem of the *characterization* (the first level of analysis). Nevertheless, his conception addresses a "broad" perspective on induction, instead of a "narrow" view. In his judgment, a "narrow" viewpoint of induction maintains that induction is a method for reasoning that allows us to achieve universal generalizations from supportive instances (Rescher 1980, p. 2).[11]

In contrast to this characterization of induction, of a clear logical inspiration, Rescher suggest a broader characterization in order to overcome some problems. His idea is to include "all of our rational devices for reasoning from evidence in hand to objective facts about the world. Induction, thus understood, will encompass the whole of 'the scientific method' of reasoning, and in treating of the justification of induction we take in hand the validation of the processes of reasoning in the sciences" (1980, p. 2).

Within this "broad" perspective, it seems that Rescher assumes that, in principle, scientific procedures always encompass an inductive element, insofar as they are based on the observation of an experimentation of particular phenomena in order to achieve general statements. Thus, when the problem of the justification of induction is addressed, in his judgment, the validity of the typical mode of reasoning used in scientific praxis is also called into question.

In his later works, Rescher basically maintains his initial view of induction as it appears in his monograph *Induction*. In this work, induction is seen as a kind of reasoning that allow us to obtain answers to question through an optimal use of the

---

[11] As an example of the "narrow" viewpoint, Rescher (1980, p. 2, n.2) mentions J. Stuart Mill's approach to induction, of Aristotelian inspiration, that sees induction as "the operation of discovering and proving general propositions."

available information (1980, pp. 19–20). Later, he offers additional elements for a more exhaustive characterization of induction. In this regard, he notes that "induction is a mode of reasoning that moves from premises that present presumably acceptable data to conclusions that make claims whose information extends above and beyond what those premises provide for" (Rescher 2014, p. 52). Therefore, for him, the major characteristic of induction consists in its *ampliative character*.

Expressly, Rescher points out that he understands the "ampliative" character of inductive reasoning in the same way that C. S. Peirce did it: "For Peirce, 'ampliative' reasoning is synthetic in that its conclusion goes beyond ('transcends') the information stipulated in the given premises (i.e., cannot be derived from them by logical processes of deduction alone), so that it 'follows' from them only inconclusively" (Rescher 1980, p. 6). The inductive character of induction (that is, that induction provides new knowledge that is not stipulated in the premises) involves a clear difference with respect to deductive Logic.

This clear difference with deduction follows well-known lines. Because Rescher claims that "we cannot pass by any sort of inference or cognitive calculation from the 'premises' of an inductive 'argument' to its 'conclusion' because (*ex hypothesi*) this would be a deductive *non sequitur*—the conclusion (in the very nature of the case) asserts something regarding which its premises are altogether silent. Clearly the paradigm mode of *inference*—of actually deriving a conclusion from the premises—is actual deduction, and this paradigm does not fit induction smoothly (...) With inductive reasoning there is always an epistemic (or conjectural) gap between the premises and the conclusion" (Rescher 2014, p. 56).[12]

Regarding the comparison between kinds of inferences from a logical point of view, it is usual to note that inductive inferences are *contingent*, whereas deductive inferences are *necessary* (Vickers 2014). These features have repercussions in scientific prediction. They affect its very possibility and reliability. This aspect can be seen with regard to one of the modes of inductive reasoning, which consists in the transition from the information about the past developments to the achievement of statements oriented towards the future. In this way, induction could be required to predict the possible future.[13]

These considerations support Rescher's rejection of what he calls a "narrow" view on induction, in the line of authors such as J. Stuart Mill and many others who insisted on the logical perspective on induction.[14] Because, since inductive reasoning is ampliative—in contrast to deductive inference—then it cannot be just a kind

---

[12] In this regard, besides the Peicean influence, Rescher quotes William Whewhell with approval. For Whewhell, deduction "descends steadily and methodically, step by step: Induction mounts by a leap which is out of the reach of method [or, at any rate, mechanical routine]. She bounds to the top of the stairs at once...," Whewell (1858, p. 114). Text quoted in Rescher (2014, pp. 55–56).

[13] "Deductive inference can never support contingent judgments such as meteorological forecasts, nor can deduction alone explain the breakdown of one's car, discover the genotype of a new virus, or reconstruct fourteenth century trade routes," Vickers (2014, sec. 1).

[14] Within the wide bibliography on this matter, the relation between inductive Logic and probability can be highlighted. See, for example, Black (1984a, b), Galavotti (2011), Gillies (2000), and Hájek and Hall (2002).

of inference that gets universal generalizations from particular cases. Thus, besides the reasoning that goes from particular to general, Rescher admits that there is a variety of modes of inductive reasoning: from the past to the future, from the sample to the population, from the instance to the type, etc. (Rescher 2014, p. 52).

So he combines two aspects in his characterization of induction. On the one hand, there is the logical basis, which has to do with induction as an ampliative mode of reasoning that can follow different ways (particular to general, past to future, etc.); and, on the other, there is the epistemological dimension, which has primacy in Rescher's approach. This is due to the ampliative character of inductive reasoning, which involves a cognitive gap, insofar as it goes further than the available information in order to provide new knowledge. In this way, he associates induction with its pragmatic usage, insofar as it is a procedure of question answering: "induction is at bottom an *erotetic* (question-answering) rather than an *inferential* (conclusion-deriving) procedure" (Rescher 2014, p. 59).

3.2.2   Justification of Induction: The Pragmatic Preference

Since inductive inferences are "ampliative" (i.e., they give new cognitive content that extends the cognitive content provided by the available information), there is always certain risk of error (Vickers, sec. 1). This leads to the second aforementioned level of the problem of induction, which has to do with its *justification*, because, in contrast to deduction, inductive reasoning can lead to false conclusions on the basis of true premises.[15] On this issue, David Hume's proposals have greatly influenced philosophical reflection on the validation of the inductive inferences (Hume 2007, book 1, part 3, sec. VI, pp. 61–65; see, in this regard, Lange 2011).

This issue of the justification of induction can be addressed, in principle, from two different perspectives: the strictly logical perspective—in the sense of providing a proof of validity of an inductive conclusion—and the epistemological perspective, which appeals to the content of knowledge provided by the inductive reasoning. From the logical perspective, in the line of the Humean criticism of induction, it might be considered that it is impossible to justify the conclusion of the inductive reasoning as true. This means that induction is not demonstrative. So it is accepted that it is not possible to provide a proof in terms of truth as justification of an inductive inference.

Once the logical path is abandoned with regard to the problem of the *justification* of induction, Rescher adopts a basically epistemological perspective (the second aforementioned possibility) with methodological repercussions. In his approach

---

[15] "In valid deduction we are in the fortunate position of having premises that provide *conclusive* grounds for our conclusions: we have situations of fully supportive pro-information. Induction effectively inverts this proceeding, resolving the questions we face correlatively with the minimum of contraindications. We seek to minimize the as-yet-visible risks in the inevitably risky venture of cognitive gap filling," Rescher (2014, p. 53).

there is a clear pragmatic component. This is a solution which is coherent with his *characterization* of induction, where the epistemological dimension has primacy over the strictly logical considerations.[16] Moreover, he thinks that the problem of the justification of induction is in no case a matter which could be solved on the basis of Logic. For this reason, he is inclined to "search for validation of induction on the basis of practical considerations."[17]

First, Rescher acknowledges the problem that is posed by the Humean account with regard to induction: "Hume may be taken to have shown, with all the lucidity that philosophical arguments admit of, that there simply can be no 'justification of induction' by way of a demonstrative proof" (1998, p. 64). He considers that the Scottish empiricist was right when he maintained that it is not possible to demonstrate that a conclusion of an inductive inference is true. Thus, he assumes the intrinsic difficulty that the achievement of a generalization of universal character from experience has, and which is something temporal and episodic.[18]

But, secondly, Rescher notes that "it is senseless to make demands or impose conditions that cannot in the very nature of things be satisfied, and absurd to require a demonstration whose accomplishment is manifestly impracticable" (1998, p. 64). For this reason, Humean objection to induction does not lead Rescher to reject the validity of the inductive reasoning, since it is possible to provide a pragmatic justification of induction. Moreover, he accepts that it is possible to justify induction on the basis of epistemological reasons, which have methodological repercussions. Thus, the inference oriented towards something general is not the most important thing, but the ability of induction to solve questions in those contexts where the available information is limited or imperfect.[19]

This kind of pragmatic justification suggested by Rescher is—my judgment— coherent with his characterization of induction. This is because, to the extent that he maintains that "induction is not really a mode of *inference*, strictly speaking, but rather one of *estimation*" (Rescher 2014, p. 63), justification of induction cannot proceed on the basis of the rules of Logic.[20] In turn, he also avoids a strictly empiricist

---

[16] On his own approach to induction, Rescher writes that "it sees induction not as a characteristic mode of drawing conclusions, but *as an estimation technique*, a methodology for obtaining answers to our factual questions through optimal exploitation of the information at our disposal" (1980, p. 20).

[17] Rescher, *Personal communication*, 19.8.2014.

[18] Rescher, *Personal communication*, 19.8.2014.

[19] Rescher, *Personal communication*, 19.8.2014.

[20] In this regard, it is advisable to distinguish between "Logic," as the study of the inferential principles and rules with a well-defined rigor, and "logic," that can consist of processes of reasoning which seek an estimation oriented towards the truth. When Rescher thinks of "inductive logic," he has usually in mind the second sense of "logic" (as truth estimation), so he assumes that inductive logic does not have the rigor that deductive logical rules have.

The advisability of distinguishing Logic—as a rigorous formal science—from logic—understood as an approach that seeks general patterns without proving in a formal way that what is obtained is true—has been pointed out by W. J. Gonzalez in order to study the work of philosophers such as Karl Popper, who sometimes uses the first sense (when he appeals to formal Logic as the basis of methodological approaches of general character), whereas on other occasions he

approach, which in principle decontextualizes inductive reasoning from the situation. Instead, Rescher maintains that we must take into account the context and purpose of inductive reasoning.[21]

From this epistemological-pragmatic perspective, the importance of induction for scientific prediction is emphasized. Insofar as Rescher understands induction as a tool for the enlargement of knowledge, this involves a clear nexus between induction and the anticipation of the future, at least at a general level: "It is undeniably possible to look upon an induction as an argument: a process of drawing a general conclusion of inherently future applicability from evidence regarding the past" (Rescher 2012b, p. 135).[22] This is because "to evidentiate our predictive claims about the future, we have no alternative but to look to the past-&-present" (Rescher 1998, p. 86). This implies that, in order to predict the possible future, the mode of reasoning that goes from past to future is an option (and it is a role that seems to be usual in certain time projections of available data).

3.3 Importance of Induction for Scientific Prediction

If induction is seen as a kind of reasoning from past to future, then there is an inductive component that goes with scientific prediction: "All modes of rational prediction call for scanning the data at hand in order to seek out established temporal patterns, and then set about projecting such patterns into the future in the most efficient way possible. For sensible prediction is always a matter of rational economy, of adapting our expectations of the future to the occurrence structures of the past in the most simple and economical way that the epistemic circumstances of the case admit" (Rescher 1998, p. 86). This implies assuming the "economics of research," which is part of the economics of science proposed by Rescher.

As a matter of fact, this pragmatist viewpoint based on the economics of research is central in Rescher's approach to induction. In his conception, inductive reasoning oriented to prediction has a crucial role. In effect, one of the modes of induction that he mentions and accepts is related to the transition from past to future. So inductive reasoning can have a predictive orientation. Furthermore, he thinks that "it is, of course, true, and *trivially* true, that any authentic *generalization* must apply to future cases" (Rescher 2012b, p. 135).

---

uses the second sense, which is wider (for example, logic is used in some approaches regarding the social sciences). These differences between the two possible senses of this term can be seen in Gonzalez (2004b).

[21] "Induction is, in the final analysis, a venture in practical/purposive rather than in strictly theoretical/illuminative reasoning," Rescher (2014, p. 63).

[22] When Rescher analyses scientific prediction, he usually considers the realm of basic science, where inductive prediction can be a test for scientific theories. But the role of generic inductive prediction can be also addressed in the context of applied science (as the previous step to prescription) and in the case of the application of science (for example, in business, where it is usual to make prediction on the basis of the trends observed in the past.)

3.3.1   Role in the "Context of Discovery"

Two different levels can be distinguished when the importance of induction to scientific prediction is addressed. One has to do with the context of *discovery*, where induction can be seen as a process that allows us to achieve predictive statements; whereas the other falls within the context of *justification*, where its task has to do with the validation of the predictive processes and their results.[23] In turn, within this second level, two possible dimensions should be considered: (a) justification of induction as a *process* to predict scientifically (that is, to what extent induction is a predictive "procedure" or a predictive "method")[24]; and (b) the role of induction for the justification of the *results*, that is, the content of prediction as such or the predictive theories.

Rescher takes into account both the "context of discovery" and the "context of justification" when he develops his approach to induction. He is aware of the importance of induction to predict scientifically the possible future. In effect, he considers that "on the standard 'inductive' model of scientific method, the predictions of science are generated by logico-mathematical derivations that apply general theories to situation-specific facts so as to preindicate future observations" (Rescher 1998, p. 161). Thus, inductive process can lead to the *discovery* of new predictions, insofar as it is possible to anticipate the possible future on the basis of the available information about the past and present occurrences.

Furthermore, Rescher generally maintains that prediction is the result of an inference from the information about the things that already happened. Thus, an argument oriented towards prediction usually fits in the type of reasoning that goes from the past to the future. This seems clear if the preconditions of rational prediction are considered. These conditions are the following: (a) data availability; (b) pattern discernability; and (c) pattern stability—the patterns observed continue into the future (Rescher 1998, pp. 86–88).[25] In this regard, induction can have a role with regard to the *goals* of scientific research when it seeks to predict the possible future.

However, the patterns can change. Consequently, the development of phenomena in the future can be different from the patterns followed in the past. Thus, there is uncertainty with regard to the future, so we have to acknowledge the limits to inductive prediction. For this reason, Rescher thinks that we have to deal with Hume's

---

[23] Javier Echeverría (1995) has suggested enlarging the traditional distinction between "context of discovery" and "context of justification." In his approach four contexts of scientific activity can be distinguished: context of education, context of innovation, context of evaluation, and context of application. He addresses this with regard to the axiology of research. So he insists on the axiological features of each of these contexts.

In this volume, the relation between scientific prediction and induction is addressed from a logical-methodological perspective. So the main interest is here in how induction can be a process to obtain scientific predictions (the "context of discovery") and how it is possible to validate predictive processes and results through induction (the "context of justification").

[24] On this distinction between "predictive procedures" and "predictive methods," see Gonzalez (2015a, ch. 10, pp. 251–284).

[25] There is a deeper treatment on this topic in Chap. 5, Sect. 5.4, of this monograph.

criticism of induction, which has clear consequences for the prediction obtained on the basis of the available information about something that has already happened. *De facto*, Rescher is mainly interested in the *justification* of the inductive process that is oriented toward question-answering regarding the possible future, which has repercussions on the problem of inductive predictions' reliability.

3.3.2   Task in the "Context of Justification"

As Rescher notes, "for Hume, the predictive aspect of all attempts to use *available* evidence to establish *general* conclusions calls for 'the transferring of past experience to the future,' a process he saw as predicated on the assumption 'that the future resembles the past'" (1998, p. 64).[26] In Hume's judgment, since experience is the only rational justification to admit this assumption, "there is no rational warrant for our inductive predictions. Its only basis lies in a habit-established psychological expectation" (Rescher 1998, p. 64).

But it seems clear that, contrary to what Hume maintained, Rescher does not think that inductive prediction requires the assumption that future invariably resembles the past. Instead of this position, he accepts that there can be changes in phenomena, because the predicted phenomena may be irregular or, at least, different from the previous ones. Even more, he explicitly recognizes that obstacles such as choice, chance or chaos are common for predicting the future. From this point of view, phenomena may follow unprecedented patterns, so there are clear limits to the predictions obtained through inductive reasoning (Rescher 1998, p. 134).

Thus, it is not possible to justify inductive prediction by merely accepting the *regularity* of phenomena, since the patterns can vary. For this reason, even when Rescher accepts Hume's criticism of inductive reasoning, he thinks that a change of perspective is required to deal with the problem of the *validation* of induction. This change of perspective—which is coherent with his characterization of induction as a mode of reasoning oriented towards question-answering—consists of putting the strictly logical viewpoint aside and adopting an epistemological approach, which has methodological repercussions. He sees the problem from a pragmatic viewpoint, which involves a contextual character.

In effect, Rescher addresses two problems related to the context of justification—the justification of induction as a *process* for predicting in a scientific way and its role for the justification of the *results*—within a framework that is clearly pragmatic. As a process used to predict the future, he thinks that there are two reasons to use induction in scientific practice: (1) utility, insofar as we know that induction may succeed as a predictive process; and (2) improvement, insofar as induction offers a better prospect of success than other alternatives (Rescher 1998, p. 64).

---

[26] According to Hume, "there can be no *demonstrative* arguments to prove, *that those instances, of which we have had no experience, resemble those, of which we have had experience*" (2007, book 1, part 3, sec. VI, p. 62).

In the basis of these reasons offered by Rescher, it seems to me that there are several issues at stake that he does not explicitly state. (i) Induction has a role in the context of discovery insofar as it allows us to obtain predictions. Thus, it is relevant with regard to the *aims* of the research, since the kind of inductive inference that goes from past to future leads us to obtain predictive hypotheses. (ii) Induction also has a role in the context of justification. It appears at two successive levels, which have to do with processes and results. (a) Regarding *processes*, there is the problem of the justification of the use of induction as a procedure for predicting the future in a scientific way. (b) With regard to *results*, justification deals with the statements about the future (the inductive predictions). This involves the problems related to the quantity and quality of the observations and experiments (to what extent they have evidential value with regard to the predictions obtained).

The solution that Rescher gives to the problems related to *justification*, which deal with the processes and the results, is contained, in my judgment, in his account of *methodological pragmatism*. In effect, "Rescher argues that a method of inquiry whose use systematically meets with success is to be seen as truth-indicative. False belief may sometimes lead to success, but it could hardly be supposed to do so on a routine basis" (Sankey 2006, p. 126). According to this methodological pragmatism, we can "monitor our acceptance of theses via the methods that substantiate them, and then validate these methods by pragmatic tests—specifically considerating how well we fare *in applying and implementing its professed claims in matters of prediction and control*" (Rescher 2000, p. 96).

First, the validity of the predictive statements of inductive character is assessed on the basis the *methods* that substantiate those statements. Secondly, the validity of the methods is evaluated with regard to their ability to give successful predictions (i.e., statements regarding the future). Therefore, according to Rescher, the justification of the results involves taking into account the processes that lead to those results. If those processes have been systematically successful with regard to prediction, then it can be said that the result is valid.

It seems clear that Rescher's methodological pragmatism itself involves an inductive element, since it is rooted in an inference that goes from past to the future. In effect, results are considered valid if the method used has achieved—in the past— a systematic success in matters related to prediction and the control of phenomena. Thus, on the basis of past experience, we conclude that the method at issue will also lead to reliable results in the future. From this perspective, the justification of induction—with regard to both processes and results—is a pragmatic issue, and the most important thing is the ability to have predictive success.

This pragmatic perspective leads Rescher to criticize explicitly Reichenbach's approach to induction (Rescher 1998, p. 258, n. 92), because, for Reichenbach, the conclusion of an inductive reasoning can be accepted to the extent that it is our best posit or wager with regard to the future: "What we obtain is a wager; and it is the best wager we can lay because it corresponds to a procedure the applicability of which is the necessary condition of the possibility of predictions. To fulfill the conditions sufficient for the attainment of true predictions does not lie in our power; let us be glad that we are able to fulfill at least the conditions necessary for the realization of this intrinsic aim of science" (Reichenbach 1949, p. 357).

Although there is no objection to accepting, in principle, that "an absolute reliability of the predictions cannot be warranted" (Reichenbach 1949, p. 86), there are reasons to reject that "it would be illusory to imagine that terms 'true' or 'false' ever expressed anything else than high or low probability values" (Reichenbach 1936, p. 156. See Gonzalez 1995). In effect, prediction fallibility is a feature that Rescher and Reichenbach's accounts have in common; but Reichenbach's notion of "prediction" is a quite weak characterization of scientific prediction, insofar as he associates the statements about the future with posits or wagers.

Rescher expressly objects to the approach of Reichenbach to prediction in terms of a wager: "Hans Reichenbach (…) saw the matter [the problem of induction] not as one of establishing knowledge but rather as making a bet" (1998, p. 258, n. 92). In contrast to the conception of the logical empiricists of the Berlin School, which associates prediction with induction and relates induction to the theory of probability in frequentist terms, Rescher insists on the epistemological component of induction, which he considers from a genuinely pragmatic perspective because, in his judgment, "the validation of an inductive knowledge of the future must in the end be founded on the non- or preinductive basis of essentially *practical* reasoning. What we have here is not a theoretical demonstration of a thesis but a pragmatic validation of a cognitive praxis" (Rescher 1998, p. 65).

Therefore, the problem of induction is not a matter of making a "wager," according to Reichenbach's expression. Instead, we have to proceed in a rational way in science, according to the available information, which is usually imperfect. For Rescher, to predict is a rational activity, which is oriented towards answering in the best possible way the questions we have about the world's future developments. From this pragmatic perspective, which deals with problem solving, the justification of the use of induction as a predictive procedure is ultimately rooted in the capability of those procedures developed on the basis of induction to obtain successful predictions.

3.4   From Logical Symmetry to Asymmetry between Explanation and Prediction

Besides the roles of deductive and inductive logic for prediction, a major problem when scientific prediction is seen from a logical point of view is the issue of the similarities or differences with respect to scientific explanation. In this regard, the thesis of the *logical symmetry* between explanation and prediction has been widely discussed. According to this thesis, explaining and predicting are like processes from a logical perspective. After its initial formulation by Carl Gustav Hempel and Paul Oppenheim (1948), this thesis had a considerable influence on other authors, such as Adolf Grünbaum (1962). The thesis of the asymmetry is against this logical account. It maintains that explanation and prediction are not equal from a structural or logical viewpoint.

Rescher actively participated in this controversy very early on. He did it as a critic of the symmetry thesis and as a defender of the logical asymmetry between explanation and prediction. In order to do this, he developed a proposal that combines logical, epistemological, methodological, and ontological elements of scientific explanation and prediction. Through these elements, he maintains that there is a "significant disanalogy" between both scientific processes.[27] Within this framework, he suggests his own alternative to the Hempel and Oppenheim's thesis: the *harmony thesis*. It has a pragmatic orientation, since it takes into account the nexus between the processes of explanation and prediction in scientific practice (Rescher 1998, pp. 167–169).

3.4.1   A Significant "Disanalogy:" Explanation of Past and Prediction of Future

Within the logical features of scientific prediction, a major problem has to do with the debate on the logical symmetry or asymmetry between "explanation" and "prediction." This controversy started in 1948, when Carl Gustav Hempel and Paul Oppenheim published their well-known paper "Studies in the Logic of Explanation,"[28] in which they propose the thesis of the logical symmetry between explanation and prediction, which maintains that "to explain" and "to predict" are symmetrical processes. This involves that they are logically equal, mainly due to the use of scientific laws, since they are valid both for the past and for the future.

These philosophers think of scientific explanation according to a deductive-nomological model, where explanation is made on the basis of laws and is supported by a deductive logical structure. In their judgment, the deductive-nomological patterns of explanation involve a logical symmetry between explanation and prediction.[29] Thus, Hempel and Oppenheim expressly maintain that "whatever will be said in this article concerning the logical characteristics of explanation or prediction will be applicable to either, even if only one of them should be mentioned" (1948, p. 138).

From this perspective, the difference between explaining and predicting is of temporal nature with a pragmatic dimension: explanation deals with past phenomena,

---

[27] Rescher's criticism appeared for the first time in a paper published in 1958. One year later, he again defended the logical asymmetry between explanation and prediction. He did this in a paper published with O. Helmer. See Rescher and Helmer (1959).

[28] The paper is in the realm of the logical empiricism that was dominant in the philosophy of science of the United States in that time.

[29] Later, Hempel would develop the inductive-statistical model of explanation. Cf. Hempel (1962). In this case there is—for Hempel—a logical symmetry between the inductive-statistical model of scientific explanation and the probabilistic prediction, cf. Salmon (1993, pp. 231–232).

whereas prediction is oriented towards phenomena that have not happened yet. This enunciation of prediction expressly implies that prediction is prior in time to the predicted phenomena. But it is also possible to think of predictions of phenomena that already exist in the present but that we do not know yet. Thus, to differentiate between explanation and prediction, they emphasize the temporal factor: "If E is given, i.e. if we know that the phenomenon described by E has occurred, and a suitable set of statements $C_1, C_2, \ldots, C_k, L_1, L_2, \ldots L_r$, is provided afterwards, we speak of an explanation of the phenomenon in question. If the latter statements are given and E is derived prior to the occurrence of the phenomenon it describes, we speak of a prediction" (Hempel and Oppenheim 1948, p. 138).

On this logical basis, which has methodological projection, the symmetry thesis implies that every scientific explanation can serve as a prediction. Moreover, for Hempel and Oppenheim, it "is this potential predictive force which gives scientific explanation its importance: only to the extent that we are able to explain empirical facts can we attain the major objective of scientific research, namely not merely to record the phenomena of our experience, but to learn from them, by basing upon them theoretical generalizations which enable us to anticipate new occurrences and to control, at least to some extent, the changes in our environment" (1948, p. 138).

But, according to Hempel and Oppenheim, every scientific prediction can serve as a scientific explanation (either within a deductive-nomological model or an inductive-statistical one) under suitable circumstances (Gonzalez 2010, p. 216). For Salmon, this is the most problematic aspect of this conception. He considers that scientific prediction is a statement about the future.[30] Thus, "as such, a prediction could not be an explanation, for an explanation, according to Peter [Hempel], is an argument. The most that could be maintained is that legitimate scientific predictions are *the conclusions of* arguments that conform to the schemas of D-N or I-S explanation" (Salmon 1993, p. 232).

This logical-methodological position on the structural symmetry between "explanation" and "prediction" was influential for several authors. Among them, Adolf Grünbaum should be highlighted. However, as Wesley C. Salmon notices, Hempel and Oppenheim's paper was practically unnoticed for a decade (Salmon 1998, p. 68). Thus, in the late fifties, Rescher (1958) was one of the first authors to criticize the symmetry thesis. Later on, in 1998, he gave shape to his own alternative to the proposal of Hempel and Oppenheim. In that time, he proposes the harmony thesis between "explanation" and "prediction" (Rescher 1998, pp. 167–169).

It should be emphasized that, in 1958, when the symmetry thesis was widely accepted, Rescher's account was critical of it: he went further than the option in favor of a simple temporal difference between explanation and prediction. He emphatically claims that "it cannot be maintained that explanation and prediction are identical from the standpoint of their logical structure" (Rescher 1958, p. 289. See also van Fraassen 2009). In his judgment, "rather than being the single point of minor difference between explanation and prediction, this temporal asymmetry is of far-reaching and fundamental import" (1958, p. 286). In fact, for him, the different

---

[30] This idea was proposed in Scheffler (1957).

temporal orientation of explanation and prediction leads to important differences between them with regard to their logical structure.

In scientific explanation, conclusion is firmly supported by premises. Meanwhile, the degree of probability associated to prediction is usually much lower (Rescher 1998, p. 166). In the paper of 1959 that Rescher published together with O. Helmer, he insists on this asymmetry between explanation and prediction: "An explanation must *establish* its conclusion, showing that there is a strong warrant why the fact to be explained—rather than some possible alternative—obtains. On the other hand, the conclusion of a (reasoned) prediction need not be well established in this sense; it suffices that it be rendered *more tenable than comparable alternatives*. Here then is an important distinction in logical strength between explanations and predictions: An explanation, though it need not logically rule out alternatives altogether, must beyond reasonable doubt establish its hypothesis as *more credible than its negation*. Of a prediction, on the other hand, we need to require only that it establish its hypothesis simply as *more credible than any comparable alternative*" (1959, p. 32).[31]

Once this logical basis has been established, Rescher maintains that there is not a merely temporal asymmetry between explanation and prediction, but a significant disanalogy between them. The "disanalogy" is based on the underlying difference between explaining and predicting, which is of an ontological kind, because explanation is about phenomena or events that have already happened, whereas prediction is oriented towards phenomena of events that we expect in the future. This feature leads him to acknowledge that the relation between informativeness and security is different in both cases. Thus, the more detailed the explanation is, the more secure it usually is; whereas, in principle, prediction is less secure as its informative content increases (Rescher 1998, p. 257, n. 90).[32]

Nevertheless, besides this logical basis for the distinction, Rescher considers that there are also epistemological, methodological, and ontological differences between explanation and prediction. (i) Epistemologically, explanation has a causal linkage that is clearer than in the case of prediction, for which there can be, in principle, alternatives. (ii) Methodologically, prediction and explanation are also different, since there are processes that we can explain, but we cannot predict them. (iii) Ontologically, past facts are different form future developments, which are still open (Rescher 1998, pp. 165–166).

Consequently, his criticism of the symmetry thesis combines logical, epistemological, methodological, and ontological elements (Gonzalez 2010, p. 264). This is something that was suggested initially in 1958, so he was one of the first philosophers

---

[31] Rescher and Helmer acknowledge that there are exceptions to this common pattern: "Of course prediction may, as in astronomy, be as firmly based in fact and as tightly articulated in reasoning as any explanation. But this is not a general requirement to which predictions *must* conform. A doctor's prognosis, for example, does not have astronomical certitude, yet practical considerations render it immensely useful as a guide in our conduct because it is far superior to reliance on guesswork or on pure chance alone as a decision making device," (1959, p. 32).

[32] "Informative content" or "informativeness" is understood here as the level of detail achieved by the prediction; that is, its precision.

who rejects the proposal—that was dominant then—on the logical symmetry between explanation and prediction. But, besides the criticism to the characterization of this thesis by Hempel and Oppenheim, Rescher came to develop his own alternative to the symmetry thesis. He calls it *harmony thesis* between explanation and prediction, since it emphasizes the idea of complementarity. The epistemological and, above all, methodological nexus between scientific explanations and predictions are highlighted in this harmony thesis.[33]

3.4.2   Rescher's Proposal on the Nexus between Explanation and Prediction

After rejecting the symmetry thesis—which had a great influence on the Received View—Rescher considers what the relations between scientific explanation and prediction are. He thinks that they are different processes from a logical point of view. In addition, he offers epistemological, methodological, and ontological reasons that support the asymmetry between explanation and prediction. He develops his own proposal from the acknowledgement of the logical asymmetry between them. Thus, the "harmony thesis between explanation and prediction" maintains that, even when to explain and to predict are asymmetrical processes, they are closely interrelated (Rescher 1998, pp. 167–169).

However, this proposal of Rescher is not, strictly speaking, a properly *logical* account of the relations between explanation and prediction. It is of a preferentially *methodological* character, insofar as it sees the problem in terms of the advancement of knowledge. In effect, his starting point is the following question: "Yet what is to be said about the relative priority of prediction versus explanation in science once one abandons the supposed equivalence at issue in the Hempel-Oppenheim thesis of logical symmetry?" (Rescher 1998, p. 167). This means that, after considering the debate on the logical symmetry or asymmetry between explanation and prediction, Rescher deals with the problem in new terms, which are oriented toward the methodological relevance of prediction and explanation.

From this perspective, the debate on the logical symmetry or asymmetry between prediction and explanation connects with the controversy about which has more relevance in evaluating scientific contents: the accommodation to what is already known or the prediction about novel facts (see Sect. 8.3.2 in this book). In this regard, Rescher's approach is certainly predictivist to the extent that he considers that "prediction is the very touchstone of science in that it affords our best and most effective test for the adequacy of our scientific endeavors" (1998, p. 161). It is, however, a *weak* or *moderate predictivism*.[34] In effect, he explicitly criticizes the

---

[33] The "harmony thesis" is already outlined in his book on scientific explanation: "The key thing in scientific understanding is the capacity to exploit a *knowledge of laws* to structure our understanding of the past and to guide our expectations for the future," Rescher (1970, p. 135).

[34] On the distinction between the strong versions of predictivism and the weak predictivism, see Hitchcock and Sober (2004) and Barnes (2008).

instrumentalist approaches to prediction,[35] according to which "prediction is *all* that matters and thereby constitutes the alpha and omega of science" (1998, p. 164).

Within the framework of his pragmatic conception, where the advancement of knowledge is oriented towards aims, Rescher thinks that to explain and to predict are not symmetrical or equivalent processes. He sees them as a part of a whole, so they are processes that should be *coordinated*, within a systemic approach to science (Rescher 1998, pp. 167–169). For this reason, there cannot be—in his judgment—an absolute instrumental priority of prediction with regard to explanation. Thus, he considers that "theories that do not yield predictions are sterile, and predictions—however successful—that lack a theoretical backing are for that very reason cognitively unsatisfactory" (1998, p. 167).[36]

After acknowledging a methodological difference, with preference for prediction, Rescher suggests the "harmony thesis" between explanation and prediction. It involves a functional complementarity: "scientific adequacy [...] involves a complex negotiation in which *both* prediction *and* explanation play a symbiotic and mutually supportive role" (Rescher 1998, p. 167). He summarizes this thesis in three principles: "1. To qualify as well established, our explanatory theories must have a track record of contributing to predictive success. 2. To qualify as credible, our predictions must be based upon theories that militate for these particular predictions over and against other possibilities. 3. Our explanatory theories should be embedded in a wider explanatory framework that makes it possible to understand why they enjoy their predictive successes" (1998, p. 168).[37]

According to this perspective, an adequate scientific *understanding* of phenomena involves the capacity to explain and predict those phenomena, at least in principle. Thus, on the one hand, scientific theories oriented toward explanation should—in Rescher's judgment—yield predictions. This fits with his methodological pragmatism, whose final goal is say how to evaluate the truth or verisimilitude of scientific knowledge. Faced with this goal of scientific activity, the capacity of theories to yield successful predictions prevails. Moreover, he maintains that "explanatory theories that yield no predictive advantages are [...] deficient. For in the final analysis only their role in providing for correct predictions can validate theories as adequate" (1998, p. 167).

---

[35] In this regard, his criticism to the predictivist thesis of Milton Friedman can be highlighted. See Rescher (1998, pp. 109 and 194–196).

[36] The first part of this proposal should be qualified, because there are examples of scientific theories that are really influential and that are not oriented towards prediction. Thus, Charles Darwin did not orient his theory of evolution towards the elaboration of predictions, but his theory was not sterile regarding the phenomena at stake. Certainly, the Darwinian evolutionary approach might generate predictions of new species, but there is no evidence that this was the aim of the author of *The Origin of Species*.

[37] Even when Rescher's proposal certainly makes sense, it could be qualified. It is possible to think of explicative theories of historical character that do not aim to make predictive contributions, at least in a direct way. There can also be considered predictive theories with correct predictions that do not yet have a well-developed explicative theory.

Meanwhile, scientific prediction should be made on the basis of reasons: science is not interested in predictive success without rational basis.[38] Thus, "with any *cogent* prediction [...] one should be able to provide a validating rationale as to why that prediction is acceptable (a rationale that need not necessarily qualify as an explanation of the phenomenon being predicted)" (Rescher 1998, p. 167). As predictive success can be not good enough for science, he manifestly rejects methodological instrumentalism. In this way, it is necessary to give the reasons that support a prediction in order to consider it as a *scientific* prediction. In this regard, Rescher thinks that "explanatory theories are best situated to yield effective predictions in a systematic and reliable manner" (1998, p. 167).

Thus, Rescher clearly rejects that explanation and prediction are symmetrical or equivalent processes. In addition, he does this on the basis of reasons that show their asymmetry in diverse realms: semantic,[39] logical, epistemological, methodological, and ontological. However, he thinks that they are coordinated process, so science must aim to achieve a harmony between explanation and prediction. In this way, he also rejects the instrumentalist predictivism and opts for a moderate version of predictivism, which is, in my judgment, a more adequate position to scientific practice.

3.5 The Temporality Factor

Salmon has also offered a quite interesting review of the symmetry thesis where the logical elements have more weight than in the case of Rescher (Salmon 1993). In fact, Salmon directly objects Grünbaum's approach to symmetry, which gives special relevance to the logical dimension of the problem from a characterization of the scientific laws. Thus, the temporality factor—the temporal anisotropy between explanation and prediction—is emphasized (Grünbaum 1962).

This issue of temporality connects with the problems posed by the notion of "retrodiction." In effect, two logical issues can be considered here. On the one hand, there is the problem of the possible equivalence of "retrodiction" and scientific

---

[38] For example, an astrologer may predict successfully, but he does not offer genuine scientific knowledge.

[39] Semantic reasons are not explicit in Rescher's criticism to the symmetry thesis. But it is implicit that there are differences between explanation and prediction from the point of view of language to the extent that he characterizes scientific prediction as a *statement*. It seems clear that the referent of a statement about the future can be different from what exists now or what existed in the past (that is the realm of explanation). The semantic differences have logical repercussions with regard to the problem of the symmetry.

However, this view of prediction as a statement is not clear in the whole set of Rescher's publications on this issue. Thus, in a paper of 1963, he characterizes prediction as an *argument*: "A *potential prediction* of the supposed fact that a system *will* exhibit the characteristic Q at time t is an argument whose conclusion is the statement that the system exhibits Q at t," Rescher (1963, p. 329).

explanation, which arises when the possibility of explaining on the basis of subsequent conditions to *explicandum* (the fact that is explained) is accepted. And, on the other hand, there is the question of the logical equivalence between retrodiction and prediction, where retrodiction is usually understood as a genuine "prediction of past."

3.5.1   Is there just an Anisotropy between Explanation and Prediction?

Following the thread of the logical-methodological proposal of Adolf Grünbaum this question is posed as a key for solving this problem. Grünbaum has been one of the major defenders of the symmetry thesis between explanation and prediction. In his judgment, many of the objections to this thesis (among them, those by Rescher) are due to an inadequate understanding of Hempel's proposal on this matter (Grünbaum 1962). For this reason, he considers that the first step is to shed light on the symmetry thesis as it was formulated by Hempel and Oppenheim in their well-known paper of 1948.

According to Grünbaum's interpretation of the symmetry thesis, we have that, in order to account for an *explanandum*-event (E), we need some particular conditions C_i ($i = 1, 2, ... n$) and the relevant laws. Particular conditions may be earlier or later than the *explanandum* in both explanation and prediction. Therefore, Hempel's criterion for distinguishing explanation and prediction is the temporal factor. When E is in the past with respect to the scientist, we have an explanation; and, when E is in the future, we have a prediction (Grümbaum 1962, p. 156). In effect, the difference between prediction and explanation is, for Hempel and Oppenheim, a *temporal anisotropy* with regard to the subject who explains or predicts the concrete fact. Thus, in scientific explanation, the fact has already happened, whereas in prediction the fact has not yet occurred (Hempel and Oppenheim 1948, p. 138).

On the basis of this characterization, Grünbaum tries to answer Rescher's objections with regard to the symmetry thesis. In his judgment, two different types of asymmetries between explanation and prediction should be distinguished: (i) epistemological asymmetry, and (ii) logical asymmetry. Epistemological asymmetry has to do with the "assertibility" of the *explanandum*, whereas the logical asymmetry has to do with the "inferability" of the *explanandum*. In this way, Grünbaum thinks that on the basis of the epistemological asymmetry between explanation and prediction it is not possible to maintain that there is also a logical asymmetry between them (see Grünbaum 1962, p. 159)..

Certainly, Rescher's criticism of the symmetry thesis takes into account several aspects, and the epistemological asymmetry between explanation and prediction is one among them, because, in his approach, science is a *system*, so the diverse elements that conforms science are interrelated. Consequently, besides the logical features, there are semantic, epistemological, methodological, and ontological aspects that support the asymmetry between explanation and prediction (Rescher 1998, pp. 165–166). From this perspective, it seems to me that Grünbaum is right when he claims that, in Rescher's criticism to symmetry thesis, the logical elements are in the background; whereas—for Hempel and Grünbaum—it is a proposal on the *logical* relations between scientific explanations and predictions.

But, in my judgment, Grünbaum is wrong when he claims that Rescher rejects the symmetry thesis on the basis of the epistemological asymmetry between explanation and prediction. Rather, there is a broader approach at stake: Rescher shows that there are several asymmetries between both processes. These differences have to do with the logical, epistemological, methodological, and ontological realms (where it is possible to add the semantic field). In this way, he offers a broad view of the relations between explanation and prediction. In addition, he gives reasons that lead to reject the symmetry between them in the different realms that can be considered.

It should be highlighted that Salmon directly criticizes Hempel's symmetry thesis on the basis of the structural elements. He does this in a book in honor of Grünbaum and due to their friendship, Salmon writes his paper as a letter and includes kind remarks about him (Salmon 1993). Certainly, Grünbaum was one of the major defenders of the symmetry thesis. Salmon is well aware of that and his starting point is the analysis of the two parts of Hempel's symmetry thesis: (i) every scientific explanation can, under suitable circumstances, serve as a prediction; and (ii) every scientific prediction can, under suitable circumstances, serve as an explanation.[40]

In the first part of the argument, Salmon considers that "*seems* clearly hold for D-N explanations" (1993, p. 231). In this way, he notes that, within a deductive-nomological model of scientific explanation, an argument that provides an explanation of a past event might also serve to anticipate it in the future (for example, the knowledge used to explain a solar eclipse can also serve to predict it). Thus, "if the explanatory facts had been at our disposal before the occurrence of the fact to be explained (the explanadum) we would have been able to predict that fact, for we would have been in possession of true premises from which it follows deductively" (Salmon 1993, p. 231).

Meanwhile, the second part of the Hempelian argument is more problematic, as Peter Achinstein admits,[41] because there is a logical difference: Salmon sees scientific prediction as a *statement*, whereas for Hempel scientific explanation is an *argument*. From this perspective, prediction cannot be an explanation, but it has its own characteristics. Thus, the most that can be maintained is that a prediction can be the conclusion of an argument that conforms to either a deductive-nomological model

---

[40] An analysis of Salmon criticism of Hempel's proposal is in Gonzalez (2002, pp. 18–19). See also Gonzalez (2010, pp. 216–217, 2015a, pp. 48–50).

[41] "This part of the symmetry thesis is usually thought to be [the] more dubious of the two. To begin with, it might be said, a prediction is simply a statement that something will occur (or perhaps that something has occurred that has not been verified)," Achinstein (2000, p. 168).

of explanation or to an inductive-probabilistic model (Salmon 1993, p. 232).[42] In addition, Salmon considers that this is the position adopted by Grünbaum in his defense of the symmetry thesis (Salmon 1993, p. 232).

It also happens that—in Grünbaum's judgment—Hempel's symmetry thesis has consequences for the notion of retrodiction, because, to the extent that the temporal anisotropy is with regard to the subject who explains or predicts, he considers that it has no relevance whether the initial conditions are antecedent conditions (previous to the fact that we want to explain) or subsequent conditions (later than the fact that we want to explain). In this way, Grünbaum maintains that both a retrodiction and a prediction can be an H-prediction and both a prediction and a retrodiction can be an H-explanation (Grünbaum 1962, p. 156).

Because Salmon focuses on examples from physics, a central problem in this regard has to do with the reversibility or irreversibility of the processes of nature and their relation with causal processes. On this issue, Salmon points out a series of considerations: (1) Objectively, the universe possesses a temporal anisotropy based on irreversible physical processes. (2) The fundamental laws of nature are symmetric from a temporal point of view. (3) The temporal anisotropy of the universe is something *de facto* not *de jure*; that is, it is based on matters of fact, not on temporally asymmetric physical laws. (4) Although many laws of nature are temporally symmetric, there are *de facto* in nature processes that are irreversible (Salmon 1993, pp. 241–242).

These are considerations that Salmon attributes to Reichenbach, and he thinks that his colleague Grünbaum would agree with them (Salmon 1993, p. 241). Thus, Salmon analyzes the consequences of these claims about the "direction of time" for the problem of the logical symmetry between explanation and prediction. In his judgment, the temporal anisotropy applies to reversible physical systems and to irreversible ones: "We distinguish, consequently, earlier and later where lunar and planetary motions are concerned just as we do with regard to ice cubes melting in glasses of ginger ale" (1993, p. 242).

This leads to three *anisotropies* that go together: (i) the anisotropy of time, (ii) the anisotropy of causation, and (iii) the anisotropy of scientific explanation (Gonzalez 2010, p. 217), because, on the one hand, causes normally precede the effects; and, on the other, if we want to explain a phenomenon or event it is not good enough to know its effects—subsequent to the phenomenon—we have to consider

---

[42] However, for Achinstein, to acknowledge that prediction is a statement does not imply an objection to this part of the symmetry thesis, because "Hempel is concerned not just with predictive statements (ones that say something will occur), but with predictive *arguments* or *inferences*, that is, with cases in which some prediction is made from, or on the basis of, something," Achinstein (2000, pp. 168–169).

He considers that a fairer objection is provided by the following example: "Suppose a drug company tests a drug on one thousand patients with symptoms S and discovers that in eight hundred cases the symptoms are relieved, while no one in a control group not taking the drug had relief. This might provide a very sound scientific basis for the prediction that the drug will be effective approximately 80 percent of the time. Yet the explanation for the drug's general effectiveness is not that it was effective in the test cases" (2000, p. 169).

the causes that are, in principle, prior to the phenomenon or event that we want to explain.[43]

3.5.2 Are Retrodiction and Prediction Logically Equivalent?

According to Grünbaum's interpretation of Hempel and Oppenheim's symmetry thesis, a retrodiction can be about past or future phenomena, so it could be equivalent to H-explanation or to H-prediction, respectively (Grünbaum 1962, p. 156). The difference is that, in a retrodiction, information used to make the inference is *subsequent* to the explained or predicted phenomenon; that is, the conditions used are *subsequent* conditions instead of *antecedent* conditions.

Salmon gives an example in which a prediction in the sense of Hempel (H-prediction) can be a retrodiction: "If, for example, the Astronomer Royal had, in 1917, established the state that the sun-moon-earth system would assume in 1921, and had (still in 1917) derived the occurrence of the 1919 eclipse from the 1921 conditions, his inference would have been both a retrodiction and an H-prediction" (1993, p. 235). This notion of "retrodiction" as equivalent to "prediction" is not especially problematic, since the information used deals with events of the past or the present and is about eclipses, which have a well-known regularity, and it is, *de facto*, "bi-directional."

On this issue, Rescher's account is the following: "any sort of rational prediction (...) will accordingly require informative input material that indicates that three conditions are satisfied: 1. that relevant information about the past-&-present can be obtained in an adequately timely, accurate, and reliable way, 2. that this body of data exhibits discernible patterns, and 3. that the patterns so exhibited are stable, so that this structural feature manifests a consistency that also continues into the future" (Rescher 1998, p. 86).

It could be considered that, in Salmon's example above, these preconditions of rational prediction pointed out by Rescher are met in both cases: first, in the prediction about the state of the solar system in 1921; and, second, in the prediction of the eclipse of 1919 using a prediction made previously (the state of the solar system in a more distant future). However, this is certainly not the usual curse of action in scientific practice.

In effect, as Salmon points out, this usage of the term *retrodiction* "may seem a bit odd from the standpoint of the ordinary usage" (1993, p. 235), because, frequently, "retrodiction" is understood as a "prediction of past" (for example, an

---

[43] The temporal precedence of the cause with regard to the effect is something clear in Salmon's approach: "Time and causality go hand in hand. The anisotropy of time is deeply connected to the anisotropy of causality. Causes come before their effects, not after them. Now, if one agrees that causality is an indispensable component of scientific explanations of *particular events*, it is natural to suppose that the anisotropy of time and causality would be reflected in an anisotropy of scientific explanation," Salmon (1993, pp. 242–243).

inference in the realm of archeology or regarding human evolution), instead of being a "prediction of future" made on the basis of subsequent conditions to the predicted event (for example, to anticipate in 1917 an eclipse that will occur in 1919 on the basis of the knowledge about the solar system's state in 1921).

Another piece of information that can clarify this discussion is the notion of "retrodiction" as "prediction of past," which appears in authors such as Stephen Toulmin, who maintained that *prediction* is an "assertion about the occurrence of a particular sort of event—whether in the past, present, or future" (Toulmin 1961, p. 31). In this regard, Wenceslao J. Gonzalez points out that prediction has a vague meaning in Toulmin "as a consequence of the sense of prediction as 'testable implication', whose reference could be in the past, present or future" (2013b, p. 355).

It should be emphasized that, faced with this option with regard to temporality, Rescher does not accept this possible usage of the term "retrodiction" as prediction of past. Expressly, he claims that "a *potential prediction* of the supposed fact that a system *will* exhibit the characteristic Q at time t is an argument whose conclusion is the statement that the system exhibits Q at t" (1963, p. 329). Meanwhile, "a *potential retrodiction* of the purported fact that a system has exhibited the characteristic Q at time t is an argument whose conclusion is the statement that the system exhibits Q at t" (1963, p. 329). Thus, on the basis of the language used, there is a clear difference between a prediction and a retrodiction (see Chap. 2, Sect. 2.3.3).

Another possibility about this problem is offered by the term "postdiction." Following the publication of *Experience and Prediction*, it has been considered that the major innovation of Hans Reichenbach regarding the problem of prediction was the introduction of the term "postdictability" as opposed to "predictability" (see, in this regard, Gonzalez 2010, pp. 48–49).[44] With this notion of "postdictability" the temporal factor is emphasized, so what is "pre-dicted" is something previous to the phenomenon at issue; whereas what is "post-dicted" occurs once the phenomenon has happened. However, "prediction" and "retrodiction" are, in my judgment, different notions (Gonzalez 1995, pp. 37–58; especially, pp. 53–54). The differences between them can be clearly seen in Reichenbach's proposal on "postdictability."

After Reichenbach, Imre Lakatos also used the term "postdiction," but he did this in a way that "prediction" also encompasses post-diction: "I use 'prediction' in a wide sense that includes 'postdiction'" (Lakatos 1970, p. 32n), which in his approach is a sort of "retrodiction." Thus, for Lakatos, prediction can be about past occurrences, instead of being always oriented toward the future (Gonzalez 2001, 2014). But it happens again that there are reasons for the differences between prediction and retrodiction, both in basic science and in applied science (Gonzalez 2015b).

Regarding basic science, the anticipation of the possible future is not the same as the enunciation of a past, which is something that has already happened and therefore will not change. Human future is open, whereas the past is already closed (at least, with regard to the existence of the phenomena). This can be seen in normal scientific practice. Concerning applied research (as well as in the subsequent phase of application of science) the difference with retrodiction is clear due to the

---

[44] The notion of "postdictability" appears for the first time in his (1944, p. 13).

connections between prediction and prescription. Thus, prediction offers knowledge about the possible future that is needed to solve the specific problem posed. It is a role that cannot encompass "retrodiction," "insofar a statement about the past does not offer us, in principle, any relevant information about the future whose problems we try to solve by means of the prescription" (Gonzalez 2015b, p. 139).

Lakatos' focus of attention is on the methodologically positive character of prediction. Reichenbach, however, was not interested in the controversy about the logical symmetry or asymmetry between explanation and prediction. Nevertheless, Hempel and Oppenheim used Reichenbach's proposal as a source that supports the symmetry thesis. In their judgment, "the logical similarity of explanation and prediction, and the fact that one is directed toward past occurrences, the other toward future ones, is well expressed in the terms 'post-dictability' and 'predictability' used by Reichenbach" (Hempel and Oppenheim 1948, p. 138, n. 2a).

But it is possible to think that Hempel and Oppenheim failed in their interpretation of Reichenbach's proposal. They did this—in my judgment—in two different senses: (a) insofar as they considered that "post-diction" can be equivalent to "explanation;" and (b) when they consider that Reichenbach laid the foundations of the defense of the logical symmetry between "explanation" and "prediction." It seems to me that, when the most notable features of Reichenbach's conception of prediction are analyzed, this interpretation by Hempel and Oppenheim has problems.

The key is that, for Reichenbach, the relation between the statements that deal with past phenomena and the statements about future developments does not rest on a possible logical symmetry, but on the probabilistic character of scientific knowledge. In his judgment, there is "a close connection between the weights of propositions concerning past events and predictions: their weights enter into the calculations of predictional values of future events which are in causal connection with the past event" (Reichenbach 1949, p. 27).

Therefore, with the introduction of the term "postdictability," Reichenbach "is emphasizing the limits of science in accordance with his probabilistic view" (Gonzalez 2010, p. 48). This is because, in his judgment, "the same limitation follows for the determination of past data in terms of given observations, and that we therefore must also speak of a limitation of *postdictability*" (Reichenbach 1944, p. 13). So Reichenbach is emphasizing that—as happens with regard to prediction—"all relations between observational data are restricted to statistical relations" (1944, p. 49).

In *The Direction of Time*, Reichenbach also dealt with the problems of postdictability. He pointed out there that there is an important difference between postdiction and prediction, which is due to the sort of information and knowledge required in each case: "*Predictions* require a knowledge of the total cause; *postdictions*, or statements about past events, can be based on partial effects, on records" (Reichenbach 1956, p. 22). Thus, he highlighted the divergent points between "predictability" and "postdictability," which basically has to do with the epistemological and methodological realms. So he went beyond the mere temporal anisotropy between them. On this basis, it is possible to maintain that—in his approach—there is not a logical equivalence between "postdiction" and "prediction."

To sum up, when prediction is analyzed from the logic of science, which deals with the structural features of scientific theories, there are several questions at stake. First, the problem of the limits of deductivism to prediction must be considered. In this regard, the elements offered by Rescher and Salmon's criticism emphasize the insufficiency of deductivism to solve the problems posed by scientific prediction. Secondly, there is the problem of *induction*, which Rescher tries to solve in a positive way. Thus, he thinks that it is possible to justify the validity of the inductive reasoning for scientific prediction on the basis of pragmatic considerations. This involves a change of perspective, since the strictly logical perspective is rejected in favor of an epistemological approach to induction that has methodological repercussions.

Thirdly, there are the problems that arise when prediction is compared with scientific explanation in terms of their logical structure. In this regard, it seems clear that there is a *logical asymmetry* between explanation and prediction, so a structural equivalence between both intellectual processes can be rejected. Finally, the perspective of *temporality* gives more elements in favor of the asymmetry thesis, at the same time that it offers an adequate framework to address the analysis of the logical differences between prediction and retrodiction. These differences can be seen in basic science and, especially, in applied science.

References

Achinstein, P. 2000. The symmetry thesis. In *Science, explanation, and rationality. The philosophy of Carl G. Hempel*, ed. J.H. Fetzer, 167–185. Oxford: Oxford University Press.

Barnes, E.C. 2008. *The paradox of predictivism*. Cambridge: Cambridge University Press.

Black, M. 1984a. Inducción. Spanish translation by Pascual Casañ. In *Inducción y probabilidad*, M. Black, 33–83. Madrid: Cátedra.

———. 1984b. Probabilidad. Spanish translation by Rafael Beneyto, in *Inducción y probabilidad*, M. Black, 85–148. Madrid: Cátedra.

Douglas, H.E. 2009. Reintroducing prediction to explanation. *Philosophy of Science* 76 (4): 444–463.

Echeverría, J. 1995. El pluralismo axiológico de la Ciencia. *Isegoría* 12: 44–79.

Galavotti, M. C. 2011. The modern epistemic interpretations of probability: Logicism and subjectivism. In *Handbook of the history of logic. Vol. 10: Inductive logic*, ed. D.M. Gabbay, S. Hartmann, and J. Woods, 153–203. Oxford: Elsevier.

Gillies, D. 2000. *Philosophical theories of probability*. London: Routledge.

Gonzalez, W.J. 1995. Reichenbach's concept of prediction. *International Studies in the Philosophy of Science* 9 (1): 37–58.

———. 2001. Lakatos's approach on prediction and novel facts. *Theoria* 16 (3): 499–518.

———. 2002. Caracterización de la "explicación científica" y tipos de explicaciones científicas. In *Diversidad de la explicación científica*, ed. W.J. Gonzalez, 13–49. Barcelona: Ariel.

———. 2004a. The many faces of Popper's methodological approach to prediction. In *Karl Popper: Critical appraisals*, ed. Ph. Catton and G. Macdonald, 78–98. London: Routledge.

———. 2004b. La evolución del Pensamiento de Popper. In *Karl Popper: Revisión de su legado*, ed. W.J. Gonzalez, 23–194. Madrid: Unión Editorial.

———. 2010. *La predicción científica. Concepciones filosófico-metodológicas desde H. Reichenbach a N. Rescher.* Barcelona: Montesinos.

———. 2013a. The roles of scientific creativity and technological innovation in the context of complexity of science. In *Creativity, innovation, and complexity in science*, ed. W.J. Gonzalez, 11–40. A Coruña: Netbiblo.

———. 2013b. Scientific prediction in the beginning of the "Historical Turn:" Stephen Toulmin and Thomas Kuhn. *Open Journal of Philosophy* 3: 351–357.

———. 2014. The evolution of Lakatos's repercussion on the methodology of economics. *HOPOS: The Journal of the International Society for the History of Philosophy of Science* 4 (1): 1–25.

———. 2015a. *Philosophico-methodological analysis of prediction and its role in economics.* Dordrecht: Springer.

———. 2015b. Prediction and prescription in biological systems: The role of technology for measurement and transformation. In *The future of scientific practice: 'Bio-Techno-Logos'*, ed. M. Bertolaso, 133–146 and 209–213. London: Pickering and Chatto.

Grünbaum, A. 1962. Temporally-asymmetric principles, parity between explanation and prediction, and mechanism versus teleology. *Philosophy of Science* 29 (2): 146–170.

Hájek, A., and A. Hall. 2002. Induction and Probability. In *The Blackwell guide to the philosophy of science*, ed. P. Machamer and M. Silberstein, 149–172. Oxford: Blackwell.

Hempel, C.G. 1962. Deductive-nomological vs. statistical explanation. In *Minnesota studies in the philosophy of science, vol. 3, Scientific explanation, space, and time*, ed. H. Feigl and G. Maxwell, 98–169. Minneapolis: University of Minnesota Press.

———. 1981. Turns in the evolution of the problem of induction. *Synthese* 46 (3): 389–404.

Hempel, C., and P. Oppenheim. 1948. Studies in the logic of explanation. *Philosophy of Science* 15: 135–175.

Hitchcock, Ch., and E. Sober. 2004. Prediction versus accommodation and the risk of overfitting. *British Journal for the Philosophy of Science* 55 (1): 1–34.

Hilpinen, R. 2000. Hempel on the problem of induction. In *Science, explanation, and rationality. The philosophy of Carl G. Hempel*, ed. J.H. Fetzer, 91–107. Oxford: Oxford University Press.

Hume, D. 2007. In *A treatise of human nature*, ed. D.F. Norton and Mary J. Norton, vol. 1. Oxford: Oxford University Press.

Lakatos, I. 1970. Falsification and the methodology of scientific research programs. In *Criticism and the growth of knowledge*, ed. I. Lakatos and A. Musgrave, 91–196. Cambridge: Cambridge University Press. Reprinted in Lakatos, I. 1978. *The methodology of scientific research programmes. Philosophical papers*, vol. 1, 8–101. Cambridge: Cambridge University Press.

Lange, M. 2011. Hume and the problem of induction. In *Handbook of the history of logic. Vol. 10: Inductive logic*, ed. D.M. Gabbay, S. Hartmann, and J. Woods, 43–91. Oxford: Elsevier.

Martínez Solano, J. F. 2005. *El problema de la verdad en K. R. Popper: Reconstrucción histórico-sistemática.* A Coruña: Netbiblo.

Nepomuceno, A. 2014. Scientific models of abduction: The role of non classical logic. In *Bas van Fraassen's Approach to Representation and Models in Science*, ed. W.J. Gonzalez, 121–141. Dordrecht: Springer.

Niiniluoto, I., and R. Tuomela. 1973. *Theoretical concepts and hypothetico-inductive inference.* Dordrecht: Reidel.

Popper, K.R. 1959. The propensity interpretation of probability. *The British Journal for the Philosophy of Science* 10 (37): 25–42.

———. 1972. Conjectural knowledge: My solution of the problem of induction. In *Objective knowledge. An evolutionary approach*, ed. K.R. Popper, 1–31. Oxford: Clarendom Press (5th revised edition, 1979).

———. 1974. Replies to my critics. In *The philosophy of Karl Popper*, ed. P.A. Schilpp, 961–1197. La Salle: Open Court.

―――. 1976. *Unended quest. An intellectual autobiography*. London: Fontana/Collins (enlarged ed. 1992. London: Routledge. Reprinted in 2002).

Reichenbach, H. 1936. Logistic empiricism in Germany and the present state of its problems. *Journal of Philosophy* 33 (6): 141–160.

―――. 1944. *Philosophical foundations of quantum mechanics*. Berkeley: University of California Press.

―――. 1949. *Experience and prediction. An analysis of the foundations and the structure of knowledge*. Chicago: The University of Chicago Press (1st ed. 1938).

―――. 1956. *The direction of time*, ed. by M. Reichenbach. Berkeley: University of California Press.

―――. 1968. Predictive knowledge. In *The rise of scientific philosophy*, H. Reichenbach, 229–249. Berkeley: University of California Press (1st ed. 1951).

―――. 1978. Induction and probability. Remarks on Karl Popper's *The logic of scientific discovery*. In *Selected writings*, vol. 2, H. Reichenbach, 372–387. Dordrecht: Reidel (Edited by M. Reichenbach and R. Cohen; translated by E. Schneewind). Originally published as Reichenbach, H. 1935. Über Induktion und Wahrscheinlichkeit. Bemerkungen zu Karl Poppers *Logik der Forschung*. *Erkenntnis* 5(4): 267–284.

Rescher, N. 1958. On prediction and explanation. *British Journal for the Philosophy of Science* 8 (32): 281–290.

―――. 1963. Discrete state systems, Markov chains, and problems in the theory of scientific explanation and prediction. *Philosophy of Science* 30 (4): 325–345.

―――. 1970. *Scientific explanation*. New York: The Free Press.

―――. 1977. *Methodological pragmatism. A systems-theoretic approach to the theory of knowledge*. Oxford: Blackwell.

―――. 1980. *Induction*. Oxford: Basil Blackwell.

―――. 1995. Predictive incapacity and rational decision. *European Review* 3 (4): 327–332. Collected in Rescher, N. 2003. *Sensible decisions. Issues of rational decision in personal choice and public policy*, 39–47. Lanham: Rowman and Littlefield.

―――. 1998. *Predicting the future. An introduction to the theory of forecasting*. New York: State University of New York Press.

―――. 2000. *Realistic pragmatism*. Albany: State University of New York Press.

―――. 2012a. Pragmatism at the crossroads. In *Pragmatism. The restoration of its scientific roots*, N. Rescher, 1–19. N. Brunswick: Transaction Publishers.

―――. 2012b. A pragmatic justification of induction. In *Pragmatism. The restoration of its scientific roots*, N. Rescher, 133–149. N. Brunswick: Transaction Publishers.

―――. 2014. *The pragmatic vision. Themes in philosophical pragmatism*. Lanham: Rowman and Littlefield.

Rescher, N., and O. Helmer. 1959. On the epistemology of the inexact sciences. *Management Sciences* 6: 25–52.

Salmon, W.C. 1981. Rational prediction. *British Journal for the Philosophy of Science* 32: 115–125.

―――. 1991. Hans Reichenbach's vindication of induction. *Erkenntnis* 35 (1): 99–122.

―――. 1993. On the alleged temporal anisotropy of explanation. A letter to Professor Adolf Grünbaum from his friend and colleague. In *Philosophical problems of the internal and external worlds. Essays on the philosophy of Adolf Grünbaum*, ed. J. Earman, A. Janis, G. Massey, and N. Rescher, 229–248. Pittsburgh: University of Pittsburgh Press.

―――. 1998. *Causality and explanation*. New York: Oxford University Press.

Sankey, H. 2006. Why is it rational to believe scientific theories are true? In *Rationality and reality. Conversations with Alan Musgrave*, ed. C. Cheyne and J. Worrall, 109–132. Dordrecht: Springer.

Scheffler, I. 1957. Explanation, prediction, and abstraction. *British Journal for the Philosophy of Science* 7: 293–309.

Sintonen, M. 2004. Reasoning to Hypotheses: Where do Questions Come? *Foundation of Science* 9: 249–266.

Toulmin, S. 1961. *Foresight and understanding*. Bloomington: Indiana University Press.
van Fraassen, B.C. 2009. Rescher on explanation and prediction. In *Rescher studies: A collection of essays on the philosophical work of Nicholas Rescher presented to him on the occasion of his 80th birthday*, ed. R. Almeder, 339–361. Heusenstamm: Ontos Verlag.
Vickers, J. 2014. The problem of induction. In *The Stanford Encyclopedia of* Philosophy, ed. E. Zalta., *http://plato.stanford.edu/archives/fall2014/entries/induction-problem/*. Accessed 10 Dec 2014.
Whewell, W. 1858. *Novum Organon Renovatum*. London: J. W. Parker and Son.

Part II
Predictive Knowledge and Predictive Processes in Rescher's Methodological Pragmatism

Chapter 4
Epistemological Factors of Scientific Prediction

Abstract This chapter addresses the pragmatic idealistic approach to rationality and the problems that arise due to the relations between rationality and the epistemological factors of scientific prediction. In this regard, five steps are followed: (1) Rescher's theory of rational choice—according to which human being's abilities for reason are bounded—is analyzed. (2) The main types of rationality (cognitive, practical, and evaluative) are addressed, and their relations with scientific prediction are considered. (3) The attention goes to the fallibilism that characterizes Rescher's epistemological approach, which is also related to predictive knowledge. (4) The epistemological limits to predictability are analyzed, where the problem of uncertainty can be highlighted. Another problem considered is how these limits can affect the decisions made on the basis of knowledge about the future. (5) The problem of risk is analyzed in relation to cognitive rationality, as a framework for addressing the nexus between risk and scientific prediction.

Keywords Rationality • Scientific prediction • Fallibilism • Epistemological limits • Uncertainty • Risk

Initially, scientific prediction can be characterized as a kind of prediction based on reasons. This feature connects the epistemological field of prediction with the general realm of human rationality, which leads to the specific realm of scientific rationality. In this regard, Nicholas Rescher's epistemology of scientific prediction has links with his approach to rationality.[1] He accepts a normative conception of rationality and gives primacy to practical reason, which, in turn, leads to an evaluative rationality. In his judgment, rationality is not simply a matter of adjustment of means to ends, because the very ends should be subject to rational deliberation (see,

---

[1] Rescher has addressed the problem of rationality in many publications. But, among them, Rescher (1988) should be highlighted. Among his papers devoted to this topic, of note are Rescher (2004, 2006a, b).

© Springer International Publishing AG 2017   101
A. Guillán, *Pragmatic Idealism and Scientific Prediction*, European Studies in Philosophy of Science 8, DOI 10.1007/978-3-319-63043-4_4

for example, Rescher 1999a). His approach is in terms of *practicable* rationality, which has some similarities with Herbert A. Simon's bounded rationality.[2]

This present chapter addresses Rescher's approach to rationality and the problems that arise due to its relation with predictive knowledge. Although he insists on the primacy of practice (Rescher 1973, Marsonet 1996), knowledge is his starting point, so the cognitive realm is crucial in his conception of human rationality, in general, and scientific rationality, in particular (see Rescher 1992a). Thus, first, his theory of rational choice is analyzed. According to this holistic theory, human being's abilities for reason are bounded. Secondly, the main types of rationality (cognitive, practical, and evaluative) are addressed. These types of rationality, as Rescher admits (1988, pp. 2–3), can be detected in I. Kant. In addition, their relation with scientific prediction is considered, insofar as prediction is, for Rescher, the result of a rational process.

Thirdly, the attention goes to the fallibilism that characterizes Rescher's epistemological approach, which is also related to predictive knowledge. Regarding this issue, two questions arise: (a) the problem of determining the truth of the predictive statements, and (b) the knowledge of the variables that are important for the prediction. Fourthly, the epistemological limits to predictability are analyzed, where the problem of uncertainty can be highlighted. A further problem considered is how these limits can affect the decisions made on the basis of knowledge about the future. Finally, the problem of risk is analyzed in relation with cognitive rationality, as a framework for addressing the nexus between risk and scientific prediction.

4.1 A Holistic Theory of Rationality

Rescher suggests an approach to rationality that gives primacy to the practical dimension, so it is not a "pure reason" (Gonzalez 1999, pp. 24–27). Thus, he distinguishes the realm of the intelligence from that of rationality, where a diversity of possibilities arises. Initially, rationality has two sectors: the theoretical sphere and the practical area. The latter has primacy in his approach, insofar as the pragmatic dimension prevails. But he admits that human rationality is subject to limits,[3] which are noticeable in the cognitive realm. Thus, he rejects the notion of "maximization," which involves having the whole range of possibilities in a context, and suggests the concept of "optimization," which means the acknowledgement of limitations, so it is oriented towards the *best we can do* in each concrete context (Rescher 1987).

---

[2] On the notion of *bounded rationality*, see the papers compiled in Simon (1982a, b, 1997a). It might be considered that Rescher is in tune with Simon with regard to the *bounded* character of human rationality, in general, and scientific rationality, in particular. However, Rescher's approach to rationality is broader, to the extent that he accepts an evaluative rationality or rationality of ends. Meanwhile, for Simon, rationality is mainly instrumental (that is, it is a rationality of means).

[3] As rational activity, prediction is a task subjected to limits; among them, there are epistemological limits.

His approach to rationality—above all, in the cognitive domain—has a certain parallelism with Simon's proposal, insofar as he assumes *de facto* an approach of bounded rationality. But their philosophical roots are different: Simon is an empiricist. He learnt this from Rudolf Carnap's positivism (Simon 1991).[4] Furthermore, the notion of "optimization" in Rescher is different from the concept of "satisficing" proposed by Simon (Rescher 1987). The difference between both approaches is mainly rooted in two issues: (I) the normative character of Rescher's proposal on rationality, in contrast to the preferential descriptive character of Simon's approach; and (II) Rescher's attention to the evaluative realm of rationality, which does not appear in Simon's proposal.

4.1.1 Intelligence and Rationality

Because Rescher considers that the intellectual field is very important in addressing human matters, he sees intelligence and rationality as "two sides of the same coin" (Rescher 1992b, p. 4). In his judgment, "the structure of rationality is a matter of system, not of sequence" (Rescher 1988, p. 124). In this way, intelligence is an instrument of rationality. He considers that "to behave rationally is to make use of one's intelligence to figure out the best thing to do in the circumstances" (1992b, p. 4). Thus, he uses the concept of "intelligence" in order to characterize what human rationality consists of.

But, sometimes, Rescher uses both notions interchangeably; for example, when he notices the difference between the ability to being rational and the exercise of that ability: "there is a crucial distinction between having the capacity for intelligent (rational) action and exercising this capacity intelligently (rationally)" (1992b, 10). In effect, one thing is being able to do something in an intelligent way and another thing is achieving that exercise in practical life. However, he admits that rationality is broader than the simple intelligence, since it encompasses several realms. Thus, when he characterizes rationality as the intelligent pursuit of the appropriate ends, he notices that "'intellligence' indicates cognition, 'pursuit' action, and 'appropriateness' evaluation" (Rescher 2003a, p. 188). So rationality encompasses three different areas: cognition, action, and evaluation; while intelligence is mainly related to cognition.

The distinction between "intelligence" and "reason" seems clearer when Rescher addresses the faculties demanded by rationality. He notices that the rational subject must have five faculties: imagination, information-processing, evaluation, selection—informed choice, and agency (Rescher 1988, p. 11). Both informed choice and agency—which he sees as the capacity to implement choices—indicate that rationality is not merely intellectual, since it involves the capacity to choose. Thus, free-will is required for rationality (Rescher 2009a, 2011).

---

[4] Carnap was one of his professors at the University of Chicago. On Simon's position, see Gonzalez (2003b).

So, he highlights that "intelligence" is different from "rationality." So, in his judgment, the capacity to implement choices "separates rational agency from mere intelligence as such, thus setting *persons* apart from mere *intelligences*" (Rescher 1988, p. 11). This issue is related to the notion of "artificial intelligence," since an artificial device can solve given problems, but only people can decide what problems they want to solve. Consequently, rationality requires intelligence ("the ability to acquire knowledge") and free will, which is "the capacity for decision and action in the light of evaluation on the basis of information" (Rescher 1992b, p. 10).

Consequently, for Rescher, intelligence is a necessary condition for rationality, which is developed in the different human domains, since the rational agent must be also an intelligent subject. But intelligence as such is not good enough to guarantee rational behavior, because rationality "consists in the intelligent pursuit of appropriate objectives" (Rescher 2006b, p. 79). To the extent that it consists in a "pursuit," rationality is related to the realm of the human action. It is an "intelligent" pursuit, since is supported by knowledge about the issues that are important in order to achieve the goals of the action developed. Therefore, knowledge, action, and evaluation are the three realms that shape rationality.

In contrast to "intelligence," which involves—in principle—immediacy, rationality involves certain mediation with a discursive component. Specifically, rationality encompasses deliberation, either in the field of knowledge, in the area of action, or in the realm of evaluation. Meanwhile, intelligence allows the grasp of something in the way of intuition. It is situated then in the realm of something "immediate;" while rationality belongs to the field of something "mediated." Thus, intelligence is related to the capacity to achieve knowledge, but the deliberation about the consistency and coherence of the knowledge achieved—theoretical, practical, or about ends—belongs to the realm of rationality.

Additionally, human rationality—with its different forms—can have a historical component. This feature is present in Rescher's approach when he notices that "the methods we use in cultivating rationality change in the light of the experience we have with them" (Rescher 1988, p. 14). Thus, although he considers rationality itself as something stable—to the extent that it consists in the effective and efficient action in order to achieve valid ends—Rescher notices that rationality involves changes with regard to the circumstances and conditions of the subject. There is a component of variability in this adaptation to the context that, in my judgment, can become genuinely historical.

4.1.2   The Realm of Rationality: Knowledge, Actions, and Values

Within this framework, a salient point is that Rescher rejects a mere instrumental conception of rationality.[5] So, in his judgment, rationality cannot be reduced to a simple matter of adjustment of means to ends, because the very ends should be subject to

---

[5] In fact, he uses the notion "practical rationality," which is broader than the simple instrumental rationality.

rational deliberation. Precisely, evaluative rationality makes his approach (which is *pragmatic*) different from an *instrumental* conception. Moreover, he maintains that rationality is *holistic*: "cognitive, pragmatic, and evaluative rationality constitute a unified and indissoluble whole in which all three of these resources are inseparably co-present" (Rescher 1988, p. vii).[6] Usually, he gives primacy to practical reason over theoretical reason, because the justification of both the ends and the beliefs is always—in his judgment—related to the realm of action (Rescher 2004, pp. 43–44). Thus, he considers that the rules of reason are "regulative principles" that guide human action on the basis of efficacy and efficiency. In this way, "reason is eminently practical—it wants what works (is efficient and effective)" (Rescher 1988, p. 17).

On this basis, he rejects clear frontiers between the theoretical and practical realms of rationality. In effect, he considers that each one of the three realms of rational deliberation—theoretical, practical, and evaluative—is supported by the others (Rescher 1988, ch. 8, pp. 119–132). Then, rational deliberation has the character of a system: an interdependence network. His approach to rationality is holistic, insofar as he sees rationality as "a unified and indissoluble whole in which all three of these resources [cognitive, practical, and evaluative] are inseparably co-present. Good reasons for believing, for evaluating, and for acting go together to make up a seamless and indivisible whole" (Rescher 1988, p. vii).

From the theoretical and practical reflection, rationality works in three realms, since it deliberates about knowledge, actions, and ends. But, in some sense, this is a conceptual distinction, because each one of those realms involves the others in the usual practice. In the case of scientific rationality, this approach entails that scientific rationality is not a simple adjustment of means to ends, because the ends should be evaluated. In Rescher's approach, rational action leads us to the achievement of preferable ends in an effective and efficient way, on the basis of well-grounded knowledge and valid values.

The acknowledgement of a rationality that deals with the ends of the human activity involves—in my judgment—an improvement with regard to narrower conceptions about human rationality, in general, and scientific rationality, in particular; for example, the approaches that give primacy to substantive rationality (focused on results)[7] and the proposals of procedural rationality (focused on processes).[8] In effect, besides the results and the processes, Rescher takes into account the rationality of the ends.

This approach has clear consequences for prediction, above all, from the epistemological, methodological, and axiological perspectives. *Epistemologically*, scientific prediction can be seen as a cognitive content, insofar as it provides knowledge

---

[6] On Rescher's approach to rationality, see Moutafakis 2007.

[7] When rationality is characterized as "substantive," then the level of the results has primacy. This is the prevailing approach in mainstream economics. Regarding scientific prediction, substantive rationality involves the primacy of the predictive success. Therefore, what matters is the result of the prediction: the predictive success, instead of the process. See Friedman (1953).

[8] Simon distinguishes his approach of "procedural rationality" (which is developed from psychology) from "substantive rationality," which has its origins in economics. Cf. Simon (1976) and Gonzalez (2015, ch. 8, pp. 203–228).

about the possible future. From this perspective, cognitive rationality has a role regarding predictive knowledge. From a *methodological* point of view, prediction is the result of a rational process directed to predicting scientifically the possible future. Therefore, not only is the result important, but also the predictive models must meet some requirements, such as the realism of the assumptions.

But Rescher's conception is not reduced to an approach to scientific rationality as a procedural rationality, because the ends sought must be also rational. Thus, *axiologically*, his holistic view of rationality also has repercussions for scientific prediction. It implies that prediction has value, first, with regard to the aims on science; and, secondly, regarding the processes and the potential results. In this way, the holistic framework that he proposes for human rationality, in general, and scientific rationality, in particular, is related to a broad approach to scientific prediction, where there are three levels at stake: the aims of scientific research, the processes oriented towards obtaining those aims, and the results eventually achieved.[9]

4.1.3 An Approach of Limited Rationality and the Viewpoint of Bounded Rationality

Certainly, Rescher considers that human rational behavior is subject to limits. *De facto*, he assumes a view of the human agent with bounded rationality: "Rationality does not make demands beyond the limits of what is genuinely possible for us—it does not require accomplishments beyond the limits of the possible" (Rescher 1988, p. 8). This leads us to relate this approach to rationality, which is in the framework of his pragmatic idealism, with two other conceptions: on the one hand, the standard economic theory of rational choice (see Hausman 1992, ch. 1, pp. 13–27); and, on the other, Simon's proposal of bounded rationality (Simon 1972).

Rescher criticizes both views from several angles, but above all his criticism is made insofar as they do not take into account the evaluative dimension of rationality (Rescher 1988, pp. 95 and 107–132). Regarding the standard theory of rational choice, he also opposes the notion of maximization of the expected subjective utility. Instead, he proposes the concept of "optimization" (Rescher 1987), which takes into account the limits of the subjects and the importance of the context, so his approach has a certain parallelism with Simon's "bounded rationality."

Like Simon, Rescher rejects the concept of "maximization" of the standard theory of rational choice; that is, the theory accepted by mainstream economics (Rescher 1988, pp. 107–132). This theory has a normative character that is accepted in the realm of neoclassical microeconomics. In its standard formulation, theory seeks to characterize how rational agents should make their choices, when they behave as rational agents. In order to do this, agents must obey the axioms demanded by the theory: transitivity, completeness, context-independence, and continuity

---

[9] The aims, processes, and results are the three successive stages in the internal dynamics of scientific research. Cf. Gonzalez (2012a, pp. 7–30; especially, pp. 8–9).

(Hausman 1992, pp. 15–19). The axioms of the theory introduce a normative element, since they establish the characteristics of the agents' preference orders. If preferences are ordered in accordance with these requirements, a utility function can be elaborated with only one maximum. In this way, a rational agent is the agent who makes decisions by maximizing his utility function.

Since the middle of the twentieth century, standard theory has to face criticism and alternative proposals, which have been suggested both with regard to the general level of human choices and to the specific realm of scientific choices. David Houghton, in his paper on "Reasonable Doubts about Rational Choice" (1995), offers a clear exposition of some of the problems of the standard theory. Among them, he highlights three: (a) the impossibility of making perfectly informed decisions; (b) the lack of empirical support to the conception of the rational subject as utility maximizer;[10] and (c) the lack of attention to social norms or rules for acting.[11]

Rescher's criticism is focused on the first two points. Regarding the first point—the impossibility of making completely informed decisions—he considers that human beings generally do not have all the important information for decision-making. In this way, "the rational resolution of problems is context-sensitive to the information in hand" (Rescher 1988, p. 22). For this reason, context is crucial for a conception of rationality with a practical component. In this way, he rejects the axiom of the standard theory about the context-independent preference order.

In effect, Rescher maintains that "rationality as such is something fixed—its nature is constant. But while rationality itself is something stable, the course of action it requires of us changes with circumstances and conditions" (Rescher 1988, p. 15). Those circumstances and conditions involve that it is not always possible to make perfectly informed decisions. But it is possible to improve and increase the information at hand. Moreover, this is a demand of rationality.[12]

Regarding the second point—the idea of rational agent as utility maximizer—Rescher thinks that the notion of "utility" in rational choice theory is "a mere fiction—sometimes useful" (Rescher 1987, p. 64). Since his starting point is a holistic conception of rationality, which has an evaluative dimension, he sees the notion of "maximization" as a narrow approach to rational choice. In fact, in his judgment, "rational choice is a matter not of one-dimensional *maximization*, but of the structurally diversified *optimization* that calls for harmonizing a complex profile of diversified goods and goals" (1987, p. 55).

The conception of rationality in terms of "utility maximization" involves assuming that "utility" is measurable and that the rational agent seeks to maximize it. This

---

[10] In this regard, Hausman notes that "to argue that utility theory is a good theory of how people actually behave because it is also a theory of how they ought to behave seems like the argument that people do not cheat on their taxes because they morally ought not to do so," Hausman (1992, p. 218).

[11] Regarding the problems of the standard theory, see also Sen (1994) and Boudon (1998).

[12] In this regard, Rescher maintains that "rationality is not just a passive matter of making good use of the materials one has on hand —in cognitive matters, say, the evidence in view. It is also a matter of actively seeking to enhance these materials: in the cognitive case, by developing new evidential resources that enable one to amplify and to test one's conclusions" (1988, p. 8).

feature means that the different values and ends of the human action can be evaluated on the basis of a common measure. In Rescher's judgment, this claim is problematic, because the realm of the human values is complex and varied (Rescher 1987, p. 58). Therefore, it is problematic to claim that the different choices of the agents can be characterized in terms of "potentiating utility." He considers that, regarding this issue, "economists incline to proceed strictly in terms of preferences, since these seem to be a common denominator in people's choices" (Rescher 1988, p. 109).

But, for Rescher, preferences are not good enough to characterize rationality. What is important is not merely what is preferred, but *what is preferable*, according to valid values and ends. For this reason, "once the link between utility and value is broken, the link between utility maximization and rational choice is also severed" (Rescher 1988, p. 111). Moreover, the very notion of "maximization" is problematic, since generally human beings cannot achieve what is the best, but what in principle can be seen as the *best possible* taking into account the context.

Consequently, Rescher does not think of human rationality as expected subjective utility maximization, but as an optimization. His approach at this point is *realistic*, since it takes into account the limitations of the knowing subject, as well as the informational restrictions to which the agents can be subjected.[13] Therefore, rationality does not demand that the agents act according to *the* rational decision from an ideal perspective, but according to *the best* decision, taking into account that neither our resources nor our capacities are unlimited.

Simon—who made a proposal on universal basis supported in the case of economics—has especially insisted on the limited character of human rationality (see Gonzalez 2007, pp. 41–69; especially, pp. 59–63). In his judgment, "there is today a very strong tradition of *a priorism* in economics, or what might be called deductionism. There is a very strong tradition of accepting the utility maximization hypothesis and then seeing, often with the aid of very powerful and elegant mathematical tools, what kind of conclusions you can draw from those premises, preferably by mathematical means. And there are even some economists who think that the theory is analytic and not refutable. I find it a rather curious point of view that a theory which purports to be about the real world should, somehow or other, follow from unrefutable premises and therefore not be subject to empirical test" (Simon 1997b, p. 407).

Although Rescher's main interest is not focused on the case of economics, a similar criticism to the "standard" conception in economics is in his work, but his reasons for the objections are different. Certainly, his proposal is also normative in origin, as it is the standard theory of rational choice. But his view is a normative theory with regard to the type of rationality that human beings, in fact, can achieve, and the justification of this type of rationality is—in his judgment—pragmatic. Thus, a rational agent is the agent who seeks the optimum understood as "the best available," instead of being the "maximum."

From different philosophical perspectives, Simon and Rescher offer alternative proposals to the standard theory of rational choice. Based on an empiricist support,

---

[13] Rescher insists especially on characterizing his own approach to rationality as a realist conception (1992b, p. 6).

Simon presents a realistic and psychological characterization of the limits to human rationality (see Simon 1976, p. 144); that is, he offers a description of human rationality "as 'bounded' and 'satisfacing'—instead of being unlimited and maximizing" (Gonzalez 2003b, p. 11). His thematic axis is an empiricist approach oriented towards the observation of human behavior.

Meanwhile, although Rescher develops a holistic approach to rationality, his framework is more pragmatic than Simon's. So, "a rational person proceeds on the basis of the grounds that are available to him (which may well also be imperfect)" (Rescher 1988, p. 7). Thus, generally, maximization is not available. So in his approach rationality demands and "optimization" according to the circumstances, which is different from "maximization:" it does not involve a possible maximum, but the best we can obtain (Rescher 1987). But "satisficing" is not good enough, because Rescher's normative approach to rationality also involves the rational deliberation about the ends of human action. This goes beyond mere "preferences."

4.1.4 Practicable Rationality

Our capacity to reason is limited and, for Rescher, we can only exercise it within the framework of *practicable rationality*. Practicable rationality is characterized as the rationality "geared to resolutions that are rationally appropriate with *everything* relevant taken into account that we can effectively manage to take account of in the prevailing circumstances—that are *optimal as best we can manage to tell*" (Rescher 1988, p. 28). This practicable rationality is, then, different from an *ideal rationality*, "which is geared to resolutions that are rationally appropriate with (absolutely) *everything* relevant taken into account—that are *optimal pure and simple*" (1988, p. 28).

Rescher is concerned about the limits of rationality both in the theoretical realm and in the practical dimension. His approach is in terms of practicable rationality, so he is aware that we do not always have all the relevant information and, in addition, the means at our disposal can be also limited. In this way, rationality only demands "that we do the best we can manage with the means in hand" (Rescher 1992b, p. 9). Thus, rational agent is required to use the limited resources he has in each concrete situation in order to guide his choices in the best possible way (that is, in an *optimal* way).

Thus, the "predicament of reason" appears. It involves that human rationality (i) aims at the absolute best; but (ii) it settles for the best that is available (Rescher 1988, p. 30). At this point, it is important to note that "*rationality* is 'information-sensitive': exactly what qualifies as the most rational resolution of a particular problem of belief, action, or evaluation depends on the precise content of our data about the situation at issue. And this dependency so functions that a 'mere addition' to our information can transform the optimality situation radically" (Rescher 1988, p. 24). Faced with these features, human beings are rational within a context: they only can do the best in the concrete circumstances of a certain context.

However, the acknowledgment of the inherent limitations of human rationality does not lead Rescher to reject a view of rationality as a matter of *idealization*.

Thus, an idealization is possible (a) because we justify possible courses of action, and (b) because the rational solution that we can achieve is that which, under similar circumstances, any other human agent can identify as the rational solution (Marsonet 1996, pp. 63–69). In this way, his approach to human rationality is not only of a normative character, but it also encompasses a descriptive element. Thus, although the rational solution to every problem—theoretical or practical—is context-dependent, it is universal at the same time. So, given exactly the same circumstances, the rational choice is the same for every agent.

To the extent that they agree with the idea that rationality is with regard to real subjects, Simon and Rescher are close in their approaches, but there is an explicit pragmatic component in Rescher, who sees the problem from practice: "being rational consists in the disposition to make good reasons constitute the motives for what one does. Since this is something we can achieve only within limits, one must regard perfect rationality as an idealization and acknowledge we humans are 'rational animals' because of our *capacity* for reason, and certainly not because of our achievement of perfected rationality" (Rescher 1988, p. 10).

The normative character of Rescher's account involves a difference from Simon, insofar as "the key point is that the significance of rationality does not, ultimately, lie in its role as a *descriptive* characterization of human proceedings (in how people *do* function) but rather in its *normative* role, as an indication of how people should function in the best interests of their cognitive and practical concerns. (...) The norms of rationality—like those of morality—are in no way undermined or invalidated by the fact that people violate them" (Rescher 1988, pp. 196–197).

For Rescher, the social scientists who, like Simon, suggest a merely descriptive theory of rationality "are engaged in a futile venture and condemned from the very outset to an inappropriate view of the rational enterprise" (Rescher 1988, p. viii). Rescher's proposal is fundamentally a normative one. But, in contrast to the standard theory of the rational choice, it is not inclined to an ideal type of rationality. Instead, it is oriented towards a "practicable" rationality, which is the rationality for which agents are in fact qualified. Moreover, it takes into account the fact that human beings sometimes behave in a non-rational way (Rescher 2006a).

He offers a broader picture. There are five requisites that, for Rescher, are demanded by rationality in matters of belief, action, or evaluation: (i) *consistency* (to avoid self-contradiction); (ii) *uniformity* (to treat similar cases alike); (iii) *coherence* (to make sure that one's commitments hang together); (iv) *simplicity* (to avoid needless complications); and (v) *economy* (to be efficient in the cost-benefit relation) [Rescher 1988, p. 16].

Those requirements give rationality a systemic character, insofar as they are "an organic (or systematic) unity of procedure, serving to make sure that everything fits together in an effective and mutually supportive way" (Rescher 1988, p. 16). Moreover, those requirements are flexible to some extent, as can be seen in his treatment of the problem of inconsistency in the case of cognitive rationality (Rescher 1988, pp. 73–91). In his approach, they are ideals that should be sought, cultivated, and valued; but their absence would not collapse every rational endeavor from the beginning.

Besides the emphasis on the normative character of his approach (with a broader picture), the rationality of ends is another important point that separates Rescher's approach to rationality from Simon's bounded rationality.[14] In effect, although Simon mentions the ends in some papers (Simon 1995), he does not contemplate in rigor the problem of the rationality of the ends chosen: there is no evaluation of ends in his approach. He is only focused on the rationality of the beliefs of the agents and the means selected in order to achieve certain ends, so the ends are assumed as "given." Meanwhile, for Rescher, "rational choice is a matter of opting not for what is preferred, but for what is preferable" (1988, p. 112). Thus, besides the evaluation of ends, he considers that rationality in terms of "satisfaction of preferences"— which is Simon's proposal—is the result of a biased approach to rationality. Moreover, rationality is not about what people merely prefer, but about what *is preferable* in accordance with human interests and values.

This feature is, in my judgment, Rescher's main contribution to the problem of rational choice and scientific rationality. Thus, by introducing the question of the rationality of ends (what is *preferable*) rationality broadens its realm to encompass the evaluation of the ends according to values. In this way, the rational agent not only has to justify his beliefs, but also the preferability of his options. Thus, Rescher suggests a holistic view of rationality, according to which rationality not only deliberates about the processes (the selection of the best means), but also about the result, insofar as it orientates the action towards the achievement of the best goals (Rescher 2004).

4.2 Scientific Prediction and the Main Types of Rationality

In order to analyze the epistemological factors of prediction in Rescher, the three types of rationality at stake must be taken into account. In his holistic approach to rationality, he notes that "philosophical tradition since Kant sees three major contexts of choice, those of *belief*, of accepting or endorsing theses or claims, of *action*, of what overt acts to perform, and of *evaluation*, of what to value or disvalue. These [contexts] represent the spheres of cognitive, practical, and evaluative reason, respectively" (Rescher 1988, pp. 2–3).

From those thematic realms, there are, in his judgment, three types of rationality according to the object of rational deliberation: (1) *cognitive rationality*, which addresses what should be believed or accepted in the realm of knowledge, either theoretical or empirical; (2) *practical or instrumental rationality*, which is about what actions should be performed; and (3) *evaluative rationality*, which focuses on what goals and ends should be preferred or valued (Rescher 1988, p. 3).

---

[14]A comparison between both approaches is in Gonzalez (2003c). The expressly acceptance of a rationality of ends can especially affect the sciences of design, since their activity is oriented towards concrete aims. This is a matter that Rescher does not develop, because he is mainly interested in the natural science and, sometimes, in the social sciences. Meanwhile, he does not expressly address the realm of the sciences of the artificial. On this matter, see Gonzalez (2008).

This approach to rationality suggested by Rescher is connected to his proposal about the epistemological factors of scientific prediction. In effect, in his judgment, predicting is a *rational activity*, so scientific predictions are characterized by being reasoned predictions.[15] Thus, they are statements about the future which are supported by theoretical or empirical reasons that justify the anticipation of the possible future. On this basis, scientific prediction can be related to the three realms of rationality that Rescher contemplates: cognitive, practical, and evaluative (Gonzalez 2010, pp. 264–265).

4.2.1 Cognitive Rationality

Initially, Rescher considers that rationality involves a *cognitive* or *epistemic* dimension. Cognitive rationality deals with what should be accepted in the realm of knowledge and, therefore, what belongs to the theoretical realm of rationality. On the one hand, this cognitive rationality demands answers to questions posed about the world; and, on the other, it requires that those answers could be justified. Thus, cognitive rationality requires increasing the available information in order to increase and improve our knowledge about the world. In addition, it allows us to justify the beliefs accepted as valid. The justification of beliefs can be of two types: discursive or presumptive (Rescher 1988, pp. 49–50).

A belief is justified *discursively* when the justification is obtained on the basis of another previously established belief. From this perspective, cognitive rationality seeks homogeneity in information-processing, since "there must be justified beliefs as inputs to arrive at justified beliefs as outputs" (1988, p. 49). Meanwhile, the *presumptive* justification does not proceed through other previously accepted beliefs, but in a direct way through a "presumption." Thus, a belief is justified in this way when "there is a *standing presumption* in its favour and no pre-established (rationally justified) reason that stands in the way of its acceptance" (1988, p. 50).

Rescher suggests a series of indicators that are "presumptions of reliability." On the basis of these "presumptions" a belief or beliefs can be justified *presumptively*. They are general principles such as the following: (a) believe the evidence provided by the senses; (b) accept the declarations of other people (in the absence of any counter-indications); (c) trust in the reliability of the cognitive aids and instruments used in the research (for example, telescopes or reference works); and d) accept, in principle, the declarations of established experts (Rescher 1988, p. 52).

Regarding prediction, scientific rationality (in its epistemic dimension) follows the same process. Thus, on the basis of the available knowledge, the aim is to obtain answers to substantive questions about future developments. Moreover, the answers given (that is, the predictions) must be justified. The presumptions of reliability are also relevant to prediction, since predictive statements are oriented towards a possible future and, therefore, it is not possible to test them now. Furthermore, prediction also has a role for the justification of hypotheses and theories. This role is usual in basic science, where prediction can be used as test for theories.

---

[15] Rescher rules out, in this way, the thesis of D.H. Mellor (1975).

Both with regard to scientific knowledge, in general, and predictive knowledge, in particular, the procedure of cognitive rationality is supported by economic principles. Thus, in Rescher's approach, the practical dimension has primacy over the theoretical component, because—in his judgment—reason is eminently practical: "Be it in matters of belief, action, or evaluation, its mission centers about the deliberate endeavor to maximize benefits relative to expenditures" (Rescher 1989, p. 11). Therefore, there is in science an *internal* economic dimension, which affects not only the actions, but also the cognitive content; because the cognitive dynamics of science is not cost-free. In this way, scientific rationality should be considered as connected to economic rationality.

Rescher suggests an economic-cognitive approach that demands an *epistemic optimization*: it seeks a positive balance between the required costs and the benefits eventually obtained with regard to an aim sought. Thus, he considers that the cognitive component of rationality involves an economic dimension. In effect, he maintains that "rationality and economy are inextricably interconnected. Rational inquiry is a matter of epistemic optimization, of achieving the best overall balance of cognitive benefits relative to cognitive costs" (Rescher 1989, p. 13).

On the one hand, the human process of knowing involves benefits, both theoretical (or cognitive) and practical (or applied) [Gonzalez 2003c, p. 72]; and, on the other, it involves costs (in terms of complexity, difficulty, resources, etc.). This economic-cognitive approach has methodological repercussions, because the methods are above all oriented towards the increase of knowledge. For Rescher, in the process of knowing there is an economy of means: "Concern for answering our questions in the most straightforward, most cost-effective way is a crucial aspect of cognitive rationality in its economic dimension" (Rescher 1989, p. 14).[16]

This feature means that scientific rationality requires an epistemic optimization; that is, it is oriented towards an optimal use of the resources with regard to the potential benefits of the cognitive endeavor (Rescher 1996c, p. 8). With this economic-cognitive approach, Rescher emphasizes the importance of the economic factors that are internal to the process of knowing. He insists on science as *human activity*, so scientific rationality is eminently pragmatic: "inquiry and the acquisition of information is itself a practical activity on the same footing with any other—a process that must be governed by the standard justificatory ground rules of practical reason" (Rescher 1988, p. 122).

This claim, to some extent cryptic, emphasizes the primacy of practice; that is, problem-solving as a proof of cognitive validity. On this basis, scientific prediction can be seen also in practical terms, as a rational procedure that involves an economic dimension, since prediction should be obtained in an effective and efficient way, so an epistemic optimization could be possible in the process of predicting. Therefore, there is a nexus between cognitive rationality and scientific prediction, which emphasizes the presence of economic factors that are internal to prediction.

---

[16] On the methodological repercussions of Rescher's "cognitive economy", see Gonzalez (2003c, pp. 65–96; especially, pp. 72–74).

In effect, *cognitive rationality* intervenes in the predictive task of science, since prediction is supported by knowledge about the important variables, as well as by adequate inferences (Gonzalez 2010, pp. 264–265). Certainly, predicting is not a cost-free task, since it requires a series of resources: procedures of data processing, experimentation techniques, observation means, etc. Thus, the minimization of costs should be sought in prediction.

In turn, this nexus between cognitive rationality and scientific prediction has repercussions for the problem of the epistemological limits of prediction. In effect, predictive knowledge is subject to limits, insofar as the cognitive capacities of the human beings are *limited*. For example, the presence of informational restrictions for the agents that make science is one of the factors that might limit the predictions. This kind of limits favors uncertainty, which accompanies the bounded rationality and has repercussions on scientific prediction. In effect, the capacity to compute the information is limited; and, in principle, the future has a number of different possibilities.

4.2.2 *Practical Rationality*

A general feature of Rescher's epistemological approach is that rationality is eminently practical; and, in a certain sense, it is also contextual. Practical rationality deals with the realm of action, and is about the means that are used to achieve a certain goal within a certain context. But, since he develops a holistic conception of reason, he has a critical attitude toward approaches such as Simon's view, because the latter did not assume the rationality of ends, but only of means.[17]

The practical dimension of rationality defended by Rescher requires the adjustment of means to ends; but the ends themselves must be adequate according to values that have an objective component (Rescher 2012b).[18] Thus, there are two issues that are closely related: selecting the ends of action and having values in order to choose them. In contrast to Simon, he also claims that "in the broad scheme of things, the two aspects are needed: the ends without the requirement of the means are frustrating; the means without adequate ends are unproductive and useless" (Rescher 1999a, p. 82).

Thus, Rescher gives primacy to the practical reason in his approach to rationality. But it is not a simple rationality of means to given ends, so it is not a purely instrumental rationality. In his judgment, the rational character of an action cannot be assessed without an evaluation of the adequacy of the ends of action. In this way, "both matters—the efficacy of means and the validity of goals—are essential aspects of practical rationality" (Rescher 1988, p. 93).

---

[17] On Simon's theory of rationality, see Gonzalez (2003a, c, 2007) and Bereijo (2003).

[18] On the objectivity of values in the axiological conception of Rescher; a more detailed treatment is offered in the Chap. 8 of this book.

From a pragmatic perspective, there is a practical criterion of evaluation. For Rescher, actions and beliefs are evaluated according to their efficacy and efficiency in the achievement of the goals and aims. Thus, rationality is associated to the achievement of goals or to meet a concrete need. In those cases, "a rational creature will prefer whatever method process or procedure will, other things equal, facilitate goal realization in the most effective, efficient, and economical way" (Rescher 2004, p. 44).

Rescher insists that practical rationality is mainly of an economic kind. Thus, it seeks the optimization of the benefits obtained from rational behavior in relation to the costs of the recourses. Therefore, the process—the search—prevails, instead of the product (the final result). This feature is related to approaches like the Aristotelian conception, where rationality is more focused on the processes (the decision-making) than on the result. Here, there is certain similitude with Simon, who also sees rationality more as a process than as a result.[19]

For Rescher, on the basis of the available information, we try to achieve the goal sought by selecting those means that allow us to achieve it in an efficient way. But this pragmatic feature not only has to do with the means. In effect, he considers that it is present in the three realms of rational deliberation: the field of the information (the cognitive realm), the area of the means (the practical component), and the realm of the ends (the evaluative aspect).

Since in Rescher's thought rationality is addressed from a pragmatic perspective, the very justification of the rational behavior is also of a pragmatic character: "Rationality has the perfectly rational justification that in failing to heed the dictates of reason we came up on the short end of the balance of benefits gained versus advantages foregone" (Rescher 1995, p. 29). Thus, practical rationality, which has to do mainly with the processes, seeks that those processes can be effective and efficient. In turn, this feature connects with the normative aspect of his approach, which is focused on the methods instead of being oriented towards the results, but which direct the processes towards what *ought to be*, which can have general validity.

In this regard, Rescher seeks that the process has universal or *universalizable* features. In this way, his approach rejects relativism. This can be seen at different levels where rational deliberation is oriented towards the procedures. Thus, his pragmatic approach to rationality involves some elements of general character: (i) rationality deals with the processes of truth validation; (ii) it is focused on act-recommending norms; (iii) it affects the processes to answer-determination; and, (iv) it is related with the procedures by which the endorsement of hypotheses is validated (Rescher 2004, pp. 45–46).

By insisting on the processes regarding rationality, the methodological aspect is crucial. But this feature does not imply that the methodological aspect is the only one that we should take into account. Thus, there is always an end that modulates

---

[19] Simon is more interested in what the disciplines of Aristotelian inspiration (logic, ethics, and psychology) say about rationality (where rationality is seen as a process) than in disciplines such as sociology or (mainstream) economics, where rationality is seen as result. In effect, against substantive rationality, Simon offers an approach to rationality that gives primacy to the procedural component. Cf. Simon (2000).

the processes, so the different processes are oriented towards the achievement of a result. In this way, human actions always have a teleological dimension. So, among the different procedures, a rational creature would choose the procedure that leads to the end sought in the most effective and efficient way.

This means that, if the processes are not self-sufficient in epistemological and methodological terms, then the rationality of the very ends sought must be taken into account. In his judgment, evaluating different courses of action is not good enough when these are about "given" ends. So the very choice of the ends must be evaluated and legitimated from a pragmatic perspective, and this is a task that should be carried out from rationality.

Within this approach, there is a nexus between practical rationality and prediction. In effect, for Rescher, to predict is mainly a *process* that is oriented towards an end: it is an activity that seeks to answer our meaningful questions about the future. In this way, practical rationality intervenes in prediction by selecting the adequate process of prediction, which must be effective and efficient in relation to the end sought. Thus, the choice of a concrete method or procedure in order to solve a predictive question depends, to a large extent, on the kind of phenomenon that we want to predict.[20] Consequently, he considers that "the comparative efficacy of predictive processes is (…) in the end an empirical matter" (Rescher 1998, pp. 111–112).

If practical rationality is eventually determined on an empirical basis, then his rationality is not only normative, but also descriptive. Moreover, Rescher gives primacy to practice in his approach to rationality, and practice is also crucial in his approach to scientific prediction. In effect, in his judgment, "to act, to plan, to survive, we must anticipate the future, and the past is the only guide to it that we have" (1998, p. 65). This approach involves an empirical and a descriptive component: "practical reasoning serves as the basis for the justification of the *inference* towards the future from past experience" (Gonzalez 2010, p. 265). Once again, his "pragmatic idealism" includes elements that are not "idealistic," but rather "realist." In my judgment, this is due to his acceptance of the realist notions of "fact" and "objectivity" (see Rescher 1992c).

4.2.3 Evaluative Rationality

Rationality also deals with what should be valued regarding the ends, according to the real interests of the human being. This task accompanies the other two already mentioned: (1) to deliberate about what should be believed on the basis of the available information; and (2) to decide what means should be used in order to achieve an end in an effective and efficient way. Thus, both rationality, in general, and scientific and technological rationality, in particular, are not reduced only to the means that should be used in order to achieve given ends. Rationality should consider the very ends sought. About this issue, Rescher explicitly shows his disagreement with Simon.

---

[20] On the distinction between "predictive procedures" and "predictive methods," see Gonzalez (2015, pp. 251–284) and Chap. 6 of this book.

For Simon, "reason is wholly instrumental. It cannot tell us where to go; at best it can tell us how to get there. It is a gun for hire that can be employed in the service of whatever goals we have, good or bad" (Simon 1983, pp. 7–8). In Rescher's judgment, this type of approaches blur the frontiers between what is rational and what is non-rational, insofar as they do not take into account the nexus between rationality and what is advisable and intelligent (Rescher 1988, p. 95). Thus, he avoids the term "instrumental" to refer to the rationality of means. Instead, he uses the notion of "practical rationality," since he considers that it is connected with a rationality of ends, which deliberates about the ends of human action.

Therefore, in contrast to Simon's approach, which is clearly of instrumental rationality and empirically based, Rescher considers that "rationality is thus a two-side, Janus-faced conception. On the side of means, it reflects a pragmatic concern for efficient process, while on the side of the 'appropriateness of ends' it reflects a value-geared concern for product. (Moreover, the acceptability of the means themselves also enters in.)" (Rescher 1988, p. 6). In that case, the issues regarding the efficacy and efficiency of means are only a part of rational action. Thus, in order to be considered as fully rational, human action should take into account the *value* of the ends it seeks.[21]

In this approach, the adequacy of ends—if they are appropriate or not as ends of human action—is an issue that belongs to the evaluative order of rational deliberation (Rescher 1988, p. 93). He assumes this evaluative component of rationality. Rescher considers that this requires justification; that is, that rationality not only deals with matters of fact, but also with matters of *value*. This is because, in his judgment, the Humean tradition of separation between reason and values is still widely accepted.

According to Rescher, the rejection of the objectivity of values and, therefore, the dismissal of the rationality of ends, is due to some confusion with the terms "taste" and "value" (Rescher 2000, p. 170). Tastes are about what people prefer, while values have to do with what should be considered as *preferable*. Thus, although "'There's no disputing about tastes' may be true, (...) 'There's no disputing about values' certainly is not. Values too can be altogether objective, in that value claims admit of rational support through impersonally cogent considerations" (Rescher 2003b, p. 31). Rationality also requires an objective assessment regarding value matters: rationality "asks for an estimation of *preferability*, more than the pure expression of a preference" (Rescher 1999a, p. 85).

Thus, Rescher admits the *objectivity* of values, which in his approach goes hand in hand with the normative character of the rationality of ends. This issue is connected with "self-interest," that is, with one's own welfare and well-being. In his judgment, "self-interest" can be interpreted in three different ways: (a) what someone *wants*; (b) what somebody *thinks* is good for him or for her on the basis of the

---

[21] Both the axiology of research and the ethics of science in Rescher are clearly teleological, insofar as he gives primacy to the perspective of human activity as oriented towards ends. On this issue, see Chaps. 8 and 9 of this monograph.

available information; and (c) what somebody objectively *ought* to want (what is, in fact, beneficial for him or for her) [Rescher 2006b, p. 79].

Usually, the third option is beyond our possibilities. But the first option is not good enough, since rationality also has a normative character. Thus, rationality requires a strengthened version of the "self-interest," as is the case in the second interpretation. Thus, self-interest consists in "what someone has *good reason* (in the prevailing state of his information) to think to be truly beneficial to him" (Rescher 2006b, p. 80).

This notion of the self-interest can be understood from either a narrow or a wide viewpoint. Rescher defends a wide notion of self-interest, according to which the self-interest of one individual also encompasses, to some extent, the other's interest (2006b, p. 80). In my judgment, this feature broadens self-interest to include reciprocity, active cooperation, and solidarity, which are three stages in interpersonal relationships.[22]

On this wide orientation, Rescher maintains that those values that implement the best interest of people are adequate values. Like the cognitive and the practical rationality, the evaluative rationality is also universal in the following sense: what I should want or prefer is "what the reasonable (impartial, well-informed, well-intentioned, understanding) bystander would think that I ought to want on the basis of what is 'in my best interests'" (Rescher 1988, p. 102). Therefore, a theory of rational choice not should be reduced to what people in fact prefer, but it should encompass *what they should prefer* according to their real interests.

Therefore, the rational agent is that who is ready to go from her preference to her interests, by subjecting her desires to rational deliberation. This feature involves an objective judgment. Through this objective judgment, it is possible to decide if our actual ends are rational or not. In this way, the valid or adequate ends are those that result in the best interest of the agents; for example, those ends related to the satisfaction of universally shared needs, such as health or affection; the ends related to the particular role one plays as relative of someone, as professional, etc.; and those ends that have to do with what one simply happens to want.

In order to consider a certain interest as appropriate, it has to meet the requirement of being connected with some universal interest (Rescher 1988, p. 101, 2000, p. 178). Thus, some concrete interest of one agent can be considered as valid if it is possible to subordinate it to a universal interest. With this position, Rescher's approach satisfactorily combines two different aspects: (I) the fact of the existence of numerous interests, which are contemplated by the agents in order to select those that will guide their actions; and (II) the existence of a universal principle of rationality, which allows us to determine the validity of those interests in an objective way.

If the claims about value matters are not beyond the realm of rationality, then they can be objective. To the extent that values are not simple tastes, rational evaluation follows the same principle as rational action in general. This feature leads to the following principle: "*proceed in the same way than a rational or reasonable*

---

[22] On recent studies about altruism, which is present in the cases mentioned, see the monographic number of *Economics and Philosophy*, which offers papers of Philip Kitcher and others in the case of economics. Cf. Kitcher (2010) and Schefczyk and Peacock (2010).

personal would proceed in those circumstances" (Rescher 1999a, p. 75). Therefore, Rescher admits the realist notion of "objectivity" as a crucial factor in his defense of the evaluative realm of rationality.

However, within a system of pragmatic idealism, the ultimate justification of the evaluative dimension of human rationality does not rest on the realist principle of objectivity; instead, it can be seen in the capacity of values to guide our actions and make them meaningful. Thus, "the pragmatic aspect of values lies in the fact that they provide a thought tool that we require in order to achieve a *satisfying* life. By contrast, the idealistic aspect of values lies in the fact that they alone enable us to achieve a *meaningful* life" (Rescher 1994, p. 248).

The idealistic aspect of evaluation can be clearly seen when Rescher character-izes values as "indispensable thought-tools" (1994, p. 249). Human knowledge has a clear teleological dimension (an orientation towards ends) and it is never passive. In his approach, values constitute an indispensable requirement in order to orientate human thought towards the ends sought. In this way, the acceptance of the rational-ity of ends has a clear repercussion in his conception of scientific activity as modu-lated by values. Among them, Rescher gives primacy to the cognitive values.

When rationality is seen in its three dimensions—cognitive, practical, and evalu-ative—human beings can go beyond what in fact *is* to evaluate what *ought to be*. Thus, "value is concerned not only with what does happen but with what might happen, and not just 'realistically' but even 'by the wildest stretch of the imagina-tion'" (Rescher 1994, p. 246). Therefore, as rational beings, we compare different possibilities for the direction of our actions.

In order to accomplish this task, evaluative rationality is crucial. In effect, Rescher considers that human beings are characterized by their capacity to choose on the basis of rational evaluation. For this reason, he maintains that "comparative evaluation is also an unavoidable requisite of the human condition. The burden of choice—and thereby of reason-guided evaluation—is one of the definitive features of the human condition" (Rescher 2006a, p. 5).

4.2.4 *Rational Coordinates and Scientific Prediction*

Within these coordinates regarding rationality, where evaluative rationality presup-poses the existence of cognitive rationality and practical rationality, scientific pre-diction can have a crucial role when different courses of action are evaluated, in order to choose among them. In fact, prediction makes it possible to anticipate future events and developments, so it provides knowledge that can have an important role in order to *orientate human actions*. This can be seen, above all, in three realms: (i) applied science, (ii) the application of science, and (iii) the use of predictive knowl-edge in everyday life. Usually, Rescher focuses on the application of science and the everyday life realms, while he hardly takes into account applied science.

First, prediction is required in applied sciences in order to make prescriptions directed towards solving a concrete problem. Second, prediction also has a role regarding the ends of the application of science, where scientific knowledge is used by agents in institutional and professional contexts (for example, in hospitals or in problems concerning policy). In this case, prediction can serve as a support in decision-making (Gonzalez 2015). Third, agents might use knowledge about the future provided by scientific predictions as a guide for action in daily contexts (for example, this is what commonly happens with meteorological predictions).

Furthermore, the predictive activity itself is a *rational activity* that is goal-oriented, so it involves an evaluative dimension. This can be seen at two successive levels, which become clear when prediction is seen as a human activity: *prediction as a value* and *the values of prediction*. On the one hand, the role of prediction *as a value* of science can be addressed (that is, insofar as it is a goal sought in order to promote scientific progress). From this perspective, prediction has a role when the aims of scientific research are selected, which are prior to the selection of the processes and the achievement of results. And, on the other hand, the question of the *values of prediction* can be considered; that is, what are the desirable characteristics that should accompany the predictive statements[23]?

Consequently, evaluative rationality intervenes in scientific activity oriented towards anticipating the possible future, insofar as prediction is, for Rescher, one of the main goals of science. When a prediction is achieved, other more specific ends can be sought (for example, to serve as a test for a theory in basic science or as a previous step to prescription in applied science). But prediction itself must be subjected to evaluation. In this regard, Rescher notes that the values that scientific prediction should have, such as correctness, accuracy, precision, etc. (Rescher 1998, pp. 119–125). First, those values modulate the ends sought and, second, they also modulate the processes oriented towards those ends. Thereafter, they also have a role for the evaluation of the result finally obtained and its possible consequences (see Sects. 8.4 and 8.5 of this monograph).

4.3   Fallibilism and Predictive Knowledge

Besides the pragmatic approach to rationality, which emphasizes the presence of cognitive and instrumental limitations, Rescher maintains a fallibilistic view of scientific knowledge. He defends that every piece of knowledge we can achieve is always subject to revision, since it is impossible to achieve perfection in science (Rescher 1999b, pp. 145–165). However, he rejects both relativistic and skeptical proposals. He considers that the information we have now is the *best possible* and that the desire to enlarge and improve it motivates research, promoting new advancements.

---

[23] From this perspective, Rescher's epistemology is clearly related with his axiology of science. The axiological aspects of scientific prediction are developed from both perspectives—prediction as a value and the values of prediction—in Chap. 8 of this book.

Within this fallibilistic epistemological framework, successful scientific prediction is connected with truth in the epistemological realm and with the progress of science in the methodological realm. Thus, prediction is a statement that provides knowledge about the possible future that can be true. Furthermore, it is a *test* for scientific theories in basic science, so it has a high confirmatory value. In this way, successful prediction allows us to evaluate the content of truth of the scientific theories. Moreover, prediction contributes to problem-solving in applied research. This feature is related to the progress of science, which in Rescher's judgment "is mostly strikingly and decisively manifested on its technological side. Science is marked by an ever-expanding predictive and physical control over nature" (Rescher 1999b, p. 39).

4.3.1   Scientific Prediction and the Problem of Truth

Rescher's approach regarding the truth of predictive statements and their role as *test* for theories (basic science) and as a previous step to prescription (applied science) is developed within a framework of fallibilism.[24] For him, fallibilism maintains that "our theoretical scientific knowledge claims are always vulnerable: they must always be staked tentatively because the prospect that further inquiry and discovery will lead to their modification or replacement can never be eliminated" (Rescher 1995, p. 72).

Therefore, on this basis, it is not possible to have certainty regarding the cognitive content of current science. In effect, scientific knowledge is the result of a rational process of research and, for Rescher, rationality is a matter of optimization, instead of maximization. Thus, he maintains that science only can achieve what, in some concrete circumstances, is the best thing available. In this case, the rational solution to a problem is not, in principle, the best solution in absolute terms; but only the best solution *possible* given a certain context. The limited and contextual character of rationality, in general, is also characteristic of scientific rationality, in particular.

Moreover, in terms of knowledge optimization, in science "the answers we give to our questions are literally the best we can provide" (Rescher 1999b, p. 30). For this reason, it is not possible to be certain that a scientific theory or statement is true; but we can claim that it is the best answer we can offer in order to solve a concrete problem. Therefore, every theory or statement is vulnerable, insofar as it can turn out to be false. With these reflections, Rescher does not call into question the *possibility* of achieving true knowledge; but he acknowledges that it is *probable* that future developments will replace or change the knowledge we have now.

Thus, Rescher rejects a skeptic or relativist approach regarding knowledge, since he admits clearly the possibility of achieving true statements. In this regard, he distinguishes "*our truth*" (what we think that is true *now*) and "*the truth*" (what is in

---

[24] Rescher attention is usually focused on basic science, and he rarely pays attention to the problems that more specifically have to do with the applied sciences. For this reason, he highlights above all the role of prediction as scientific test.

fact true) [1999b, p. 36]. For him, this is a "convenient fiction," as a hypothesis that allows us to make advancements in the process of research.[25] Consequently, it should be acknowledged that every belief we accept can turn out to be false. Moreover, many of our beliefs will be falsified (Rescher 1999b, p. 34).

In order to justify this epistemological approach, Rescher contemplates the history of science. The historical dimension of science implies that this activity is "one transitory state of things in an ongoing process" (Rescher 1999b, p. 38). In this sense, he thinks that "the clearest induction from the history of science is that science is always mistaken" (1999b, p. 36). However, it is not a "pessimistic induction" in the style of Larry Laudan (1981),[26] since in Rescher's approach this induction is connected with an account of science according to which science is progressive regarding knowledge: "later science is better science—that is, better warranted science" (Rescher 1999b, p. 40). In this way, the main lesson is that it is always possible to improve the available knowledge.

Rescher establishes the progressive character of science on the basis of the continuous improvement and increase of the available knowledge (Rescher 1978). But the way of evaluating this progress is within a framework of methodological pragmatism (Rescher 1977), according to which the predictive success and the control over nature are the best available indicators regarding scientific progress.[27] In this realm, prediction connects with his epistemological fallibilism. In my judgment, this connection can be seen in four different directions: (1) insofar as prediction provides knowledge about the future that is fallible; (2) because prediction is a test for the adequacy of theories, so it has a relevant role in the justification of the accepted beliefs; (3) since prediction is a guide for prescription (applied science) or for both individual and social agents' decision-making (application of science); and (4) prediction as an indicator of scientific progress, so the advancement of science can be evaluated on the basis of predictive success.

1. Rescher considers that scientific prediction can have an objective content. Thus, he insists on the rational basis of prediction and the realism of the assumptions. However, every predictive statement is revisable. From an approach of "bounded rationality," the task of predicting is also a task subjected to limits, both cognitive and practical. But accepting that our intent of anticipating the future is subject to limits does not mean that, for this very reason, it is doomed to failure. Therefore, he maintains that we should persist "in using the resources of reason to doing the best we can in the recognition that while overall this is going to prove to be quite a lot, it will never be nearly as much as we would ideally like" (Rescher 1998, p. 222).

---

[25] "We have no alternative to proceeding on the working hypothesis that in scientific matters *our* truth is *the* truth. But we must also recognize that that is simply not so —that the working hypothesis in question is no more than just that, a convenient fiction," Rescher (1999b, p. 36).

[26] It has been noted that the pessimistic induction can even lead to semantic antirealism. Cf. Frost-Arnold (2011).

[27] Rescher's methodological pragmatism is analyzed in Chap. 5 of this book.

Therefore, scientific prediction is a rational prediction. It consists in a statement about the future that is obtained as the result of processes in which cognitive, practical, and evaluative rationality intervene. Thus, the rational bases—theoretical or empirical—of prediction justify the anticipation of the possible future. Moreover, as the content of the prediction is the result of a rational process, it can be objective. In this way, the use of scientific prediction as a test for theories is also justified.

2. In effect, scientific prediction can be used as a test for scientific theories. However, within a fallibilistic epistemological approach, prediction cannot provide a definitive proof of the validity of theories. For Rescher, prediction is the best we have, but "the scientific bearing of predictive success is not demonstrative but merely evidential, in this domain even our best confirmed theories are no more than reasonable but also provisional estimates of the truth" (Rescher 1998, p. 171).

3. Furthermore, the reliability of the predictive statements in applied science has repercussions for prescription. On the one hand, prediction can turn out to be false. This has a negative incidence on prescriptions, which are designed by using the knowledge about the possible future. And, on the other hand, the fallible character of scientific prediction requires—in my judgment—adopting a wary attitude when the action is designed on the basis of predictive knowledge. This also happens in the context of the application of science, where prediction can serve as a basis for the agents' decision-making. Thus, the fallible character of the cognitive content of scientific prediction makes the decision-making of the agents (both individual and social) difficult when they try to solve practical problems of acting.

4. Within this fallibilistic framework, Rescher's attention usually shifts towards the realm of basic science. In this regard, he thinks that scientific prediction has an important role for scientific progress. So, as a test for theories, prediction is also an important indicator of scientific progress (Rescher 1998, p. 164). Thus, although his proposal is clearly pragmatic, it is not an instrumentalist approach, since scientific progress is not directly connected with problem-solving. In this regard, the Kantian component of this thought prevails, so scientific prediction is above all a cognitive content, which also has methodological importance.

4.3.2 Knowledge of the Variables of Prediction

Although Rescher usually focuses on the role of prediction in basic science, scientific prediction has several roles, according to the type of activity where prediction is achieved or used (Gonzalez 2010, p. 11). (I) Prediction can be used in basic science (either natural sciences or social sciences) as a test for establishing the scientific character of the hypotheses and theories. (II) Within the realm of applied science (pharmacology, medicine, economics, etc.), prediction is usually the previous step to prescription. Thus, the anticipation of the possible future is required in order to guide the action towards the solution of concrete problems. (III) Prediction also has a role in the application of science, where it can serve as a support for the procedures of decision-making (Gonzalez 2010, p. 11).[28]

Above all, Rescher's attention goes to the role of prediction as a test for theories, since his main interest is usually in the realm of basic science. In order to serve as a test, the knowledge provided by the predictive statement must be objective knowledge. He admits the objectivity of the knowledge about the future, which can be true. In his approach, the objectivity of prediction is related to its rational character. So, in his judgment, the rational basis of prediction is a necessary condition in order to have a genuine scientific prediction (see, in this regard, Chap. 2, Sect. 2.3).

Since prediction can be used as a test for theories in basic science, the standards for the acceptance of a prediction in science should be more demanding. From this perspective, the knowledge about the relevant variables for prediction is crucial, since it allows us to justify an inference of the future from the available data about the past and present. In fact, a distinction has been proposed in economics between "foresight," "prediction," and "forecast," on the basis of their different degree of control of the variables (Fernández Valbuena 1990). Thus, "foresight" is the most secure kind of prediction, and "forecast" is the least, because it is a prediction with a margin of error (cf. Sect. 2.4.2).

This conceptual distinction is not considered by Rescher, although he insists on the importance of the information about the relevant variables in order to get predictive success. In effect, he notes that prediction is *sensitive* to the information available. For this reason, "with prediction, as elsewhere, we must be careful not to identify automatically the vastly extensive (ontological) realities that make for the actuality of what is predicted with the comparatively modest (cognitive) considerations that furnish a prediction's evidential warrant" (Rescher 1998, p. 59). This means that the more variables we know and the better our knowledge about these variables is, the more secure the prediction will be.

However, for Rescher, scientific perfection is not always attainable. On the one hand, there is always risk of error; and, on the other, the available information can be incomplete. In this regard, he notices that prediction is inherently risky, so "when we make claims about the future, things can almost invariably go awry" (Rescher 1998, p. 59). But we must take the risk, because the attainment of knowledge about the future events and developments is important for the human beings. Thus, the best perspective of success involves the rational bases of prediction, which are objective and, therefore, reliable.

He suggests that prediction must be the result of a rational process supported by proofs (either theoretical or empirical); because, although the future is observationally and physically inaccessible, it is not inaccessible in a cognitive way. Access to the future is possible through prediction, "even though it always involves an intrinsically risky, error-liable epistemic leap from information regarding the past-&-present to claims regarding the yet unrealized future" (Rescher 1998, p. 54). The

---

[28] The roles of prediction in scientific activity, as well as the differences between basic science, applied science, and the application of science, are developed further in Sect. 5.2.

perspective of error cannot be completely eliminated, but the reliability of a prediction will depend on the quality of the information it is based on. In this way, in the rational process of prediction, knowledge about the variables—number and quality—is crucial for predictive success.

4.4 Epistemological Limits to Predictability

Due to the primacy of the epistemological realm in Rescher's philosophical approach, the epistemological limits to scientific prediction are an especially important issue. Expressly, he maintains that "epistemological limits on prediction exist insofar as the future is *cognitively inaccessible*—either because we cannot secure the needed data, or because it is impossible for us to discover the operative laws, or even possibly because the requisite inferences and/or calculations involve complexities that outrun the reach of our capabilities" (1998, p. 134). Thus, he assumes, *de facto*, an approach of bounded rationality as a framework for the analysis of the epistemological limits that affect predictability. It is another feature that separates Rescher's philosophy of science from the traditional philosophical idealism.

4.4.1 The Problem of Uncertainty

In Rescher's thought, the epistemological realm has primacy over the ontological dimension (Gonzalez 1999); because concepts and categories are what allow us to articulate the reality. Thus, science is *our science*: it is the result of an activity carried through by human beings with limited capacities, within a context where the available information is usually also limited. Consequently, Rescher notes that "the limits of our experience set limits to our science" (1999b, p. 216). Now then, this experience is articulated from our concepts and categories, our mental structure.

Insofar as human experience is limited, Rescher sees uncertainty as the main obstacle to scientific prediction in the epistemological realm.[29] His approach insists on this point: "the circumstances of our existence are such that many of our decisions—and many of the most important ones—have to be made under conditions of unavoidable uncertainty" (Rescher 2003b, p. 33). This feature especially affects prediction. In effect, since prediction is usually oriented towards a *possible future*, uncertainty can be a clear obstacle to predictability.

Therefore, uncertainty can limit scientific prediction according to various degrees: (i) uncertainty can imply unpredictability (that is, the impossibility of obtaining a prediction); (ii) it is possible that uncertainty entails non-predictability with regard to a concrete issue (the current impossibility to state a prediction);[30] and

---

[29] The analysis of uncertainty as a limit to scientific prediction is also in Guillán (2014).

[30] See Gonzalez (2010, p. 289). This distinction between "unpredictability" and "non-predictability" was dealt with in Sect. 2.5 of this book. See also Eagle (2005).

(iii) uncertainty can make it difficult to achieve an ideal degree of exactness and precision. In this case, it might only be possible to obtain a generic prediction, instead of a specific prediction.[31]

Regarding how to address the problem of predictive uncertainty, cognitive or epistemic rationality is the starting point. But Rescher does not endorse an approach to rationality based on the idea of maximization, as some mainstream economists do.[32] He considers that human rationality is a "bounded rationality" or, at least, a limited rationality.[33] In his judgment, rationality is bound to a circumstantial optimization (the best thing that can be done in a concrete situation), instead of being associated with something absolute or maximization (Rescher 1987, 1988). This has to do with human beings' environmental conditions or social milieu, which are usually affected by uncertainty.

Uncertainty is one of the aspects that go hand in hand with bounded rationality. On the one hand, it is not usual for us to have *all* the relevant information; and, on the other, human ability to compute information is also limited (Rescher 2003a, ch. 11, pp. 187–206). Hence, in Rescher's approach, rationality and uncertainty are closely related. This feature has direct repercussions on scientific prediction, insofar as it is the result of a rational activity. Moreover, this rational activity is oriented to a future that, in principle, has a number of possibilities.

In Rescher's conception, rational prediction is the result of a process. It commonly involves several aspects: (1) prediction is obtained where there is the relevant information about past and present events; (2) the paths reflected in this body of data for prediction are discerned; and (3) the patterns detected in past and present phenomena are stable, to some extent, so they continue into the future (Rescher 1998, p. 86). If we consider these preconditions for rational prediction, it seems clear that uncertainty is, *de facto*, one of the main limits to predictability (see Sect. 5.4. of this monograph).

Furthermore, Rescher notices that "uncertainty produced by sheer ignorance is clearly the most obvious obstacle to prediction" (1998, p. 135). Thus, although the phenomenon that we are trying to predict is, or may be, a regular one, failure is still possible when all the relevant information about its operation is not available. Obtaining the relevant information is a *necessary condition* to predictability. However, too often it is difficult, or even impossible, for human beings to gain access to information (at least, to the relevant information).

---

[31] Regarding what we can or cannot predict in science, Rescher thinks that the best source of information we have comes from science itself: it is not an external issue. Moreover, only science itself can inform us about the achievable degree of precision for scientific prediction. This depends on circumstances such as the scope—short, medium or long run —, the available technology, etc. Additionally, it also depends on the question we want to ask. In principle, the more concrete the question is, the more complicated it will be to answer it accurately. Rescher, *Personal communication*, 17.6.2014.

[32] On rationality as maximization and the alternative of bounded rationality, see Gonzalez (1997).

[33] On the similarities and differences between Simon's bounded rationality and Rescher's limited rationality, see sec. 4.1.3.

   This aspect leads Rescher to accept that "the limits of one's information set unavoidable limits to one's predictive capacities" (1998, p. 135). Nevertheless, it is necessary to distinguish between uncertainty and ignorance. *Ignorance* means the complete lack of information about a concrete issue, whereas *uncertainty* has to do with the characteristic of indecisiveness. Thus, "with uncertainty we know (or think we know) what the range of possibilities is: it is based on (presumed) knowledge of the possibility range for correct. Accordingly we can generally grapple with uncertainty by means of probabilities—at least in favorable circumstances" (Rescher 2009c, p. 57).

   It should be noticed that, even if it were possible to deal with uncertainty by means of probabilistic knowledge, this may not be good enough to guarantee a scientific prediction. Available information can be insufficient, making the predictive task difficult: "possibilities rest on actualities that require information to project possibilities, although unfortunately, misinformation will also come into play" (Rescher 2009c, p. 58). Thus, uncertainty might affect the probabilistic calculus, because the available information might be insufficient or wrong.

   But Rescher insists that prediction cannot be reduced to a probabilistic statement. In his judgment, to make a prediction is something more than assigning probability to phenomena's occurrence. In fact, he thinks that "the probability of a prediction thus affords an index of its acceptability—a measure of the extent to which rational confidence in its realization is warranted in the prevailing circumstances" (Rescher 1998, p. 44). Consequently, claiming the occurrence of something (that is, making a prediction) is different from claiming that something is going to happen with certain probability. From this viewpoint, probability can be important to predict the degree of uncertainty of the forecast obtained.

   Since scientific prediction is the result of an intellectual rational activity, it should be made on the basis of available knowledge and the control of the variables that are relevant with respect to the phenomenon at issue. It should also be based on the appropriate inferences. This will lead to reducing the uncertainty associated with future phenomena, and to the conclusion of reliable and correct predictions. Thus, probability can be a tool that helps to overcome the obstacles related to uncertainty, but probabilistic knowledge alone is not good enough to obtain a prediction.

   Furthermore, the difficulty of achieving appropriate knowledge about the studied phenomena has repercussions on the temporal dimension, which affects the scope of prediction. Rescher distinguishes between long-run forecast and short-run forecast, depending on the temporal distance of the predicted phenomenon with respect to the present moment (Rescher 1998, pp. 76–78). Thus, to the extent that we cannot predict what we cannot conceive, uncertainty will increase as we try to predict a more distant future (however, this is not a general rule: some phenomena can be easier to predict in the long run than in the short run).[34]

   According to Rescher's viewpoint, predicting is similar to trying to see through the fog: "very little can be seen at a distance—and that little with but little clarity" (1998, p. 76). Undoubtedly, the problem of complexity can increase uncertainty. Even if it is known that something is possible, this sometimes might be insufficient to

---

[34] This is the case of the desertification of part of the Earth.

make a prediction. The more complex the studied phenomenon is, the more problematic it will be, in principle, to obtain a reliable prediction about it (Gonzalez 2012b).

There is also another aspect that adds difficulty to scientific prediction: uncertainty is related with predictive scatter. When we are dealing with a limited body of information, it is possible that various competing theories arise. This can lead to conflicting predictions. Hence, "the prospect of conflicting predictions has to be accepted as a pervasively recurrent phenomenon" (Rescher 1998, p. 135). Conflicting predictions concern both the natural sciences and, to a greater extent, the social sciences (especially, in economics).

When the question at issue is the prediction about the development of future science, uncertainty is also present. In fact, Rescher maintains that future science is unpredictable. He is thinking on the Kantian "principle of question proliferation," that intervenes here, making knowledge of future science impossible (at least, in a specific way). According to this principle, each answer given to solve a problem makes new questions arise, which, in turn, need an answer. Furthermore, we cannot predict what questions we will ask in the future, because we cannot anticipate what the answers to currently open questions will be. In this case, the available information about past developments does not justify an inference about the future advancements (Rescher 2012a).

According to Rescher, scientific progress is basically of a conceptual nature. Thus, scientific research advances trough conceptual creativity (see Guillán 2016, pp. 144–146). Consequently, "the questions we can pose are limited by our conceptual horizons" (Rescher 1995, p. 76). In this way, prediction about future science is only possible at a generic level. This is because there is a cognitive indetermination here: the more detailed and precise the prediction is, the less confidence we can attribute to it. It is possible, for example, to predict with certainty that scientific means for observation and experimentation will improve in the future, but it is not possible to anticipate what these improvements will be.

This allows us to infer that uncertainty has more weigh in specific predictions than in generic predictions. This is so because there is a relation of balance between informativeness (detail, precision, etc.) and security, both in scientific knowledge, in general, and in scientific prediction, in particular. Thus, as the degree of detail of the prediction increases, the uncertainty with respect to its reliability also increases. This is an especially relevant issue, since the utility of the prediction lies in its informativeness.

In Rescher's words, "an ironic but critically important feature of scientific inquiry is that the unforeseeable tends to be of special significance just because of its unpredictability. The more important the innovation, the less predictable it is, because its very unpredictability is a key index of importance" (Rescher 2009b, p. 15).[35] This

---

[35] Rescher does not distinguish clearly between creativity and innovation. In this regard, it has been noted that it is usual to distinguish the concept of "creativity," as it appears in a scientific context, from the notion of "innovation," which connects with the technological realm (where there are often modifications of existing realities rather than something completely new or original). Cf. Neira (2012, p. 217).

increases the problem of uncertainty because, generally, those things we do not know are, for that very reason, the most important for us.

On this basis, it is not advisable to obtain predictive security by losing informativeness, since the value of prediction lies in its content. Hence, it has to aspire to accuracy and precision. However, when it is impossible to obtain a specific prediction due to uncertainty, at least a generic prediction may be possible. In this case, uncertainty does not entail unpredictability, but it acts as an obstacle with respect to the achievable degree of accuracy and precision.

Comparatively, according to the type of phenomena that prediction is about, it is possible to claim, in my judgment, that prediction generates fewer difficulties in the natural sciences than in the social sciences. This is due to the higher level of complexity that social phenomena can have (Gonzalez 2011, pp. 319–321), where uncertainty also has more weight than natural phenomena.[36] Usually, Rescher's approach focuses on natural sciences (mainly, physics), so he leaves open the question of uncertainty with respect to social sciences (and he certainly does not pay special attention to prediction in the sciences of the artificial).

In principle, social sciences have to do with agents' actions and choices, and the decisions are taken in changeable social and historical settings. These structural and dynamic factors involve an additional source of complexity for prediction in these sciences. Too often, Rescher focuses on the problem of prediction with regard to the limits of the information, that is, on the internal aspects of science. In the specific case of scientific prediction, uncertainty has to do then with the future environment, so that available information does not allow us to predict or, at least, makes it difficult for predictions to have the desired degree of accuracy and precision.

In social sciences, in general, and in economics, in particular, there is an additional source of uncertainty. It is *decision-making*, which is related with the problem of rational decision. In economics, the problem of uncertainty is not only related with the future environment, but also has to be considered with regard to the agents' decision-making, which is usually carried out in circumstances of uncertainty.[37] Consequently, when prediction is about economic agents' decision-making, it is not usual to obtain a "foresight" (the securest kind of prediction), but a mere "forecast," which always involves a margin of error and, hence, intrinsically involves uncertainty.

In this context, it is possible to highlight the relevance of the methodological role of uncertainty, especially in the case of economic prediction (Gonzalez 2012a, p. 91). When the result of economic forecasts are known, "the corresponding fore-

---

[36] According to Rescher, social phenomena are more difficult to predict than natural phenomena. He considers that there is a very simple reason for this: social prediction deals, in principle, with people's acts and choices. How people consciously and deliberatively act depends not on the reality of the world, but on what people think about that reality (i.e., it depends on beliefs, ideas, expectations, etc. that are immensely variable). By contrast, natural prediction is about natural processes, which are an objective matter, and depend on the state of affairs of the world. Rescher, *Personal communication*, 10.6.2014.

[37] About uncertainty as an obstacle to prediction in social sciences, in general, and in economic science, in particular, see Simon (2000).

cast errors and the anticipated forecast uncertainty can be used to evaluate the models from which the forecasts were generated" (Ericsson 2002, p. 19). Thus, uncertainty has a methodological role that has repercussions on the use of forecast as a test to assess the appropriateness of predictive models.

Therefore, it is possible to state that, with regard to uncertainty, there are more questions than those contemplated by Rescher. In his approach, which is primarily oriented to the natural sciences, uncertainty is the main epistemological obstacle to predictability. It has to do with lack of knowledge about the regularities of phenomena, which is due to epistemic failures in obtaining or computing the information. But, besides the epistemological dimension, there are, in my judgment, a methodological and an ontological aspect which can be seen clearly when the problem of prediction in economics is analyzed.

Methodologically, the problem arises about whether or not prediction is a necessary condition or test to have science. This is a question that has been debated above all in economics, where the problem of the lack of reliability of economic predictions has special relevance.[38] Uncertainty influences this question, so it is possible to ask whether it is good enough to obtain forecasts (predictions that involve uncertainty) instead of a genuine foresight.

This methodological problem related with prediction has ontological roots. *De facto*, in the social sciences and the sciences of the artificial, "it sits on the *complexity* of human activity involved in the social environment. This complexity contributes to the frequent lack of reliability of economic predictions, which has its roots in the object of study of this science: economic reality is a social and artificial undertaking, which is commonly mutable, as a consequence of its dependence on the human activity that develops historically" (Gonzalez 2012a, p. 92). From this point of view, it is possible to consider uncertainty as a source of complexity in economic activity, so it is an obstacle of special relevance to prediction in economics.

However, the epistemological dimension is fundamental, because uncertainty has to do, in principle, with the lack of knowledge about the regularities, which is due to the lack of information. Consequently, it might be the case that uncertainty implies "unpredictability", i.e., the complete impossibility of predicting. But, commonly, uncertainty accompanies scientific predictions, so prediction is possible under conditions of uncertainty (at least, at the level of forecasts or generic predictions).

However, it must be considered that a phenomenon we cannot currently predict, because of the presence of uncertainty—or an issue that we can only predict generically, or by means of forecasts—can be predictable in the future. Furthermore, the accuracy and the degree of precision of the obtained predictions can increase. In this case, it is necessary to stress the rational basis of scientific prediction, since it is possible to overcome uncertainty if the knowledge and control of the relevant variables increase.

---

[38] About prediction as the "scientific test" of economics, see Gonzalez (2006). This paper analyzes the approaches of four Nobel Prize winners in economics: Milton Friedman, John Hicks, James Buchanan, and Herbert Simon. See also Gonzalez (1996).

4.4.2   Prediction and Decision-Making

Although Rescher focuses mainly on natural science, he also takes into account the role of prediction in social sciences, and notes limits to prediction in this field. In this regard, predicting the behavior of the agents in a social context involves the assumption that these agents make choices in a rational way. They should seek an optimal choice with regard to their context. But, even acting as rational agents, there are situations in which agents are not capable of predicting the optimal course of action (Rescher 2003b, pp. 39–47). This implies clear limits to the predictive capacity of social sciences.

An especially complex task in social science is to meet the necessary conditions for prediction, because of the obstacles inherent to their object of study for predictability. Thus, the *indeterminism* of human affairs must be taken into account, as well as the *predictive errors* related to the knowledge of the variables that intervene in the process (Gonzalez 2010, p. 273). For Rescher, the limits to prediction in social sciences "lie in the intractability of the issues, so that there is little reason to think that the relatively modest record of the past will be substantially improved upon the future" (Rescher 1998, p. 202).

Initially, social sciences have to do with the actions and choices of the agents, above all if we assume the methodological individualism. In that case, it can be noticed that both actions and choices are predictable to the extent that they are also rational. From this perspective, the behavior of the agents can be predicted if they choose the optimal option in their specific circumstances. However, agents' do not always behave rationally. Thus, arbitrary choices intervene, which imply non-rational and, therefore, not-predictable decision making (at least with the desired level of accuracy and precision). In this way, it is possible to predict that an agent will act in an arbitrary way, but we cannot predict what his or her action or choice will be in a rigorous way (Rescher 2003b, p. 39).

However, even if the agents act in a rational way, prediction in a social context is not always possible. Rescher contemplates two cases where, although the behavior of the agents is fully rational, it is impossible to formulate a reliable prediction about their actions (Rescher 2003b, pp. 40–45). First, he notes the case of informational underdetermination; and, second, he analyzes overdetermination as a limit to predictability.

Informational underdetermination can make prediction impossible; because, if we do not have information (or the information that we have is incomplete), then we cannot achieve a rational prediction about the agents' actions and choices, even if they are ideally rational agents. In addition, the analysis of overdetermination implies that it is possible to reach several rational solutions that are divergent among themselves. Thus, the agent whose actions we want to predict is in a situation where an optimal solution cannot be established, since several different choices are equally rational. The prisoner's dilemma is a clear example of this kind of obstacle.

Despite the different limits mentioned, Rescher thinks that human actions should be, in principle, predictable, insofar as they are rational actions.[39] In fact, the social system—which is certainly complex—is supported by the predictability of agents' behavior. Thus, even when he admits that "mere arbitrary *choice* can provide resolutions that lie beyond the dictates of reason" (Rescher 2003b, p. 46), he also understands that "the acts of rational agents are usually predictable" (2003b, p. 39).

This idea is related to the thesis maintained by Merrilee Salmon, when she analyzes the possibility of predicting in the realm of the social science. This philosopher considers that "if people were completely unpredictable, social life would break down altogether" (Salmon 1992, p. 406; see also Gonzalez 2005). There are many social phenomena that include regularities and, therefore, they are predictable. On this basis, predictive success in social sciences is an attainable goal. However complexity, to the extent that it involves the lack of regularities, undoubtedly makes scientific prediction more difficult, both in science, in general, and in social sciences, in particular.

4.5 Rationality and the Problem of Risk

Rescher considers that rational behavior involves an important risk: error. He thinks that there are three main categories of error, which corresponds with the three realms of rationality: (i) cognitive error, which arises when there are failures in information acquisition; (ii) practical error, which involves failures with regard to the goals of action; and (iii) axiological error, which has to do with failures in the realm of evaluation (Rescher 2007, p. 1).

Consequently, when we accept a belief, make an action, or evaluate different alternatives, there is always certain risk of error. Rescher's interest is mainly in the problem of risk in relation to cognitive rationality. His approach is, above all, a criticism of skepticism. Thus, for him, the acceptance of information as valid always involves certain risk. However, he considers that we need to calculate the risk and, in some cases, assume it, if we expect to obtain some benefit from the cognitive undertaking.

This issue is connected with scientific prediction at several successive levels. First, prediction is a statement that provides information about a potential future. Therefore, it can be considered that, by accepting a predictive statement as true, we assume a higher risk than when we accept as true information regarding past or present facts. In effect, it is not possible to test now what prediction states about the

---

[39] "The acts of rational agents are usually predictable because it is often and perhaps even usually possible to figure out on the basis of general principles what the rational thing to do is in the prevailing situation. This circumstance makes rationality into a crucial predictive resource in matters of human action. Indeed, it is on just this basis that we try to understand people, since we ordinarily credit them with being rational until such time as they prove themselves otherwise. In consequence, the actions of free agents must be substantially predictable—if they are rational, at any rate," Rescher (2003b, p. 39).

future. Secondly, prediction can anticipate a risk (a natural disaster, for example), so it can be linked with what should be done with respect to that risk. Thirdly, the prediction of a risk can be related with the problem of decision making by agents; because, once the risk is anticipated, decisions should be made about the adequate measures to avoid or, at least, minimize that risk within given circumstances.

4.5.1 Cognitive Rationality and Risk: A Criticism of Skepticism

Within an epistemological framework of fallibilism, it is assumed that every attempt to obtain knowledge—human, in general, and scientific, in particular—involves certain risk. In Rescher's judgment, "virtually all of our ventures in claiming knowledge about reality carry some risk of cognitive error in their wake: it is unavoidable companion of the enhancement of knowledge" (2009c, p. 59). Furthermore, the achievement of knowledge (human and scientific) is a human need.[40]

When we try to obtain knowledge, it is not possible for us to achieve conclusions that we can consider true beyond doubt. History of science itself leads us to think that the hypotheses and theories that we accept now will be revised in the future (Rescher 2012a). In effect, Rescher considers that "the skeptical tradition reminds us that all our claims to knowledge and truth carry some element of risk" (1988, p. 72).

Regarding the problem of risk, Rescher maintains that there are three different approaches that are related to three different types of personalities: (a) *risk avoiders*, who are people who have little or no tolerance for risk; (b) *risk seekers*, who are people extremely tolerant for risk; and (c) *risk calculators*, who proceed in a cautious way, taking risks only when the situation seems to be favorable (Rescher 1988, pp. 54–64). Regarding cognition, the first of these approaches corresponds with a skeptical position regarding knowledge, according to which no risk should be assumed, since there is no guarantee that it is possible to achieve true knowledge.

But, although risk avoiders never assume risks, the possibility of error is not completely avoided. This is because, regarding knowledge, errors can be of two different types: errors of commission and errors of omission (Rescher 2009c, p. 61). On the one hand, there can be errors of commission (when a false belief is accepted); and, on the other, there can be errors of omission (when a true belief is rejected).

Errors of *commission* can be avoided in two different ways: by giving less informative answers or by giving no answer to the questions posed. In both cases alike there are *errors of omission*, because, by offering vague and insufficient answers, the risk of errors of commission is avoided (to accept false claims); but the knowl-

---

[40] For Rescher, "the knowledge that orients our activities in this world is itself the most practical of things—a rational animal cannot feel at ease in situations of which it can make no cognitive sense. We have questions and want (nay, *need*), to have answers to them. And not just answers, but answers that cohere and fit together in an orderly way can alone satisfy a rational creature. This basic practical impetus to (coherent) information provides a fundamental imperative to cognitive intelligence" (Rescher 1988, p. 65).

edge that can be achieved in this way is unsatisfactory, because detail, accuracy, precision, etc. are values that should accompany scientific knowledge.

Risk avoiders will incur many errors of omission; while risk seekers will incur may errors of commission. For Rescher, the most adequate position is the approach of the *risk calculators*, who base their choices and actions on "sensible calculation and prudent management" (Rescher 1988, p. 59). In effect, the achievement of answers is a human need, which should be satisfied. Faced with the risks of this task, the rational thing to do is "[act] as best we can to balance the positive risks of outright loss against the negative ones of lost opportunity" (1988, p. 59).

In this way, the practical aspect of rationality and therefore the methodological component of the scientific undertaking are highlighted. So, Rescher agrees when the skeptical tradition when he claims that "*each* of our accepted beliefs *may* turn out to be false, and many of our accepted beliefs *will* turn out to be false" (Rescher 1999b, p. 34). But this feature does not imply, in his judgment, that every attempt to obtain knowledge is doomed to failure. Skepticism is directed to the content of concrete theses. For Rescher, this is a wrong approach, since the problem is of a methodological character and is related to the policies for the acceptance of theories (Rescher 1988, pp. 61–64).

In other words, we have to establish cognitive policies that can be justified and that allow us to accept as valid the concrete statements, hypotheses, and theories. This is because, at the level of the cognitive content, there is always certain risk of error. In the case of the knowledge about the future, for example, it should be accepted that both a prophecy of a clairvoyant and a scientific prediction can turn out to be false. However, unlike a prophecy, we can establish that a scientific prediction is reliable, if there are theoretical or empirical proofs that justify the inference of future.

Nevertheless, it is not possible to completely eliminate the cognitive risk. But a policy regarding risk should be accepted, in order to minimize the errors and have the maximum possible benefits, so we can meet our need of knowledge. Once again, practical rationality is crucial, because the costs and the benefits of scientific research should be correctly calculated. Thus, "the crucial fact is that inquiry, like virtually all other human endeavors, is not a cost-free enterprise. The process of getting plausible answers to our questions also involves costs and risks. Whether these costs and risks are worth incurring depends on our valuation of the potential benefit to be gained" (Rescher 2009c, p. 61).

Therefore, Rescher's approach to the problem of risk is mainly pragmatic. The presence of risks in the cognitive venture must be acknowledged. This feature should not lead us to skepticism, but to a realist position about what is possible— both in the sense of attainable and acceptable—for us. In effect, achieving knowledge is a human need that serves as a basis for action. An action can be performed on the basis of non-cognitive policies (instinct or tradition, for example), but this is not good enough. As a rational creature, a human being not only seeks to act, but also *to justify* the actions he chooses to perform (Rescher 1988, p. 67).

Consequently, if no cognitive risk is assumed, the costs are too high, since this avoids, from the very beginning, every attempt to obtain true knowledge. Also the opposite position—that tends to be extremely tolerant regarding risk—has problems, because it involves many risks of commission; that is, to accept as valid beliefs that are false. Some "error management" is necessary which allows us to reduce the probability that an error occurs and to reduce the negative consequences of the error when it occurs (Rescher 2009c, p. 62).

Ideal knowledge is far from the type of knowledge that, in fact, we can achieve. But "an analogue of the old Roman legal precept is operative here—one is never obliged beyond the limits of the possible (*ultra posse nemo obligatur*)" [Rescher 1988, p. 70]. Thus, risk is always present due to the possibility of error, but this possibility can be reduced when the value of knowledge is admitted and, consequently, the potential benefits of the cognitive venture justify the acceptance of a certain previously calculated risk. In Rescher's words, "with cognition as elsewhere rationality calls for a pragmatic balance of costs and benefits in the presence of limited resources. Here too we must strike a reasonable compromise between what is ideal and what is affordable" (2009c, p. 63).

4.5.2 Risk and Prediction

For Rescher, every attempt to obtain knowledge involves risks. In the case of the predictive knowledge this is even clearer, because prediction is about a potential future or a reality that we do not know now, instead of dealing with past or present phenomena and events. Thus, prediction can turn out to be false. For this reason, by accepting its content we assume risks of error, either of commission or of omission, assuming the risks is a question that depends on two crucial factors: on the one hand, the *reliability* of the prediction; and, on the other, the *benefits* derived from the knowledge that prediction provides.

The reliability of a statement about the future depends on its rational bases. Thus, when there are no rational bases that justify the inference from the available data, prediction is not credible. In effect, "only reasonable and substantive predictions—those which are both informative and can be rendered plausible to other people by way of substantiation—are of any cognitive interest" (Rescher 1998, p. 55). Consequently, the risk of error would be higher in a "prediction without reasons" than in a reasoned prediction, since rational prediction has an evidential basis that justifies it.

However, predictive success cannot be guaranteed in many cases (i.e., when phenomena are not deterministic). Moreover, Rescher considers that prediction is an inherently risky business (1998, p. 59). But the risk should be assumed, since the complete lack of knowledge about the future is more damaging than the possible errors derived from our attempt to know it. Thus, he insists that to obtain some knowledge about the future events and developments is a human need: "to act, to plan, to survive, we must anticipate the future" (Rescher 1998, p. 64). So, for him, every human action depends, to some extent, on information about the future.

Also, prediction can be about a matter that involves risks (the anticipation of a natural disaster, for example). In that case, prediction connects with risk management that, in turn, is related with prescription (applied science) and decision-making (application of science). Rescher defines risk as "the chancing of a negative outcome" (1983, p. 5). Thus, he considers that "risk" is an ontological rather than an epistemological issue. This distinction leads him to differentiate two aspects: on the one hand, *to run a risk*; and, on the other hand, *to take a risk*, which involves an epistemic component that is not present in the first case (Rescher 1983, pp. 6–7).

To run a risk is something that happens independently of the knowledge that there is or there will be a situation of risk. In this case, the person or the society is facing a risk that is unknown, so the situation of risk is not linked with any type of action performed by the agents. Meanwhile, when a risk *is taken*, there is a previous knowledge that some kind of harm or loss can occur. Thus, it implies a choice or choices by the individual or the society, which act on the basis of that knowledge. In this way, the anticipation of risks connects with rational choice and leads to ethical problems that have to do with prediction.

In this second case—"taking a risk"—the problem of risk management arises. Thus, although Rescher's main interest is not the realm of the applied science, his approach to risk connects with the role of prediction in this area. In fact, from this perspective, prediction is crucial, since only by the anticipation of risks—that is, by prediction—can policies be suggested, either to avoid some occurrence or to produce some result. However, he thinks that the anticipation of the possibility of a risk is easier than the identification of the risk and the determination of its magnitude or negativity (Rescher 1983, p. 18). Therefore, the evaluation of the magnitude and negativity of the risk predicted is the previous step to what is properly the risk management.

In order to evaluate the magnitude of a risk, three main questions must be valued: (i) character, which is the identification of what type of negativity is at issue: physical injury, monetary loss, etc.; (ii) extent, which is the gravity and magnitude of some risk, which involves issues such as the number of persons affected or the amplitude of the area at danger (in the case of a natural disaster, for example); and (iii) timing, which is the duration of the situation of risk (Rescher 1983, p. 19).

But there is the possibility that we cannot evaluate in a precise way the three indicators of the magnitude of a risk. Thus, even when the risk is anticipated, there is usually uncertainty. The very risk of error related to every prediction impedes that we consider a predictive statement as absolutely secure. Moreover, there is a cognitive indetermination, according to which the more informative a prediction is, the less secure it is (Rescher 1998, p. 62). In the case of a prediction about a risk (earthquakes, tsunamis, volcanic eruptions, etc.), this involves that the prediction is less secure the more we go deeply in questions such as the extent or timing of the risk.

Within this framework, Rescher notices that uncertainty is the indeterminacy of some of the characteristic elements in a situation of risk (1983, p. 94). This feature is especially important, since "impredictability and risk go hand in hand in human affairs" (Rescher 1998, p. 237). Therefore, the cognitive limitations that affect prediction should be taken into account, as well as the problem of uncertainty. In turn, these limits affect the human capacity to control future phenomena and events, since "a future we cannot foresee is a fortiori a future we cannot control" (Rescher 1998, p. 236). Therefore, avoiding or minimizing a potential risk is only possible if that risk has been anticipated. This has repercussions in the configuration of the applied sciences, where prediction is the previous step to prescription and the subsequent application of science (Gonzalez 2008, pp. 181–182).

Through risk management the aim is to have control over the future events or the repercussions that these events can have on people, society, or the environment. For Rescher, control is *"the capacity to intervene in the course of events so as to be able both to make something happens and to preclude it from happening, this result being produced in a way that is not only foreseen but intended or planned"* (Rescher 1998, p. 235). Therefore, there are two main kinds of control: negative control (to preclude something from happening) and positive control (to make something happens). Both can be found in a context of applied science.

Rescher notes that, generally, the control we can have over future events and phenomena is a negative control (1998, p. 235). This usually happens in risk-management. Thus, once a risk has been anticipated, the prescription, in principle, seeks to preclude the risk from happening. In order to do this, we need to intervene in the course of events. This requires a *causal participation* of the individuals in the course of events (Rescher 1998, pp. 235–236). But this is not always possible in applied science.

For example, we can anticipate an earthquake, but we cannot prevent it from happening. In this case, the aim of the prescription (applied science) and the planning (application of science) is to minimize the harm predicted. In this cause, the so-called "precautionary principle" is used, according to which measures of protective character should be taken if there is some indicator that a situation of risk will happen (a natural disaster, for example) or when a certain product or technological device might involve risks for health or for the environment (Gómez 2003). Thus, although it is not possible to anticipate a risk with certainty, we should act in order to prevent it.[41]

However, in principle, the prescription will be more effective if the risk is known in detail. In that case, prediction is crucial, because of the evaluation of risk; and, after that, the prescription to orientate the action will be made on the basis of the kind of knowledge about the future that we can achieve. Therefore, although, as Rescher notes, "impredictability" (or, at least, non-predictability) "and risk go hand in hand in human affairs" (1998, p. 237), two things should be sought: first, to reduce the uncertainty that accompanies the risk in order to achieve reliable predictions; and, secondly, to implement an effective and efficient risk-management in order to minimize the harms that have been anticipated.

---

[41] On this issue, see Luján and López Cerezo (2003), and Luján and Todt (2013).

References

Bereijo, A. 2003. La racionalidad en las Ciencias de lo Artificial: El enfoque de la racionalidad limitada. In *Racionalidad, historicidad y predicción en Herbert A. Simon*, ed. W.J. Gonzalez, 131–146. A Coruña: Netbiblo.

Boudon, R. 1998. Limitations of rational choice theory. *The American Journal of Sociology* 104 (3): 817–828.

Eagle, A. 2005. Randomness is unpredictability. *The British Journal for the Philosophy of Science* 56: 749–790.

Ericsson, N.R. 2002. Predictable uncertainty in economic forecasting. In *A companion to economic forecasting, Blackwell*, ed. M. Clements and D.F. Hendry, 19–44. Oxford: Blackwell.

Fernández Valbuena, S. 1990. Predicción y Economía. In *Aspectos metodológicos de la investigación científica*, ed. W.J. Gonzalez, 2nd ed., 385–405. Madrid-Murcia: Ediciones Universidad Autónoma de Madrid and Publicaciones Universidad de Murcia.

Friedman, M. 1953. The methodology of positive economics. In *Essays in positive economics*, M. Friedman, 3–43. Chicago: The University of Chicago Press (6th reprint, 1969)

Frost-Arnold, G. 2011. From the pessimistic induction to semantic antirealism. *Philosophy of Science* 78 (5): 1131–1142.

Gómez, A. 2003. El principio de precaución en la gestión internacional del riesgo medioambiental. *Política y Sociedad* 40 (3): 113–130.

Gonzalez, W.J. 1996. On the theoretical basis of prediction in economics. *Journal of Social Philosophy* 27 (3): 201–228.

———. 1997. Rationality in economics and scientific predictions: A critical reconstruction of bounded rationality and its role in economic predictions. *Poznan Studies in the Philosophy of the Sciences and the Humanities* 61: 205–232.

———. 1999. Racionalidad científica y actividad humana: Ciencia y valores en la Filosofía de Nicholas Rescher. In *Razón y valores en la Era científico-tecnológica*, N. Rescher, 11–44. Barcelona: Paidós.

———., ed. 2003a. *Racionalidad, historicidad y predicción en Herbert A. Simon*. Netbiblo: A Coruña.

———. 2003b. Herbert A. Simon: Filósofo de la Ciencia y economista (1916–2001). In *Racionalidad, historicidad y predicción en Herbert A. Simon*, ed. W.J. Gonzalez, 7–63. A Coruña: Netbiblo.

———. 2003c. Racionalidad y Economía: De la racionalidad de la Economía como Ciencia a la racionalidad de los agentes económicos. In *Racionalidad, historicidad y predicción en Herbert A. Simon*, ed. W.J. Gonzalez, 65–96. A Coruña: Netbiblo.

———. 2005. Sobre la predicción en Ciencias Sociales: Análisis de la propuesta de Merrilee Salmon. *Enrahonar* 37: 181–202.

———. 2006. Prediction as scientific test of economics. In *Contemporary perspectives in philosophy and methodology of science*, ed. W.J. Gonzalez and J. Alcolea, 83–112. A Coruña: Netbiblo.

———. 2007. Configuración de las Ciencias de Diseño como Ciencias de lo Artificial: Papel de la Inteligencia Artificial y de la racionalidad limitada. In *Las Ciencias de Diseño. Racionalidad limitada, predicción y prescripción*, ed. W.J. Gonzalez, 41–69. A Coruña: Netbiblo.

———. 2008. Rationality and prediction in the sciences of the artificial. In *Reasoning, rationality and probability*, ed. M.C. Galavotti, R. Scazzieri, and P. Suppes, 165–186. Stanford: CSLI Publications.

———. 2010. *La predicción científica. Concepciones filosófico-metodológicas desde H. Reichenbach a N. Rescher*. Barcelona: Montesinos.

———. 2011. Complexity in economics and prediction: The role of parsimonious factors. In *Explanation, prediction, and confirmation*, ed. D. Dieks, W.J. Gonzalez, S. Hartman, Th. Uebel, and M. Weber, 319–330. Dordrecht: Springer.

———. 2012a. La vertiente dinámica de las Ciencias de la Complejidad. Repercusión de la historicidad para la predicción científica en las Ciencias Diseño. In *Las Ciencias de la Complejidad: Vertiente dinámica de las Ciencias de Diseño y sobriedad de factores*, ed. W.J. Gonzalez, 73–106. A Coruña: Netbiblo.

———. 2012b. Complejidad estructural en Ciencias de Diseño y su incidencia en la predicción científica: El papel de la sobriedad de factores (*parsimonious factors*). In *Las Ciencias de la Complejidad: Vertiente dinámica de las Ciencias de Diseño y sobriedad de factores*, ed. W.J. Gonzalez, 143–167. A Coruña: Netbiblo.

———. 2015. *Philosophico-methodological analysis of prediction and its role in economics*. Dordrecht: Springer.

Guillán, A. 2014. Epistemological limits to scientific prediction: The problem of uncertainty. *The Open Journal of Philosophy* 4 (4): 510–517.

———. 2016. The limits of future knowledge: An analysis of Nicholas Rescher's epistemological approach. In *The limits of science: An analysis from "barriers" to "confines,"* Poznan Studies in the Philosophy of the Sciences and the Humanities, ed. W.J. Gonzalez, 134–149. Leiden: Brill.

Hausman, D. M. 1992. *The inexact and separate science of economics*. Cambridge: Cambridge University Press (2nd rep., 1996).

Houghton, D. 1995. Reasonable doubts about rational choice. *Philosophy* 70 (271): 53–68.

Kitcher, Ph. 2010. Varieties of altruism. *Economics and Philosophy* 26 (2): 121–148.

Laudan, L. 1981. A confutation of convergent realism. *Philosophy of Science* 48: 19–49.

Luján, J.L., and J.A. López Cerezo. 2003. La dimensión social de la Tecnología y el principio de precaución. *Política y Sociedad* 40 (3): 53–60.

Luján, J. L. and Todt, O. 2013. Precaution: Building bridges between innovation and regulation. In *Creativity, innovation, and complexity in science*, ed. W.J. Gonzalez, 173–185. A Coruña: Netbiblo.

Marsonet, M. 1996. *The primacy of practical reason. An essay on Nicholas Rescher's philosopy*. Lanham: University Press of America.

Mellor, D.H. 1975. The possibility of prediction. *Proceedings of the British Academy* 65: 207–223.

Moutafakis, N.J. 2007. *Rescher on rationality, values, and social responsibility. A philosophical portrait*. Heusenstamm: Ontos Verlag.

Neira, P. 2012. Complejidad en Ciencias de la Comunicación debido a la racionalidad: Papel de la racionalidad limitada ante la creatividad e innovación en Internet. In *Las Ciencias de la Complejidad: Vertiente dinámica de las Ciencias de Diseño y sobriedad de factores*, ed. W.J. Gonzalez, 205–230. A Coruña: Netbiblo.

Rescher, N. 1973. *The Primacy of Practice*. Oxford: Basil Blackwell.

———. 1977. *Methodological pragmatism. A systems-theoretic approach to the theory of knowledge*. Oxford: Blackwell.

———. 1978. *Scientific progress*. In *A philosophical essay on the economics of the natural science*. Oxford: Blackwell.

———. 1983. *Risk: A philosophical introduction to the theory of risk evaluation and management*. Washington, DC: University Press of America.

———. 1987. Maximization, optimization, and rationality. On reasons why rationality is not necessarily a matter of maximization. In *Ethical idealism. An inquiry into the nature and function of ideals*, N. Rescher, 55–84. Berkeley/Los Angeles: University of California Press.

———. 1988. *Rationality. A philosophical inquiry into the nature and the rationale of reason*. Oxford: Clarendon Press.

———. 1989. Knowledge and scepticism in economic perspective. In *Cognitive economy. The economic dimension of the theory of knowledge*, N. Rescher, 3–32. Pittsburgh: University of Pittsburgh Press.

———. 1992a. *A system of pragmatic idealism. Vol. I: Human knowledge in idealistic perspective*. Princeton: Princeton University Press.

————. 1992b. The light of reason. In *A system of pragmatic idealism. Vol. I: Human knowledge in idealistic perspective*, N. Rescher, 3–14. Princeton: Princeton University Press.

————. 1992c. Cognitive limits. In *A system of pragmatic idealism. Vol. I: Human knowledge in idealistic perspective*, N. Rescher, 243–254. Princeton: Princeton University Press.

————. 1994. *A System of pragmatic idealism. Vol. III: Metaphilosophical inquires*. Princeton: Princeton University Press.

————. 1995. *Satisfying reason. Studies in the theory of knowledge*. Dordrecht: Kluwer.

————. 1996c. *Priceless knowledge? Natural science in economic perspective*. New York: Rowman and Littlefield.

————. 1998. *Predicting the future. An introduction to the theory of forecasting*. New York: State University of New York Press.

————. 1999a. *Razón y valores en la Era científico-tecnológica*. Barcelona: Paidós.

————. 1999b. *The limits of science*, revised ed. Pittsburgh: University of Pittsburgh Press.

————. 2000. *Realistic pragmatism*. Albany: State University of New York Press.

————. 2003a. *Epistemology. An introduction to the theory of knowledge*. Albany: State University of New York Press.

————. 2003b. *Sensible decisions. Issues of rational decision in personal choice and public policy*. Lanham: Rowman and Littlefield.

————. 2004. Pragmatism and practical rationality. *Contemporary Pragmatism* 1 (1): 43–60.

————. 2006a. Homo optans: On the human condition and the burden of choice. In *Studies in philosophical anthropology*, N. Rescher, 1–7. Heusenstamm: Ontos Verlag.

————. 2006b. Rationality and moral obligation. In *Studies in philosophical anthropology*, N. Rescher, 79–93. Heusenstamm: Ontos Verlag.

————. 2007. *Error: On our predicament when things go wrong*. Pittsburgh: University of Pittsburgh Press.

————. 2009a. *Free will: A philosophical reappraisal*. New Brunswick: Transaction Publishers.

————. 2009b. *Unknowability. An inquiry into the limits of knowledge*. Lanham: Lexington Books.

————. 2009c. Cognitive compromise. On managing cognitive risk in the face of the imperfect/flawed. In *Epistemological studies*, N. Rescher, 57–63. Heusenstamm: Ontos Verlag.

————. 2011. Free will. In *Philosophical explorations*, N. Rescher, 61–77. Heusenstamm: Ontos Verlag.

————. 2012a. The problem of future knowledge. *Mind and Society* 11 (2): 149–163.

————. 2012b. Pragmatism and purpose. In *Pragmatism. The restoration of its scientific roots*, N. Rescher, 21–47. New Brunswick: Transaction Publishers.

Salmon, M.H. 1992. Philosophy of the social sciences. In *Introduction to the Philosophy of Science*, ed. M.H. Salmon et al., 404–425. Englewood Cliffs: Prentice Hall.

Schefczyk, M., and M. Peacock. 2010. Altruism as a thick concept. *Economics and Philosophy* 26 (2): 165–187.

Sen, A. 1994. The formulation of rational choice. *The American Economic Review* 84 (2): 385–390.

Simon, H.A. 1972. Theories of bounded rationality. In *Decision and organization*, ed. C.B. McGuire and R. Radner, 161–176. Amsterdam: North-Holland. (Reprinted in Simon, H.A. 1982. *Models of bounded rationality. Vol. 2: Behavioral economics and business organization*, ed. H.A. Salmon, 408–423. Cambridge, MA: The MIT Press).

————. 1976. From substantive to procedural rationality. In *Method and appraisal in economics*, ed. S.J. Latsis, 129–148. Cambridge: Cambridge University Press.

————. 1982a. *Models of bounded rationality. Vol. 1: Economic Analysis and Public Policy*. Cambridge, MA: The MIT Press.

————. 1982b. *Models of bounded rationality. Vol. 2: Behavioral economics and business organization*. Cambridge, MA: The MIT Press.

————. 1983. *Reason in human affairs*. Stanford: Stanford University Press.

————. 1991. *Models of my life*. New York: Basic Books (reprinted in 1996. Cambridge, MA: The MIT Press).

————. 1995. Rationality in political behavior. *Political Psychology* 16: 45–63.

———. 1997a. *Models of bounded rationality. Vol. 3: Empirically grounded economic reason.* Cambridge, MA: The MIT Press.

———. 1997b. Why economists disagree. In *Models of bounded rationality. Vol. 3: Empirically grounded economic reason*, ed. H.A. Simon, 401–420. Cambridge, MA: The MIT Press.

———. 2000. Bounded rationality in social science: Today and tomorrow. *Mind and Society* 1 (1): 25–39.

Chapter 5
Conceptual Framework of the Methodology of Prediction and Preconditions for Rational Prediction

Abstract Following a pragmatic setting, Rescher's methodology of scientific prediction is connected with his general methodological approach. Prediction has a fundamental role in it, since it is one of the main indicators of methodological efficacy. In order to clarify the conceptual framework of the methodology of prediction, this chapter follows several steps. First, the focus is on Rescher's methodological pragmatism as a framework for the analysis of scientific prediction from a methodological perspective. Second, the attention shifts to the roles of prediction in scientific activity. Third, the different groups of empirical sciences (the natural sciences, the social sciences, and the sciences of the artificial) are considered.

Thereafter, the preconditions for rational prediction are analyzed. In Rescher's judgment, these preconditions are three: data availability, pattern discernability, and pattern stability. In his approach, preconditions are necessary and sufficient conditions for predictability, so they are especially relevant in his methodological proposal about scientific prediction.

Keywords Methodological pragmatism • Scientific prediction • Basic sciences • Applied sciences • Application of science • Natural sciences • Social sciences • Sciences of the artificial • Preconditions for rational prediction

In Nicholas Rescher's conception, the methodological characters of scientific prediction are within a framework of methodological pragmatism. According to this proposal, scientific claims and theories should be evaluated according to methodological criteria (Rescher 1977, ch. 5, pp. 66–80). To do this, there is a procedure that consists of two steps: (1) the truthlikeness of scientific propositions[1] or theories can be evaluated through methods of a cognitive character, which are oriented towards the confirmation of those propositions or theories; and, then, (2) the validity of the methods used should be evaluated on the basis of practical criteria; mainly, the capacity of these methods to achieve successful predictions and control over nature.

---

[1] "Proposition" is used here with the meaning of the content expressed by a statement, a content that can be evaluated regarding truth or truthlikeness.

© Springer International Publishing AG 2017 143
A. Guillán, *Pragmatic Idealism and Scientific Prediction*, European Studies in Philosophy of Science 8, DOI 10.1007/978-3-319-63043-4_5

Within this context, Rescher's methodology of scientific prediction is connected with his general methodological approach, which is pragmatic. Prediction has a fundamental role in it, since it is one of the main indicators of methodological efficacy. In order to clarify the conceptual framework of the methodology of prediction, this chapter follows several steps. First, the study is focused on Rescher's methodological pragmatism as a framework for the analysis of scientific prediction from a methodological perspective. Second, the roles of prediction in scientific activity are considered. Third, the attention goes to the different groups of empirical sciences (the natural sciences, the social sciences, and the sciences of the artificial),[2] which have their own characteristics that have repercussions on prediction in each one of those realms.

Thereafter, the preconditions for rational prediction are analyzed. Rescher's identifies three preconditions: data availability, pattern discernability, and pattern stability (Rescher 1998a, pp. 86–87). In his approach, preconditions are necessary and sufficient conditions for predictability, so they are basic in his methodological proposal about scientific prediction. His analysis is made within the framework of methodological pragmatism, which connects with the roles of prediction in sciences, in general, and in each group of sciences, in particular.

5.1 Methodological Pragmatism as a Framework for Scientific Prediction

For Rescher, the questions about the validity of scientific knowledge can (and should) be addressed in an objective way from a methodological perspective. Thus, he seeks to clarify the process for the rational warrant of knowledge. According to his approach—which he labels *methodological pragmatism*—we can "monitor our acceptance of theses via the methods that substantiate them, and then validate these methods by pragmatic tests—specifically considering how well we fare *in applying and implementing its professed claims in matters of prediction and control*" (Rescher 2000, p. 96).

In this regard, prediction has a crucial role in the articulation of a methodological pragmatism. So, for him, the validity of theories is considered through the use of methods that lead to the confirmation or disconfirmation of these theories; and, after that, the validity of the methods is assessed in relation to their capacity to provide successful predictions. For this reason, the methodological characters of scientific prediction in his work should be analyzed from the viewpoint of methodological pragmatism. To do this, first, his characterization of this approach to pragmatism is considered; and, second, the repercussions of methodological pragmatism to scientific prediction are analyzed.

---

[2] The three types of sciences are analyzed here, although Rescher is mainly interested in the natural sciences (above all, in physics).

5.1.1 Nicholas Rescher's Characterization of Methodological Pragmatism

Certainly, nowadays pragmatism is not a homogenous philosophical doctrine.[3] There is, in effect, a great diversity of philosophico-methodological approaches within what is generically called "pragmatism."[4] So it is important to note a set of features that are characteristic of Rescher's methodological pragmatism, which allows us to establish important differences between his approach and other pragmatic proposals. Above all, these differences are related to the notions of "objectivity" and "truth." Moreover, he considers that many of the theses maintained by authors like William James, John Dewey and, later, by philosophers such as F. C. S. Schiller and Richard Rorty, are not properly pragmatic, but are the result of a "deformation" of pragmatic philosophy.[5]

On this matter, Rescher writes that "deflationary epistemologists, including such soft-line pragmatists as William James, are fearful that if we take a hard objectivistic line on the meaning of truth, then truth becomes transcendentally inaccessible and scepticism looms. And they accordingly insist that we soften up our understanding of the nature of truth. But another option is perfectly open, namely to retain the classical (hard) construction of the *meaning* of truth as actual facticity ('correspondence to fact') and to soften matters up on the epistemological/ontological side by adopting a 'realistic' view of what is *criteriologically* required for staking rationally appropriate truth claims" (2004, p. 59).

Clearly, he prefers the second option. For this reason, he criticizes Rorty's conception, according to which pragmatists "suggest that we not ask questions about the nature of Truth and Goodness" (Rorty 1982, p. xiv. Quoted in Rescher 2000, p. xi). In Rescher's judgment, the abandonment of the notions of "truth" and "value" or the attempt to replace them with other concepts, such as "utility," would finally lead to abandoning philosophy.[6] Thus, he does not think that Rorty's approach is properly a pragmatic approach. In effect, as Rescher sees it, pragmatism must admit the notions of "truth," "fact," "objectivity," and "value." Furthermore, a realist

---

[3] In this regard, see Margolis (2002), Bacon (2012), Burke (2013), and Caamaño (2013).

[4] It should be highlighted that frequently pragmatism goes hand in hand with another complementary philosophical approach, above all regarding epistemological matters. For example, Philip Kitcher combines realism with pragmatism. Cf. Gonzalez (2011c).

[5] "A noteworthy—and distinctly curious—aspect of contemporary American philosophy relates to the fate of 'pragmatism,' which has undergone a remarkable deformation from its original conception. Many—indeed most—philosophers nowadays think of pragmatism as something radically different from what was originally at issue with this conception. And, oddly enough, this latter-day sort of pragmatism is not a 'new improved version' but a markedly inferior product," Rescher (2012b), p. 1.

[6] "On such a view, pragmatism is not so much a philosophical doctrine as a position that urges the abandonment of philosophy and recommends finding something else to do instead," Rescher (2000, p. xi).

approach to these notions can be compatible with a pragmatic proposal about the rationality of human beliefs, actions, choices, and evaluations.

In this way, Rescher's philosophy is in tune with the pragmatist tradition of Charles Sanders Peirce, who sought to provide a standard of objectivity that can be used as a test for the appropriateness of our factual beliefs (Rescher 2000, p. 58). Moreover, he sees his own proposal of methodological pragmatism as a return to the Peircean roots of pragmatist thought (Rescher 2012b, pp. 10–11). Thus, he rejects other proposals, such as James' subjectivist pragmatism, Dewey's pragmatism as social and cultural constructions, and Schiller's and Rorty's relativistic approaches (Rescher 2000, ch. 1, pp. 1–56; especially, pp. 15–31 and 44–47, and ch. 2, pp. 57–80).

Within pragmatism, which he sees as a heterogeneous philosophical doctrine, Rescher distinguishes two main directions: (a) pragmatism of the left; and (b) pragmatism of the right (Rescher 2000, pp. 64–69). "Pragmatism of the left" has its origins in William James' approach, which is articulated on the basis of the preferences of the individuals. It is an account that admits cognitive pluralism and relativism, since it is orientated towards the local and personal dimensions. In this way, it gives primacy to the subjective or intersubjective components.

By contrast, "pragmatism of the right"—which has its origin in the theses maintained by Peirce and C. I. Lewis—seeks objective components. Against the simple preferences, it focuses on what is effective and efficient in the satisfaction of universal human needs. Furthermore, it considers that both the very notion of efficiency and the determination of the universal human needs are questions that can be established in an objective way. For this pragmatic approach, objectivity is crucial and it is connected with an ontological realism that allows us to preserve the notion of truth.

According to Rescher, pragmatism of the left is not properly a pragmatic approach, but an "inferior product" (2012b, p. 1). For him, "pragmatism properly understood is a positive doctrine—not one that substitutes practice for truth but one that involves practice as our best available *test* of truth" (1997, p. 34). For this reason, Rescher sees his *methodological pragmatism* as a return to the Peircean roots of the pragmatist tradition.

5.1.2   Methodological Pragmatism as an Evaluative Proposal

With this retrospective background regarding the pragmatist vision about the processes of research, methodological pragmatism—as Rescher conceives it—is a proposal about how is it possible to evaluate the truth or truthlikeness of scientific propositions or theories. There are, in his analysis, two main ways of justifying beliefs: (a) discursively; and (b) methodologically (Rescher 1977, pp. 69–70). Thus, a belief is justified discursively when other previously accepted claims are offered as reasons for the acceptance of the belief. Meanwhile, a belief is methodologically justified when considerations are used that appeal to methods; that is, to processes of research that have clear rules.

When a belief is evaluated methodologically, a procedure is used that follows two steps (Rescher 1977, p. 67): Firstly, theses are justified by the application of a method; and, secondly, the adoption of a particular method is justified on the basis of certain practical criteria, "preeminently, success in prediction and efficacy in control" (1977, p. 67). In this regard, it should be noticed that Rescher's thought, which is a system of pragmatic idealism, combines a realist notion of truth with a pragmatic approach to the evaluation of the content of truth of the scientific claims and theories (see Sect. 1.2 of this book).

Thus, Rescher considers that the truth of a proposition or theory depends on its agreement with reality, and he conceives reality as something independent from the knowing subject. In this way, the results of the research—the theses themselves—can be objective. However, as an idealist philosopher, he insists that the access of human beings to that extramental reality is always mediated by our concepts and mental categories, so scientific knowledge is the result of the interaction between the subject, the researcher, and the reality researched (which in his case is usually the natural reality) [Rescher 1990].

Therefore, within a framework of pragmatic idealism, Rescher thinks that it is not good enough to justify a belief on the basis of other beliefs that were previously accepted, since this task is only possible within a concrete conceptual scheme and, hence, it is a process that depends on the subject. Furthermore, the justification of a belief on the basis of previously accepted knowledge is a circular procedure, so there are no objective bases.

These reflections can be seen as an objection to approaches such as the Bayesian, where beliefs have a basic role (see Howson and Urbach 1989). In effect, when a belief is discursively justified, other beliefs are offered as reasons for its acceptance. In turn, these other beliefs are justified through other theses that support them, and so on (Rescher 1977, p. 70). By contrast, an *instrumental* justification[7] (i.e., properly methodological) is thought-independent, and—on the basis of the efficacy and efficiency of the processes—it avoids the circular character that inheres a discursive justification.

For Rescher, "the truth/reality connection that is operative here is certainly not a cognitively isolated issue subject to no theory-external quality controls. 'Thought externalized' objectivity is still at our disposal. For with regard to our methodological resources of truth-estimation we can indeed deploy a theory-external means of quality control, such as applicative efficacy" (Rescher 2004, p. 59). Thus, for him,

---

[7]Although Rescher uses many times the term "instrumental" regarding his methodological approach, it is certainly not an instrumentalist approach in the sense of a subordination of theory to practice in terms of importance. In his judgment, practical rationality (which allows us to decide about the rationality of the processes) also encompasses a theoretical dimension. Practice and theory are equally important. However, "it takes considerations of purposive effectiveness to provide the test-standard for the adequacy of the operative principles of human endeavor—alike in theoretical and in practical matters. Effective implementation is its pervasive standard of adequacy," Rescher (2004, pp. 43–44).

the question whether a method works (that is, if it is effective and efficient in achieving a goal) can be determined in an objective way on the basis of predictive success.

There is a search for objectivity: the pragmatism that Rescher suggests does not reduce the questions about the validity of scientific knowledge to intersubjective or merely subjective criteria. On the contrary, objective bases are sought in order to establish the truth of a claim or theory (or, at least, its truthlikeness). In order to perform this task, the effectiveness of the methods has a crucial role, since it is a matter that is independent from the subject: "methods possess an inherent objectivity and freedom from any sort of personal dependence" (Rescher 1977, p. 73). In addition, on the basis of the efficacy and efficiency of the processes, the comparative evaluation of alternative methods is possible, as it is to assess the improvements of a concrete method.

Furthermore, Rescher thinks that the efficacy of a method is not only the best criterion for its evaluation, but also the "natural" way of establishing its adequacy (1977, p. 8). In order to support these claims, he notes the two main characteristics of the scientific methods, in general, and the cognitive methods, in particular: (i) they are teleological (that is, scientific methods are *means* orientated towards the achievement of a certain goal); and (ii) they seek generality, so they can be used in successive occasions (Rescher 1977, pp. 2–5).[8]

Regarding the methodology of science, Rescher claims that an *instrumental* conception should be adopted, understood as a combination of mediation and utility, to the extent that methods "are *means for doing things of a certain sort*" (Rescher 1977, p. 3). Therefore, insofar as they are instruments, methods require an instrumental justification. In turn, this instrumental justification involves taking into account the teleological character of the processes, since a method is always oriented to the attainment of some end: "a *method*, after all, is something intrinsically purpose-relative" (Rescher 1977, p. 3). This feature implies that methods, in some sense, are contextual and can be diversified according to the different ends sought.

Due to the primacy of the practical view (Rescher 1973), the justification of a method should be, for Rescher, pragmatic. Because the methods are means in order to achieve some goal, their validity is something that directly depends on their efficacy and efficiency in the achievement of that goal. For this reason, he considers that "the pragmatists were surely right: there can be no better or more natural way of justifying a *method* than by establishing that 'it works' with respect to the specific appointed tasks that are in view for it" (1977, p. 3). Thus, the basic criterion for a rational evaluation of a method is its *success*, which he understands as efficacy and efficiency regarding the goal sought.

For Rescher, when something is oriented towards the achievement of a certain goal, for this very reason it should be subject to an evaluation that takes into account its efficacy. Thus, the adequacy of a method depends on to what extent it achieves its goal in an effective and efficient way. This connects the notion of efficacy with human rationality, since the role of economic rationality is highlighted with regard

---

[8] On the problem of the methodological universalism, see Gonzalez (2012c).

to the selection of the processes: "a rational creature will prefer whatever method process or procedure will, other things equal, facilitate goal realization in the most effective, efficient, and economical way" (Rescher 2004, p. 44).

Besides its teleological character, another characteristic of the methodology of science is its general character.[9] For this reason, what matters is not success as such, but that success regarding methods should be *systematic* (Rescher 2000, p. 84). On the basis of the general character of the scientific methods, he maintains that "one success does not validate a method" (1977, p. 5). Thus, the occasional or isolated success is something irrelevant (Rescher 2000, p. 84). In this way, his perspective regarding the methodology of science is clearly systematic. Moreover, it assumes the general character of scientific methods.

This systematic character allows Rescher to face the objection regarding the role that chance or luck can have in the success of an action or procedure. He admits that the success of a method—that is, that it achieved its goal in one occasion in an effective and efficient way—cannot be the basis to claim the validity of that method. The same happens when a method "fails" on one particular occasion. Undoubtedly, an action performed on the basis of *false* beliefs can have a positive result, as can an action performed on the basis of *truth* beliefs have a negative result.

Therefore, Rescher considers that the generality of scientific methods should be taken into account, insofar as they are characterized by being capable of implementation on numerous occasions. For this reason, he maintains that a "sensible pragmatism" would propose an instrumental justification at a generic and systematic level (Rescher 2000, p. 85). In this way, the success of a method should be evaluated taking into account its performance on numerous occasions and in the long run. In these terms, his methodological pragmatism seeks an evaluation of scientific knowledge on the basis of objective indicators. Methodological efficacy is a key factor, since it is "a matter of how things go 'across the board' generally and in the long run" (Rescher 1977, p. 5).

On the basis of the efficacy and efficiency of the processes as objective matters that are independent from the subjects, Rescher defends a procedure to decide about the truth of scientific propositions and theories that takes into account the success of methods. He does not reject a realist notion of "truth," neither does he suggest replacing it with the concepts of "efficacy" or "efficiency;" but he does consider efficacy and efficiency at the methodological level as the best criterion for the truthlikeness of scientific propositions and theories. Thus, as procedures, scientific methods depend in no way on subjective (or intersubjective) considerations (in this way, he rejects the primacy of the consensus for science),[10] but are susceptible to an

---

[9] Rescher considers that a method consists of a series of general rules or patterns in order to perform a task, so it is possible to use a method on successive occasions. However, there are different levels of generality or abstraction regarding methods, depending on the general or specific character of the matter at issue. Hence, he assumes that different levels of scales of reality (macro, meso, and micro) require different methods. Cf. Rescher, *Personal communication*, 10.6.2014.

[10] About the role of consensus in other contexts from a broad perspective, see Rescher (1993).

objective evaluation, which takes into account how they, in fact, perform their function.

Within this framework, prediction has a fundamental role. It is crucial in Rescher's methodological pragmatism; because, in order to claim that a scientific theory is true or has truthlikeness, the main criterion is ultimately its capacity to predict future phenomena with success. This use of prediction as a test for theories is because prediction, in a clear way, can be tested with experience. The success of prediction allows us to confirm theories, while its lack of success leads to the disconfirmation of the theories.[11] Thus, in his approach, the success of predictions is "the pivotal controlling factor for quality control in scientific theorizing" (Rescher 1998a, p. 161).

5.1.3 Application of Methodological Pragmatism to Scientific Prediction

The applicative efficacy of a method is—for Rescher—its capacity to provide successful predictions and effective control over phenomena. Thus, scientific prediction is very important in the framework of methodological pragmatism. He maintains that the acceptability of theories can be justified on the basis of methods. In turn, the adequacy of the methods depends on their efficacy in obtaining successful predictions and in the achievement of control over phenomena. Therefore, prediction is a test for the acceptability of theories, which, in rigor, should vary in order to adequate the different objects (natural, social, or artificial).[12]

As a test for theories, Rescher considers that prediction is also an important indicator of scientific progress. Thus, there is progress when there is an increasing success in prediction. In this regard, Ilkka Niiniluoto has highlighted the discrepancies between Rescher's methodological pragmatism and a realist approach to scientific progress. Niiniluoto writes that, "according to Rescher, science is realist 'in its intention,' but its achievements or its progress must be defined in terms of its increasing success in the control or 'physical domain over nature.' Here, the realist [philosopher] disagrees with the pragmatist [author]: this pragmatic success is, at the most, one indicator of cognitive success. Furthermore, there can be genuine cognitive success without practical applications: all science is not reducible to applied research" (Niiniluoto 1997, p. 402).

---

[11] In this regard, Rescher notes that "on the standard 'inductive' model of scientific method, the predictions of science are generated by logico-mathematical derivations that apply general theories to situation-specific facts so as to preindicate future observations. Then, insofar as the actual observations *agree* with those predictions, the theories at issue are confirmed and thereby evidentially substantiated, and insofar as they *diverge*, the theories are disconfirmed and evidentially undermined," Rescher (1998a, p. 161).

[12] Usually, Rescher does not distinguish between the subject matters of the natural sciences, the social sciences, and the sciences of the artificial. His approach seeks to be as general as possible.

However, there are several points where Rescher's approach—of methodological pragmatism—and Niiniluoto's proposal about scientific progress (open to some pragmatic components) are close. This is because Rescher's pragmatic idealism accepts realistic contributions in key concepts (truth, fact, etc.). For this reason, in my judgment, Niiniluoto's criticism does not grasp adequately Rescher's proposal about the role of prediction as an indicator of cognitive success and, therefore, of scientific progress.

In this regard, a comparison between the approaches of both authors should take into account several levels of analysis: (i) the *semantic* level, which seeks to clarify the concept of "progress;" (ii) the *methodological* dimension, which focuses on the question about what are the reliable indicators in order to claim that a theory is "progressive" with regard to an alternative theory; and (iii) the *factual* level, which analyzes when we can claim that science has made *de facto* some progress, i.e., an advancement that can be assessed in historical terms (on these aspects, see Gonzalez 1997).

Rescher and Niiniluoto agree with the characterization of scientific progress from a *semantic* perspective: they accept scientific theories as sets of propositions (expressed through statements) oriented towards the truth or, at least, towards truth-likeness. Both authors see "scientific progress" as a notion relative to the goals of science. This claim appears explicitly in Niiniluoto,[13] and it is implicit in Rescher's proposal, according to which we can defend that scientific progress is continuous on the basis of the goals of science (Rescher 1999a, ch. 10, pp. 145–165). In this case, the concept of "progress" has commonly a positive connotation, since it implies an improvement with regard to what was previously available. In this way, it is different from other terms such as "development" or "change."[14]

Regarding the *methodological* issue about which the reliable indicators of scientific progress are, Niiniluoto writes that those indicators are "cognitive factors such as truth, information, explanatory power, predictive capacity, precision, and simplicity" (1997, p. 402). Thus, he considers that Rescher's insistence on prediction as indicator of the truthlikeness of theories results in a partial view of scientific progress. Pragmatic success would be, at most, an indicator of cognitive success, but neither the only one nor the most important one (Niiniluoto 1997, p. 402).

But, in rigor, Rescher does not claim that predictive success is by itself an indicator of the truthlikeness of theories. In his judgment, "only a reciprocally interactive gearing of explanation, prediction, and control can in the final analysis provide a satisfactory standard of scientific adequacy" (Rescher 1998a, p. 165). This is because usually scientific prediction should be supported by an explicative knowledge of phenomena. Furthermore, this interrelation of explanation and prediction is the basis for establishing the barriers between scientific prediction and non-scientific prediction (Rescher 1998a, pp. 169–171). For this reason, when there is some

---

[13] "'Progress' is a normative or goal-relative—rather than purely descriptive—term," Niiniluoto (1980, p. 427). See also Niiniluoto (1984).

[14] Changes might be positive or negative as well as developments, which might be positive or negative.

improvement in the predictive capacity of science, there is also progress at the theoretical level.[15]

The main difference between both approaches—scientific critical realism and Rescher's pragmatism—is, then, rooted in the way they address the *factual* question about when we can claim that science has progressed *de facto*, which can be assessed historically. The answer, for Niiniluoto, is in the notion of "truthlikeness." Thus, we claim that science has made some progress in accordance to the relative success of scientific theories in the achievement of true or, at least, verisimilar knowledge about reality (Niiniluoto 1997, p. 402). Meanwhile, for Rescher, this *factual* issue is, actually, methodological. Thus, the capacity of the methods to achieve successful predictions or an effective control over reality is the best criterion we have to assess the theoretical adequacy of theories.

From this perspective, he does not defend a realist approach when he maintains that "science does indeed progress not, to be sure, by way of 'approaching the ultimate truth,' but by providing us with increasingly powerful instrumentalities for prediction and control" (Rescher 1999a, p. 42). However, he does not consider that there is scientific progress only in pragmatic lines. Successful prediction is not, for Rescher, the only indicator about the comparative theoretical adequacy of theories, but it is the best criterion we have. Predictive efficacy is, in his judgment, "the best available token for the explanatory adequacy of our theories" (Rescher 1998a, p. 164).

According to Niiniluoto's approach, scientific progress consists of an increasing truthlikeness (1984, ch. 5, pp. 75–110), where prediction has a role. Rescher also considers that science makes progresses in that direction. However, in his judgment, prediction should be emphasized when the question at stake is to justify that truthlikeness has, in fact, been achieved. This is because successful prediction provides an objective criterion in order to confirm or disconfirm a theory and to compare its adequacy in relation to other alternative theories. In that case, prediction appears as a result, which acquires the form of a statement and is backed up by experience.

Therefore, prediction is a key notion in Rescher's characterization of scientific progress. He addresses this issue mainly from the methodological realm, as a process oriented towards the increase of knowledge and the control over phenomena. Furthermore, the prominent role of prediction in his methodological pragmatism has also incidence in the configuration of the methodological characters of scientific prediction. This can be seen in the importance of the methodological component within the whole set of the philosophico-methodological characters of prediction (semantic, logical, epistemological, methodological, ontological, axiological, and ethical).

Moreover, in my judgment, many of the most important contributions of Rescher to the study of scientific prediction are in the realm of the methodology of prediction. Thus, one can highlight his effort to offer a methodological approach to

---

[15] This can be clearly seen in Rescher's "thesis of harmony," according to which explanation and prediction are not symmetrical processes, but are closely interrelated as crucial goals of science. See, in this regard, Sect. 3.4.2 of this book.

prediction that is exhaustive and appropriate to scientific practice. This involves taking into account the roles of prediction in scientific activity, which varies in the contexts of basic science, applied science, and the application of science. Thus, Rescher considers prediction as an aim of science. Within an approach of methodological pragmatism, he especially insists on the role of prediction as a test for theories and guide for discovery; but he also takes into account its use in the applied sciences, where prediction is usually the previous step to prescription.[16]

He also seeks to clarify the common features of the different processes of prediction, insofar as those processes are rational. In this regard, he suggests three *preconditions* for rational prediction: data availability, pattern discernability, and pattern stability (Rescher 1998a, pp. 86–87). Concurrently, he assumes a methodological pluralism regarding prediction. Thus, he offers an analysis of the diverse predictive processes, which he classifies into three groups: (1) estimative procedures; (2) discursive elemental processes; and (3) discursive scientific methods (1998a, p. 87).

Thereafter, his contribution to the analysis of the methodological characters of scientific prediction encompasses different problems at stake: (a) the roles of prediction in scientific activity; (b) the preconditions for rational prediction; and (c) the methods of prediction. Furthermore, regarding the methods, he made an effective contribution, because he was one of the creators of the Delphi predictive procedure.[17] Within the special level of the methodology of science—which takes into account the distinctive features of each science or each group of sciences (the natural sciences, the social sciences, or the sciences of the artificial)—he pays more attention to the realm of the natural sciences; although he also takes into account the role of prediction in the methodology of the social sciences. Thus, he also addresses some specific problems of economics and sociology (Rescher 1998a, ch. 11, pp. 191–208; especially, pp. 193–202).

Nevertheless, due to his methodological pragmatism, he highlights above all the methodological aspect of scientific prediction as a test for theories. In turn, this feature leads to a pragmatic approach of prediction itself, according to which prediction is, above all, an *instrument*.[18] The instrumental aspect of prediction appears insofar as he considers that prediction allows us to judge the comparative theoretical adequacy of the theories and to the extent to which prediction might serve as a guide for human action. This approach to prediction as an instrument is due to its connection with practice, but it does not involve an instrumentalist account of science.

De facto, Rescher is clearly against methodological instrumentalism. He maintains that "some philosophers take this matter of the predictive utility of good theories too far by adopting a wholly 'instrumentalistic' view of the theories of natural science as mere predictive instruments, altogether dismissing the issue of describing and explaining the world's occurrences. On this approach, prediction is *all* that

---

[16] On the role of prediction in the applied sciences, see, for example, Gonzalez (2007) and Simon (1990).

[17] On the predictive procedure Delphi, see Ayyub (2001) and Bell (2003).

[18] "Prediction, in sum, is our instrument for resolving our meaningful questions about the future, or at least of *endeavoring* to solve them in a rationally cogent manner," Rescher (1998a, p. 39).

matters and thereby constitutes the alpha and omega of science" (Rescher 1998a, p. 164).

Therefore, Rescher's methodological pragmatism involves an approach to prediction as an instrument, but it does not encompass a genuine methodological instrumentalism that subordinates scientific methods to the aim of predicting (Rescher 1998a, p. 39).[19] In this way, he highlights prediction as a fundamental component of science (this can be seen in relation to the roles that prediction plays in scientific activity) without subscribing a strong predictivist thesis, according to which prediction has a clear primacy over any other goal of science.

5.2 The Roles of Prediction in Scientific Activity

In order to analyze the roles of prediction in scientific activity, the distinction between basic science and applied science should be considered. Because basic science and applied science are different activities,[20] the uses of prediction in them can be also different. Wenceslao J. Gonzalez has highlighted that the differences between both kinds of sciences have to do with three successive levels of scientific research: (i) the goals or aims; (ii) the processes; and (iii) the results (Gonzalez 1999, p. 158, 2015, pp. 32–40). In turn, those differences between both kinds of sciences (basic and applied) have incidence on the uses of prediction in scientific activity.

From the perspective of the *aims* of the research, we have basic research when scientific activity seeks either to obtain new knowledge or to increase the knowledge already available. Thus, basic research is mainly oriented towards giving answers to questions of a cognitive character. Meanwhile, applied science seeks to achieve new knowledge with a specific purpose, which can be either to solve a concrete problem or to solve it in a more efficient way (Niiniluoto 1993, pp. 3–5).

There are also differences between both kinds of sciences regarding the *processes*. From this perspective, scientific methods in basic science are mainly oriented towards the achievement of new knowledge or the improvement of the available knowledge (both predictive and explicative), so a main feature is the search for empirical support for the hypotheses and theories. However, in applied science "the means acquire an operative character, on having had direct relation with specific ends (that means, the practical knowledge has to allow to achieve more efficient processes to solve the particular problems that have been raised)" [Bereijo 2011, p. 338].

Moreover, there are differences between basic science and applied science from the point of view of the *results*. These differences have to do, above all, with the criteria for evaluating the results obtained. In basic science, the criterion of

---

[19] In this regard, Rescher's view is quite different from M. Friedman's (1953).

[20] On the differences between basic science and applied science, see Niiniluoto (1993, 1995) and Gonzalez (2005, 2013a).

truthlikeness is the main one. Thus, the results are evaluated on the basis of the increase of the available knowledge, to the extent that truthlikeness—the main aim—is achieved. Meanwhile, in applied science, the evaluation of the results can be done following cognitive criteria (the adequacy of the knowledge in order to solve the concrete problem at stake) or according to practical parameters (efficacy and efficiency in the solution of the problem) [Bereijo 2011, p. 338].

In this way, the differences between basic science and applied science—which have to do with the aims or goals, the processes, and the results—highlight a background difference between both kinds of research: there is a pragmatic or instrumental feature that is more emphasized in applied science than in basic science, which gives primacy to the theoretical or epistemic component. In turn, these differences between basic science and applied science have repercussions on the role played by prediction in both kinds of sciences. Thus, the roles of prediction vary according to the context in which it is made: basic research or applied research.

Basic science seeks first to describe phenomena and, second, explain or predict these phenomena. If it is merely confined to describing phenomena, then there is no genuine scientific contribution. Thus, prediction is a main aim of basic research, because it provides knowledge about future events or happenings. For this reason, scientific methods can be oriented towards prediction, which also has a fundamental role in the evaluation of the results. In effect, the knowledge about the future can be used as a test for hypotheses and theories, since it provides the empirical content that is required for testing them.

But, when prediction is made in applied science, prediction is, besides being an important aim in itself, a tool for decision-making. In this way, prediction is connected with prescription, because in order to prescribe (i.e., to suggest paths of action to solve a concrete problem), it is necessary to predict. Thus, in the realm of applied science, prediction is the previous step of prescription. For this reason, prediction in applied sciences can be considered as a methodological tool: prediction about the possible future is needed in order to establish the paths that should be followed (Gonzalez 2008, p. 181).

Now, when the distinction between applied science and the application of science is considered, it seems clear that we should take into account another role of prediction, which is related to practical problem-solving (in political contexts, economic, ecological, etc.). This distinction between applied science and the *application of science* has been highlighted by Niiniluoto. In his judgment, "the former is a part of knowledge production, the latter is concerned with the use of scientific knowledge and methods for the solving of practical problems of action (e.g., in engineering or business)" (Niiniluoto 1993, p. 9). From this perspective, prediction can be the basis of decision-making in contexts of policy.[21]

Within the application of science, the use of knowledge by agents (individual or institutional) prevails. In this way, on the basis of the same applied knowledge, two agents can apply knowledge in different ways in their respective contexts. A

---

[21] On the distinction between basic science, applied science, and the application of science, see Gonzalez (2013a, pp. 11–40; especially, pp. 17–18).

prediction about the possible future has been obtained, but what prevails is an agent-relative component, since the agents make decisions according to different contexts. This feature also affects the prescriptions (Gonzalez 2013a, pp. 17–18).

Therefore, prediction encompasses several roles. (1) In basic science, prediction can be used as a test for theories, in general, and hypotheses, in particular. This use of prediction can be seen both in natural sciences (for example, physics, chemistry, or astronomy) and in social sciences (among others, economics or sociology) (2) In the case of the applied science (pharmacology, medicine, applied economics, etc.), prediction is usually the previous step to prescription. In this realm, the anticipation of the future is necessary before paths of action can be suggested in order to solve concrete problems. (3) When the problem of the application of science is considered, prediction also has a role, since it can serve as the basis of the procedures of decision-making (Gonzalez 2010, p. 11).

5.2.1 The Role of Prediction as Test in Basic Science

When Rescher considers uses of scientific prediction, he usually focuses on the role of prediction as test for theories. In this regard, he notes that prediction has mainly two uses in science: "as a test of the acceptability of theories and as a guide to discovery. No other factor shows more clearly that we are making real (rather than merely putative) progress in natural science than the successful prediction of new phenomena" (Rescher 1998a, p. 160).

Commonly, he contemplates the use of prediction as a test regarding the natural sciences (above all, physics). In general, he thinks that the predictive capacity is much higher than in the case of social sciences (he does not address expressly the sciences of the artificial). Concretely, in the case of economics, he considers that quantitative prediction—in accurate and precise terms—is usually not possible. Thus, predictive success in this science is habitually obtained at the level of generic predictions regarding trends and probabilities (Rescher 1998a, pp. 193–199; especially, p. 198).

For this reason, it is possible to maintain that there is a duality in Rescher's methodological approach to prediction and its roles in scientific activity. At the general level, he considers prediction as a reliable test for the theories; while in the special level (regarding the social sciences and, concretely, economics) it seems that predictions do not usually achieve the level of detail required in order to serve as a test for theories. To a certain extent, he seems to agree with Herbert A. Simon (1989), who maintained that we should have a *wary* attitude regarding the use of prediction as a test for economics as a science (Rescher 1998a, pp. 198–199 and p. 277, n. 264).

When Rescher thinks of prediction as a test for theories in the natural sciences, he considers the most specific case of the *surprising predictions*. In this regard, he notes that the importance of prediction is in the *cognitive novelty* that it encompasses: "After all, it would seem to be *cognitive* novelty that is the crux, and futurity as such (mere *chronological* novelty) seems immaterial. The predictive aspect is surely incidental; surprising predictions are important for confirmation, but on account of their surprisingness rather than their predictivity, seeing that it is epistemic novelty that carries the burden of the work" (1998a, p. 162).

When Rescher analyses scientific prediction, he assumes two different components of the notion of "novelty:" an ontological feature and an epistemic aspect.[22] Scientific prediction implies, in principle, novelty in the ontological sense, because it is oriented towards a possible future. But Rescher highlights the epistemic aspect. In this way, the cognitive content (and, therefore, the epistemic novelty) is the most important feature of a prediction. For this reason, when prediction is used as a test for theories, surprising predictions should have more weight: "if the [predicted] fact is something new in kind—a new phenomenon or a new type of fact that was not experienced before—then it indeed is in a position to make significant evidential contribution" (Rescher 1998a, pp. 162–163).

In this regard, Rescher considers two options for prediction, which should be taken into account when predictive success is evaluated in the context of basic science: (a) prediction of new instances of familiar phenomena; and (b) prediction of the occurrence of new phenomena, which have not been investigated before (Rescher 1998a, p. 163). In his judgment, the most important thing is not just having more elements from a quantitative perspective about something already known, but the qualitative achievement of new areas (Gonzalez 2010, p. 267). Thus, in principle, scientific prediction involves a strictly temporal or ontological factor (it is oriented towards a possible future); but surprising predictions are characterized by their *epistemic novelty*, which gives them value as a test for theories.

Moreover, this use of prediction as a test for hypotheses and theories is a central feature of Rescher's methodological pragmatism. Thus, in his judgment, theories are justified by methods for the validation or confirmation of factual statements; and, in turn, those methods must be evaluated according to practical criteria (above all, the success in prediction and the efficacy in the control of phenomena) [Rescher 2000, p. 96]. In this case, scientific methods should be oriented towards prediction, to the extent that successful prediction is "the pivotal controlling factor for quality control in scientific theorizing" (Rescher 1998a, p. 161).

From this perspective, Rescher's account—in the context of the discussion between "prediction" and "accommodation"—is certainly predictivist.[23] When empirical support is required for theories, in general, and for hypotheses, in particular, prediction is, in his view, the decisive factor. It is decisive insofar as prediction provides an *epistemic novelty* that can be compared with the future observations: "insofar as the actual observations agree with those predictions, the theories at issue are confirmed and thereby evidentially substantiated, and insofar as they diverge, the theories are disconfirmed and evidentially undermined" (Rescher 1998a, p. 161).

---

[22] There is also a third possibility: heuristic novelty. See Gonzalez (2014, pp. 1–25; especially, pp. 14–16).

[23] On the controversy about the methodological weight of "prediction" and "accommodation," see Gonzalez (2010, pp. 288–292). A defense of the predictivist position is in White (2003). Rescher's account in this regard is further developed in Chap. 8, Sect. 8.3.2 of this book.

Consequently, the most rigorous knowledge that science can provide is the knowledge obtained through the predictive success. For this reason, it is possible to claim that scientific methods should be mainly oriented towards prediction, since it is the best test we have for the scientific character of theories. However, Rescher's methodological pragmatism does not involve an instrumentalist approach of scientific methods; i.e., he does not think that scientific methods are just instruments of prediction. Moreover, in his judgment, prediction is not a necessary condition for having science.

It is important to point out that scientific prediction is fallible as a test for theories (a false theory can lead to true predictions, and even a true theory can make predictions that will be disconfirmed by future observations) [Batitsky and Domotor 2007]. This feature is acknowledged by Rescher: "the complex interweaving of fact, theory, and conjecture in scientific prediction means that even good theories sometimes yield poor predictions. And contrariwise, even where we make successful predictions this will not necessarily mean that the basis of theory from which they emerge is scientifically appropriate" (1998a, p. 169).

Moreover, prediction should not be the only aim of scientific research. For Rescher, prediction is an important aim of science; but it is one aim *among others*. Thus, besides prediction, science should be oriented towards the description, explanation, and control over reality (mainly nature). For this reason, scientific methods cannot be simple predictive instruments. However, the methods that allow us to predict have a high value, because successful prediction is the best criterion in order to evaluate the theoretical adequacy of theories.

However, according to the harmony thesis between explanation and prediction suggested by Rescher, scientific methods (in basic science) should be oriented towards the two main aims of the scientific activity: the explanation about past (and present) phenomena and the prediction about the future phenomena. Methods should lead research to theories with both explicative and predictive power, since, in Rescher's judgment, "theories that do not yield predictions are sterile, and predictions—however successful—that lack a theoretical backing are for that very reason cognitively unsatisfactory" (1998a, p. 167).[24]

In this regard, it should be noticed that some of our best theories, such as the theory of evolution, do not, properly speaking, make predictions. On this objection, Rescher maintains that "while evolutionary theory does not predict specific outcomes by way of forecasting the modifications of particular species, it does, nevertheless, provide [the content] for predictive inferences at the general level of trends and statistical tendencies" (1998a, p. 161).

From these elements, it is possible to maintain that there are theories that are only oriented towards past developments (and that, therefore, do not make predictive inferences), which can have *predictive import*, to the extent that their content can serve as a support in order to achieve statements about the future.[25] Obviously,

---

[24] This is a clear expression of his rejection of the instrumentalist predictivism without realism of the assumption, which was defended by M. Friedman (1953).

[25] On this issue, see Sect. 3.1 of this book.

Rescher is thinking of the natural sciences. But in the realm social sciences, where human events are involved, this approach to prediction as a test for hypotheses and theories is more problematic. In effect, there are sciences mainly oriented toward the explanation (for example, history). So if Rescher's approach is accepted, without the required qualifications, the scientific character of these disciplines would be questioned.

5.2.2 The Role of Prediction as a Guide in the Task of Applied Science

Besides the use of prediction as a test for theories, another important role of prediction is to serve as a guide for action within the framework of the applied sciences. In this realm, prediction is the previous step to prescription, because the anticipation of the possible future is required in order to perform the task of problem-solving. Therefore, prediction is a methodological tool in this context: knowledge of the possible future is required in order to suggest what paths of action should been followed (Gonzalez 2008, p. 181). Thus, in this context of the applied sciences, prediction is a useful tool for prescription, which is oriented towards providing information in order to solve practical problems.

Since in applied sciences the aim is the solution of concrete problems, the usual procedure is to achieve predictions in order to give prescriptions (Gonzalez 2008, pp. 181–182). In this regard, two features can be highlighted in the relation between prediction and prescription: (i) prediction is prior to prescription, because the indications about how to solve a problem (prescription) are given once the problem has been anticipated (prediction) (Gonzalez 2008, pp. 181–182); and (ii) prediction makes the prescriptive task of applied science possible, because in order to make a prescription some knowledge about the possible future is always required (Simon 1990, Gonzalez 2015, ch. 12, pp. 317–338). The first is a chronological feature with methodological incidence, while the second leads to a clearly epistemological-methodological aspect.

Regarding prediction as a guide for the task of applied science, Simon has offered a quite interesting proposal. In his judgment, the main aim of the applied sciences, in general, and of the sciences of design, in particular, is prescription. He thinks that most of the predictive models are oriented towards the prediction of phenomena that human beings cannot control (for example, the meteorological models). For this reason, the main aim of applied science is prescription, which seeks to favor the best possible adaptation to those phenomena (Simon 1990).

Unlike Simon, Rescher's main interest is not focused on the sciences of the artificial, but on the natural sciences. Moreover, he rarely takes into account the role of prediction as a guide for prescription in the realm of applied science. This is because his pragmatism goes hand in hand with a Kantian approach, so prediction is mainly a *cognitive content* valuable by itself. However, as a pragmatic philosopher, his interest in the nexus between prediction and human action should be highlighted. In this regard, he addresses the problem of the human capacity to shape the future (1998a, ch. 14, pp. 231–246; especially, pp. 232–236).

But, very often, Rescher addresses this problem in relation to human action, in general, instead of doing so with regard to scientific activity, in particular. In effect, in his judgment, "to act, to plan, to survive, we must anticipate the future" (1998a, p. 65). Thus, he does not insist on the use of prediction as a guide for the development of the scientific activity that seeks to provide solutions to practical problems (that is, the role of prediction in applied science); but his main concern is the use of prediction by agents (or groups of agents) in the everyday context.

However, Rescher makes some reflections about important background issues that have to do with the relation between prediction and prescription. First, he addresses the problem of the tractability of the future; that is, to what extent we can shape the future; and, second, he takes into account the human capacity to *control* future phenomena or events in an effective way. In this regard, he thinks that the control of phenomena is one of the main aims of the natural sciences (besides description, explanation, and prediction) [Rescher 1999c, p. 106], which clearly leads to a context of applied science.

On the first issue—the tractability of the future—he thinks that, in principle, three positions can be maintained (1998a, pp. 234–235): (1) the future is completely intractable because reality is determined, so it is not possible for us to exert any kind of influence over future events; (2) the future is completely tractable, so we can influence the future course of events without any limitation; and (3) an intermediate position, according to which we can shape the future within certain limits.

Rescher subscribes the third option. He considers that the future events can be influenced in an intentional way by human agents. However, this is something that only can be done within certain limits, which are mainly due to our capabilities to anticipate the possible future. In his judgment, "a future we cannot foresee is *a fortiori* a future we cannot control*" (Rescher 1998a, p. 236). Here the common methodological path of applied sciences is implicit, to the extent that prediction is required in order to prescribe. Certainly, prescription encompasses the problems related to the human ability to control the future phenomena.

In effect, Rescher defines control as "*the capacity to intervene in the course of events so as to be able both to make something happen and to preclude it from happening, this result being produced in a way that is not only foreseen but intended or planned.* Control thus calls for the possibility of causal participation ("intervention") in the course of event ("to make something happen or preclude it") with a power that can be exercised both positively ("to make happen") and negatively ("to preclude from happening")." (Rescher 1998a, p. 235).

Therefore, he is interested in the relation between prediction and the control of phenomena; i.e., to what extent we can intervene in phenomena in an effective way, so we can make something happen or prevent something from happening. In this regard, he thinks that our abilities are limited: "the limits of predictability set limits to control as well" (1998a, p 237). However, as Simon notes, the prescriptive task of the applied science is usually oriented to providing paths of action in order to deal with matters that we cannot control. So it is usual to seek the best possible adaptation to the foresight problems. This is the case, for example, of the models oriented toward meteorological prediction (Simon 1990).

Within applied science, prediction has to do with information useful for paths for action in order to solve specific problems, so it does not necessarily involve a control over phenomena (natural, social, or artificial). Nevertheless, instead of considering planning as a way of configuring actions in space and time, made on the basis of a previous knowledge about the future, Rescher sees "planning" as a tool to deal with "not predictability" (or even "unpredictability") (Rescher 1998a, p. 238).

Rescher thinks that "by canalizing our actions into tried and true patterns we can clearly render the future less obscure, and thus less problematic" (1998a, p. 238). In this way, "prediction as such is not altogether essential to the rational management of our affairs. Very rough prediction will often serve our planning needs perfectly well: to make adequate provision for the future we (most frequently) do not have to know its precise character in many or most cases" (1998a, p. 238).

In my judgment, such an approach highlights that Rescher is not thinking of a scientific context of problem-solving, where prediction is required in order to provide prescriptions about what should be done. He is mainly interested in the use that agents can make of this knowledge of the future provided by science, either in an individual way (the everyday behavior or in a collective way (issues related to policy). Thus, it is a proposal about the applications of science, which takes into account the possible uses of scientific prediction in the direction of practical actions. It is not properly an approach to the role of prediction in the realm of applied science.

In this regard, it seems clear that prescription should be performed on the basis of an anticipation of the future. In order to make an effective prescription, forecasts (the least secure kind of prediction) can be good enough, instead of having a genuine foresight. But prediction is always needed, since it is prior to prescription from a chronological viewpoint and, moreover, it makes prescription *possible*. This can be clearly seen in the case of the sciences of design, where the usual procedure is to anticipate possible design problems and then give the paths to solve those predicted problems (Gonzalez 2008, pp. 181–182).

Therefore, Rescher does not offer an exhaustive analysis of prediction as a guide for the problem-solving activity of the applied sciences. Instead, his attention is focused on issues related with the role of prediction in a context of application of science, so he focuses on the use of prediction as a guide for human action. However, he addresses some questions that are very important when prescription is considered, such as the *tractability* of the possible future and the human capacity to *control* future phenomena and events. In this regard, his main contribution is, in my judgment, that he highlights the *limits* of the task oriented towards shaping the future, which are derived from the limits of the predictive activity itself.

5.2.3 The Problem of the Application of Science: The Role of Prediction

Regarding the roles of prediction, an important issue is the problem of the applications of scientific prediction. This question is connected with the distinction between "applied science" and "application of science," which has been pointed out by Niiniluoto. As this philosopher notes, in the realm of applied science, research is oriented towards the solution of concrete problems, so the search for new knowledge has a specific purpose. Meanwhile, the problem of the application of science deals with the use that can be made of scientific knowledge in order to solve practical problems of action (as happens, for example, in professional practices) [Niiniluoto 1993, p. 9. See also Gonzalez 2013a, pp. 17–18].

Certainly, this is an issue that has repercussions in the roles of scientific prediction. Thus, while in applied science prediction has mainly a role as a guide for the task of problem-solving; in a context of application of science, prediction can be the basis for the decision-making of agents, either in an individual mode or in a collective way. Rescher pays special attention to this problem, due to the pragmatic character of his thought. Thus, he is interested in how agents use scientific knowledge, in general, and knowledge about the future, in particular.

Because his philosophical conception is in terms of "system," Rescher addresses the problems that have to do with decision-making, above all, in public policy matters, where scientific prediction can have an important role (Rescher 1999b. 2012c). It is a problem he sees in relation to complexity, since "the decision problems that we face in contemporary public affairs are often too complex to allow a resolution by way of rational calculation and what might be called the application of 'scientific principles'" (Rescher 2012c, p. 205). In this way, complexity has repercussions on the application of scientific predictions, so when the aim is to manage complex systems, a large number of variables that are open to the future must be taken into account.

In general, complex systems are less predictable than simple systems. For this reason, when the applications of science deal with complex systems—which is usually the case in policy—the obstacles that hinder prediction have repercussions on the human capacity to manage and control the system, as well as on our ability to plan in order to solve practical problems. Furthermore, Rescher considers that complex systems, in general, and social systems, in particular, can be unpredictable (or at least non-predictable) in the long run (2012c, p. 209).

In this case, unpredictability can be due to several factors. (1) It is possible that chaos and chance intervene, so "the course of events over the longer term in matters of social interest depend too much on subtle interactions which, while virtually indiscernible at present and negligible in the short term, can make an enormous difference to what happen over the long term" (Rescher 2012c, p. 209). (2) There can also be factors such as novelty, spontaneity, and creativity, so the patterns that a system followed in the past cannot allow us to infer its development in the future (2012c, p. 210).

The difficulties to prediction such as chaos, chance, or novelty have repercussions in the human capacity of management. Thus, "complex systems are inherently less amenable to successful comprehension, management, and control" (Rescher 2003, p. 86). In effect, in order to manage or control a system in an adequate way, first, we need some kind of knowledge about how this system will behave in the future; and, second, the success of the management depends, to a large extent, on the correct anticipation of the results and consequences of the measures suggested.

But, when the matter we want to manage is a complex one, "the eventual effects of the measures we take to address the challenges become lost in a fog of unpredictability" (Rescher 2003, p. 89). This is a problem that makes the decision-making process difficult, to the extent that various possible solutions can be suggested for the same problem. As Rescher points out, "the fact is that in a complex modern society there is often no way to get a rational grip on the consequences of public policy measures and employ 'scientific intelligence' to foretell the consequences. There are no calculable solutions here—all that we ever seem to get is a clash of 'my experts' versus 'your experts'" (2012c, p. 212).

Therefore, when predictions do not reach the desirable level of detail, the interpretations of the experts or social entities that address or manage a problem are possible. For this reason, on the same basis, such as a prediction or a set of predictions that anticipate a problem, different possible solutions might be suggested. In turn, there are difficulties for the correct anticipation of the results of the measures adopted. Moreover, regarding one solution, different experts may have opposite opinions about the results and consequences of that solution.

In view of these problems posed by the applications of science, Rescher considers that "the best that we can do is to feel our way cautiously step by step—to experiment, to try plausible measures on a small scale and see what happens, and to let experience be our guide" (2012c, p. 212). He labels this way of proceeding "political pragmatism." It is a proposal about the rational procedure in decision-making about issues of public interest; because, in his judgment, complexity of the matters of public policy implies the impossibility of trusting that the experts would offer adequate solutions.

Based on the lack of a direct transfer of the solutions suggested by the scientists, Rescher thinks that the process of decision-making in order to solve practical problems should be a "democratic" process, with collective participation (2012c, pp. 213–215). Certainly, the applications of science in many of the applied sciences (such as economics, medicine, or pharmacology) have to do with questions of public interest, to the extent that they can affect people, society, or the environment (in a positive or negative way). This can be analyzed at three successive levels: aims, processes, and results (and the consequences) of the applications of science (Gonzalez 2005, p. 26). At each of these levels, ethical values are important; above all, those related to social responsibility (which might lead to legal responsibility).[26]

It should be highlighted that, in Rescher's approach, decision-making should be collective, at least regarding issues that have social repercussion, such as the

---

[26] These ethical features are analyzed in-depth in Chap. 9.

problems of public policy. But this makes the attribution of responsibilities a difficult task, for "when things go wrong—when even our best conceived measures do not deliver on their promises and live up to expectation—in a system of genuinely participatory decision making, 'we the people' will at least have no one to blame other than ourselves" (Rescher 2012c, p. 214).

Thus, an important problem that arises here is, in my judgment, that responsibility can fade when *groups of agents* (instead of individual agents) make the decisions. This might make it difficult to talk about a "collective responsibility" in decision making. In this regard, Rescher maintains that collective responsibility is rooted in the individual members of a group, provided that the individuals act in a coordinated, intentional way—either through consensus or by delegation—in order to generate a result. He considers that only in this case can there be a genuine collective responsibility (Rescher 1998c).

Rescher rejects a naïve view of the capacity of science to solve practical problems of action. In effect, when the problems are complex, science can have difficulties in offering optimal solutions, "and this occurs not because the experts are incompetent but because the problems are intractable. They are of such complexity that scientific analysis and expert deliberations simply cannot settle matters" (Rescher 2012c, p. 209). So, he rejects a scientism of the application of science: the prediction of the possible future is not always followed by prescriptions that, in practical life, are directly applicable to provide the results sought.

It seems clear that, in this case, the existence of limits to scientific prediction due to the complexity of the reality (that is, the complex systems) should be considered. This is highlighted by Rescher when he notes the nexus between scientific prediction and the management or control of a system. Thus, when predictions are not possible or they are not reliable, management can lead to undesired or even undesirable effects. For this reason, successful predictions are needed, insofar as they can provide secure bases for acting. In view of this, the methodological dimension is, in my view, crucial, since the improvement of the predictive processes is basic in order to achieve predictive success.

5.3   From the General Realm to the Special Level

Usually, Rescher is concerned about prediction in the natural sciences (especially, in physics), but he also considers the role of prediction in the methodology of the social sciences (above all, in economics) [Rescher 1998a, ch. 11, pp. 191–208]. In this regard, he highlights that there is generally a problem of *unreliability* that affects predictions about social and human matters; while predictions about natural phenomena are, in principle, more reliable. This is a methodological problem that leads to ontological roots, since it rests on the complexity of the human activity that is developed in a social milieu (Gonzalez 2012a, p. 92).

But prediction also has an important role in the sciences of the artificial, which is a realm that Rescher does not take into account expressly. In this realm, the sciences of design seek to enlarge human possibilities in the human made field. In this way, these sciences develop an activity of a teleological character, where applied knowledge is needed to solve concrete problems. In this context of applied science, prediction and prescription have a relevant role, because the activity of design requires the anticipation of the possible problems (*prediction*) in order to suggest what should be done (*prescription*).

5.3.1 Prediction in the Natural Sciences and in the Social Sciences

Regarding scientific prediction, Rescher's approach is mainly focused on the natural sciences. However, in *Predicting the Future*, there is a chapter devoted to prediction in the social sciences, where he makes a comparison with prediction in the realm of the natural sciences. Thus, he considers that, from a methodological perspective, prediction in social sciences is more difficult than in the natural sciences. For him, "the difficulties that the predictive project encounters in [natural] science pale in comparison with those it encounters in human affairs" (1998a, p. 192).

The difficulty of predicting about social and human matters leads to an ontological dimension, because it is rooted in the characteristics of the reality (natural or social) predicted. So, when the prediction is about social phenomena, "it is the nature of the phenomenology of the domain—its volatility, instability, and susceptibility to chance and chaos—that is responsible for our predictive incapabilities here, rather than our imperfections as investigators" (Rescher 1998a, p. 202). However, those obstacles to predictability can also affect natural phenomena, so it should be considered to what extent indeterminism *especially* affects the social prediction.

Rescher's attention goes to natural reality when he analyzes the relation between *predictability* and *predetermination*. Regarding this issue, he claims "that this world of ours indeed is such a deterministic, wholly predictable Laplacean world is, in the present state of our knowledge, somewhere between implausible and false. The role of predetermination-blocking factors (chance, choice, and the like) is a real and prominent fact of life in the world as we know it" (1998a, p. 73).[27]

It can be noticed that, although there are factors such as volatility, instability, chaos, or chance that can clarify the lack of predictive success in social sciences, it does not seem that they can elucidate, by themselves, why social sciences have generally less predictive capacity than the natural sciences. In effect, as Rescher himself admits, natural sciences have to deal with this kind of ontological obstacles, so the chaotic or volatile character of some social phenomena is not the only reason for the methodological difficulties of the social prediction.

---

[27] On the discussions about determinism and freedom, see Gonzalez (2012d).

Furthermore, Rescher considers that the indeterminism that characterizes human matters does not necessarily involve "unpredictability" (i.e., the complete impossibility of predicting): "the operation of a power of free choice certainly does not mean that there *must* be unpredictability" (Rescher 2009a, p. 46). In his judgment, to the extent that the human actions and choices are *rational*, they might be also predictable. Thus, he thinks that "the acts of rational agents are usually predictable because it is often and perhaps even usually possible to figure out on the basis of general principles what the rational thing to do is in the prevailing situation" (Rescher 1995, p. 327).

Nevertheless, Rescher admits that the free choice of agents can be an obstacle for predictability in social sciences. In these sciences, predictions are generally about issues that are related with the actions and choices of rational agents that have free will, so these actions and choices are open to changes (Rescher 1998a, p. 192). Therefore, although factors such as change, chance, or chaos are not only present in the social reality, it is possible to claim that, in general, social phenomena are more instable than natural phenomena, because these factors are more pervasive in the social realm.

From this perspective, an important difference between natural phenomena and social events has to do with the *regularity* of these events or happening. This is an ontological issue, which has clear methodological repercussions, to the extent that the preconditions for rational prediction demand the discernability and stability of the patterns exhibited by phenomena (see Chap. 7, Sect. 7.1 herein). Consequently, if it is admitted that social phenomena are generally less regular and more instable than natural phenomena, then it is also possible to maintain that there are more methodological difficulties for prediction in social sciences, due to the kind of issues they deal with.

In this regard, Rescher relates the predictability of natural phenomena to the existence of laws about those phenomena. For him, "nature is predictable insofar as its phenomena exhibit discernible patterns that reveal rulish lawfulness in its operations" (1998a, p. 176). In that case, we can discern patterns in natural phenomena and those patterns have also a nomic expression. Thus, in natural sciences the inference from laws is a usual predictive method.

Certainly, when a prediction is the result of an inference from laws, it is more reliable than when other predictive procedures are used. Firstly, the presence of laws implies that there are lawful regularities in phenomena, which can serve as mechanisms of connection between the events of the past and the happenings of the future; and, secondly, nomic stability makes the secure inference of statements about the future possible. Meanwhile, regarding social phenomena, there is a higher instability, so the use of less reliable methods is more usual: estimative procedures, trend extrapolation, etc.

Regarding prediction in social sciences, one of the main difficulties is due, in my judgment, to the diversity of variables (endogenous and exogenous) that are open to the future. Furthermore, there are difficulties due to the complexity of selecting the *quantifiable* factors and the analysis of the representative interrelations among different variables in quantitative (and not just qualitative) terms (Gonzalez 2010,

p. 221). These are questions that have repercussions on the complexity and reliability of social predictions. Thus, it can be claimed that scientific-social prediction is generally more complex than prediction in the natural sciences (Gonzalez 2011b, pp. 319–321).

For Rescher, "the predictive limitations of social science are ultimately rooted in the immense complexity of the processes through which human beliefs and desires are shaped in the first place" (1998a, p. 201). But his approach to complexity in social science is, I believe, too generic, insofar as he seems to associate complexity with the presence of factors like chaos, volatility, chance, etc. (1998a, p. 202). Thus, this approach to complexity of social phenomena contradicts, to a certain extent, his own approach to the problem of complexity.

In effect, as Rescher maintains in his book *Complexity*, the study of complexity cannot be determined only by the extent to which factors such as chance, randomness, and lawful regularity are absent.[28] This feature leads him to suggest a framework of complexity with many aspects. In his analysis, there are mainly two dimensions of complexity: epistemological and ontological (to which it is possible to add a methodological facet). Furthermore, although his approach is preferentially focused on structural complexity, it is also open to dynamic complexity, which has to do with the changes of complex systems over time (see Chap. 7, Sect. 7.4 herein).

However, Rescher's analysis of the complexity of prediction in social sciences is too generic. He explicitly maintains that "the comparatively limited progress of the social sciences in matters of prediction does not lie in a want of trying, a lack of dedication or intelligence, a deficiency of method or of information-collecting methods, or in some other error of omission or commission. Instead, chance, chaos, choice—in fact, all of the bugaboos of rational prediction—play a prominent part throughout the social sphere. In this domain, where the causal phenomenology at issue is so enormously complex, volatile, and chaotic, there is only so much that can be done" (1998a, p. 202).

On the one hand, Rescher is right in considering that the difficulties in dealing with complexity in the realm of social sciences allow us to elucidate the poor predictive success in those sciences. But, on the other hand, I believe his approach in this regard is excessively general: there is not the desirable degree of detail. This is because his interest in prediction is mainly focused on the realm of the sciences of nature, so he does not develop in depth the specific issues of the methodology of prediction in the social sciences.

The methodological repercussions of the problem of complexity in the realm of social sciences have been emphasized above all in philosophy and methodology of economics. In this regard, it has been highlighted that *complexity* "contributes to the frequent lack of reliability of economic predictions, which has its roots in the object of study of this science: economic reality is a social and artificial undertaking which

---

[28] "As many writers see it, complexity is determined by the extent to which chance, randomness, and lack of lawful regularity in general is absent. But this cannot be the whole story, since law systems themselves can clearly be more or less complex," Rescher (1998b, p. 8).

is commonly mutable, as a consequence of its dependence on the human activity that develops historically" (Gonzalez 2012a, p. 92).

Certainly, economics has a dual status: it is a science of design, insofar as it has artificial elements that enlarge human possibilities; and, in addition, it is a social science, because it deals with human needs, such as food, housing, etc. (Gonzalez 2012a, p. 88). But, when Rescher analyzes the problem of prediction in economics, he only takes into account the dimension of social science, so his approach does not grasp all the issues at stake. In his judgment, the methodological difficulties to economic prediction are rooted in the changeableness of the social reality, where there is a great diversity of variables, endogenous and exogenous, which are instable or even chaotic (Rescher 1998a, pp. 193–199).

But economics is also a science of design. This feature of being a dual science (social and artificial) adds more factors of analysis to the problem of economic prediction, which Rescher does not take into account. For this reason, in my judgment, his methodological approach to scientific prediction can be completed through the analysis of the role of prediction in the sciences of the artificial. First, prediction has an important role in this group of sciences (so it is not only present in the field of the natural sciences and the social sciences); and, secondly, there are social sciences (such as economics) that are also sciences of design, so an exhaustive analysis of prediction in these sciences cannot be limited to the study of their features as social disciplines.

5.3.2 The Artificial Realm: Prediction in the Design Sciences

From an epistemological perspective, the sciences of design develop contents of the sciences of the artificial, because of the type of object they deal with (see Guillán 2013). Scientific designs belong to the realm of the "human-made" and can be considered as the result of a task of synthesis (Simon 1996, pp. 4–5), where prediction and prescription have an important role. This issue connects with an ontological dimension, which Simon highlighted when he made the distinction between artificial objects and natural items.

He made this distinction in order to establish the boundaries for the sciences of the artificial. In his judgment, there are four main differences between them: "1. Artificial things are synthesized (though not always or usually with full forethought) by human beings. 2. Artificial may imitate appearances in natural things while lacking, in one or many respects, the reality of the latter. 3. Artificial things can be characterized in terms of functions, goals, adaptation. 4. Artificial things are often discussed, particularly when they are being designed, in terms of imperatives as well as descriptives" (Simon 1996, p. 5).

Following epistemological and ontological features, design can be characterized as a human-made undertaking, which is made through a task of synthesis and brings about an artificial thing. Moreover, according to Simon, the concept of "synthesis" can be used as a synonymous with the notion of "design." In his judgment, "design, as I am using the term, means synthesis. It means conceiving of objects, of processes, or ideas for accomplishing goals, and showing how these objects, processes, or ideas can be realized" (Simon 1995, p. 246). Thus, to some extent, "design" and "synthesis" can be understood within this approach in the broad-sense of devising objects and processes, and thinking of how they can be accomplished in an effective and efficient way.

This teleological character of "design," which has a relation with the artificial domain, leads Simon to stress the link between design sciences and technology. In his judgment, "as soon as we introduce 'synthesis' as well as 'artifice,' we enter the realm of engineering. For 'synthetic' is often used in the broader sense of 'designed' or 'composed.' We speak of engineering as concerned with 'synthesis,' while science is concerned with 'analysis.' Synthetic or artificial objects and more specifically prospective artificial objects having desired properties are the central objective of engineering activity and skill. The engineer, and more generally the designer, is concerned with how things *ought* to be in order to *attain goals*, and to *function*" (Simon 1996, pp. 4–5).

De facto, both technology and design sciences share the need for designs to attain their goals (Gonzalez 2008, p. 168). Nevertheless, "'design sciences' belong to a realm that is *scientific* rather than technological, and they have a scientific rationality that is different from technological rationality" (Gonzalez 2008, p. 168). The main difference between them is neat: science seeks a variety of aims by cognitive means in order to increase our knowledge (basic science) or to solve specific practical problems (applied research) [Niiniluoto 1993, pp. 3–6, 1995]. Meanwhile, technology is oriented towards a creative transformation of reality in order to bring about new results (generally, an artifact, which is a tangible reality).[29] Thus, "science" and "technology" are different human activities, although there are connections between them, which are especially relevant in the case of design sciences.

These differences between "science" and "technology" have repercussions on the concept of design. In effect, there are differences between "designs," in general, and "scientific designs," in particular. These differences are clear insofar as scientific design involves the addition of epistemological and methodological features to the kind of design that is common in professional practices.[30] Therefore, scientific design seeks to solve concrete problems through the use of applied scientific knowledge and scientific methods (Gonzalez 2007, p. 11).

In Simon's view, "design like science is a tool for understanding as well as for acting" (1996, p. 164). This feature appears insofar as design "is concerned with how things ought to be, with devising artifacts to attain goals" (Simon 1996, p. 114). Thus, when a design is elaborated, there is a purposed aim. Consequently, it is necessary to choose the most appropriate processes, and these processes may eventually

---

[29] On the differences between science and technology, see Gonzalez (2005, pp. 11–12).

[30] Since Simon emphasizes the connections between sciences of design and technology, his approach to the notion of design is somewhat ambiguous, insofar as it does not make it possible to distinguish "design" form "scientific design." Cf. Gonzalez (2007, p. 11).

lead to previously established outcomes.[31] In this way, design is a teleological activity, and sciences of design—as sciences of synthesis—imply the enlargement of human possibilities by the use of creative designs to solve particular problems.

The teleological character of scientific design implies a link between the activity of design and the applied knowledge. In fact, sciences of design are applied sciences oriented towards problem-solving activities. In this way, the goal-oriented nature of design sciences involves a *prescriptive* component that deals with how things *ought to be* in order to attain certain goals. This implies that the elaboration of a scientific design requires both prediction and prescription. First, prediction is needed to know whether the design is feasible as well as to anticipate any possible problem; and, secondly, prescriptions are made on what should be done to achieve the goals.[32]

Therefore, in the realm of the sciences of design, since they are applied sciences, the most prominent role of the prediction is in its relation with prescription. In this case, prediction is mainly a methodological tool, because the knowledge about the possible future is required in order to establish the relevant paths for the prescription (Gonzalez 2008, p. 181). In this way, in applied sciences, in general, and in the sciences of design, in particular, the usual methodological path is to predict in order to prescribe.

This has been emphasized by Wenceslao J. Gonzalez when he analyzes the role of prediction in economics as a design science. In this regard, he notes that, "in order to carry out a design activity in science, the common path is to consider in advance whether the project is feasible (prediction), before we give indications about how to solve the problem that is foreseen (prescription). Thus, to make a prediction is, in principle, chronologically prior to establishing a prescription when the problems involved are in the realm of design science" (2008, pp. 181–182).

Explicitly, Simon has paid special attention to this problem. In fact, his approach gives primacy to prescription over prediction, because he considers that prescription is the most important goal of science, instead of just predicting what is going to happen in the future (Simon 2002). For him, prediction is prior to prescription both in science, in general, and in sciences of the artificial (and among them design sciences), in particular. Additionally, he thinks that the knowledge about the future (i.e., the prediction) is what makes the prescriptive task of science possible (Simon 1990, pp. 7–14; especially, pp. 10–12).

Within this context, a question arises about the level of accuracy and precision that predictions should achieve in order to establish effective prescriptions. On this problem, José Francisco Martínez Solano notes that, for Simon, prescription "does not involve the need for accurate prediction of the future because the main concern is to shape the future by designing it correctly, instead of predicting accurately what is going to occur" (Martínez Solano 2012, p. 248). Nevertheless, it seems clear to me that some kind of knowledge about the future is needed in order to prescribe,

---

[31] Of course, the outcomes attained could differ from the stated aims. This can happen commonly when the possible problems are not correctly anticipated.

[32] On the role of prediction and prescription in the sciences of the artificial, see Gonzalez (2008, sect. 4, pp. 179–183).

although it might be good enough to have a forecast (the least reliable type of prediction) instead of a genuine foresight.

In my judgment, a forecast might be good enough, but a genuine prediction or foresight—the most reliable kind of knowledge about the future—provides us a more rigorous knowledge in order to establish a prescription. In principle, the more accurate and precise the prediction is, the more successful the prescription could be. It is for that reason that I consider that applied sciences, in general, and the sciences of design, in particular, should be oriented towards the search for more and more reliable and informative predictions. In this way, it would be possible to choose from among alternative courses of action those which lead us to an effective and efficient solution to stated problems.

However, this aim can be difficult to attain due to the role of human creativity in the scientific activity of design (see Guillán 2013). On this issue, it seems to me that creativity can influence prediction and prescription in two different ways: (a) as an obstacle to achieve accurate predictions; and (b) as a key factor for problem-solving. According to Rescher, human creativity is a major limit to predictability. In his judgment, "human creativity and inventiveness defies predictive foresight" (Rescher 1998a, p. 149). From this perspective, creativity can be seen as a source for complexity in design sciences.[33]

But creativity can be also a key factor in the problem-solving activity of the applied sciences, where prescription has a main role. This feature is especially clear in the case of the sciences of design, insofar as "the science of design is directly connected to prescribing: design looks for courses of action whose aims are to change existing situations into preferred ones, and those processes require identification of some prescribed paths to be followed" (Gonzalez 2008, p. 182).

Certainly, this task can be accomplished through a creative act, because the role of creativity in the sciences of the artificial is related "to the *invention of forms that are to satisfy some requirements or purpose*" (Dasgupta 1994, p. 8). Therefore, creativity can be considered as a dual notion: on the one hand, it can be a source of complexity in design sciences, so it is a clear obstacle to predictability in this realm. But, on the other hand, it can be an element which helps us to overcome that complexity, insofar as it seeks new ways that serve as key elements for prescription in design sciences.

5.4 Preconditions for Rational Prediction

Epistemologically, Rescher characterizes scientific prediction as a rational prediction. Therefore, scientific predictions have rational bases that distinguish them from other non-rational attempts to anticipate the future (for example, prophecies). In his judgment, "we who do not ourselves directly observe or experience the future, and

---

[33] Creativity as an ontological limit to predictability, which affects above all the social sciences and the sciences of the artificial, is analyzed in Chap. 7, Sect. 7.3.3 of this book.

lack any self-authenticating precognitive insight into it, can only get knowledge about it by way of rational evidentiation, of 'inference' from the available data regarding accomplished facts. Prediction as we know it is a matter of thought and not perception. To us, ordinary nonclairvoyant humans, the future can only be conjectured on the basis of experience-derived information" (Rescher 1998a, p. 54).

Unlike authors like D. H. Mellor, who maintains that predictions do not need reasons (Mellor 1975), Rescher argues in favor of the rational bases of scientific prediction. In his approach, in contrast to simple prophecy, scientific predictions are credible to the extent that there are some bases (theoretical or empirical) that justify the inference from the past to the future (Rescher 1998a, pp. 53–56). Furthermore, he thinks that the predictive inference must be *necessarily* supported by information about the behavior of the phenomena in the past and in the present, since we cannot directly "observe" the future.

As the future is something that is not yet, when the problem of the characterization of the scientific prediction is addressed, the methodological orientation is crucial. In Rescher's judgment, scientific prediction is the result of a *rational process*. Thus, he admits a methodological pluralism in relation to scientific prediction: there is a variety of methods—estimative and discursive—that scientists can use in order to achieve the aim of predicting. But, as *rational* processes, the different methods share a series of characteristics.

On methodology of prediction, his attention goes in two different directions: on the one hand, he wants to make the different predictive methods explicit, which involves taking into account the distinctive features of each concrete process; and, on the other hand, he pays attention to the common features of the different predictive procedures and methods. This second line of analysis—the bases shared by the predictive processes—is developed by Rescher according to the "preconditions for rational prediction" (Gonzalez 2010, pp. 271–274).

These are a series of conditions that he considers *prior* and *necessary* for the methods oriented towards predicting, since "any sort of rational prediction—no matter how naive or how complex and sophisticated is mode of operation—will accordingly require informative input material that indicates that three conditions are satisfied" (Rescher 1998a, p. 86). In his judgment, those three preconditions for rational prediction are the following ones:

(1) Data availability, which implies that data should be obtained in an accurate, reliable, and timely way (i.e., prior to the occurrence or non-occurrence of the phenomenon predicted). (2) Pattern discernability; that is, there should be identifiable patterns in the data obtained. (3) Pattern stability, since the success of prediction depends, to a large extent, on the fact that the patterns followed by the phenomena in the past and in the present have certain stability towards the future (Rescher 1998a, p. 86). These three preconditions justify the inference of future, because "rational prediction pivots on the existence of some sort of appropriate *linkage* that connects our predictive claims with the input data that provide for their justification" (1998a, p. 87).

5.4.1 Data Availability

The first of those three preconditions—data availability—has to do mainly with two fundamental issues: (i) the access to information (to what extent we have the data required in order to make the prediction), and (ii) the quality of the data, where problems such as the accuracy or reliability of the data should be contemplated. Both problems are connected with the epistemological limits to scientific prediction, which Rescher addresses mainly with regard to the obstacles that have to do with information (Rescher 2009b, ch. 6, pp. 91–122; see also Chap. 4, Sect. 4.4 herein).

Regarding the first issue—*the access to the relevant data*—two problems can be considered which affect scientific prediction in different degrees: (a) ignorance (the lack of information), and (b) uncertainty (which initially affects prediction when some data are not available). Both cases are epistemological obstacles to scientific prediction from the perspective of the access to information.

In his analysis of *ignorance*, Rescher makes an initial distinction between two different types of ignorance: contingent ignorance and the ignorance of a necessary character. When ignorance is *contingent*, it is "grounded in operations of nature that render certain fact-determinations impossible in the circumstances, and it is sometimes grounded in the insufficiency of our information-accessing resources" (Rescher 2009b, p. 141). In the former case, ignorance has ontological roots; while in the latter the problem is epistemological. Meanwhile, the necessary ignorance is rooted in logico-conceptual considerations (2009b, pp. 140–141), so it can involve unpredictability.

Within this framework, ignorance that affects data availability in order to make predictions is a contingent ignorance of epistemic basis. According to Rescher, this type of ignorance is rooted in the inadequate character of our resources for information accessing, among which are the current means of observation (2009b, p. 141). The result of this type of ignorance is the impossibility of accessing now the relevant data in order to explain, predict, or prescribe a certain matter.

When there is ignorance with regard to the relevant data for prediction, the first precondition for rational prediction, which has to do with the *availability* of the information, is not fulfilled. In that case, the required data are not available, so prediction appears as an impossible task from the very beginning. In other words, those phenomena about which we have no knowledge, either explicative or merely descriptive, are unpredictable phenomena or, at least, non-predictable.

However, to the extent that it is a *contingent* ignorance—not a *necessary* ignorance—it can be overcome in principle (for example, to the extent that there are improvements in the means of observation or experimentation). For this reason, when the impossibility of prediction is due to ignorance, the research should be first oriented toward obtaining the required data; that is, toward guarantying the access to the relevant information for the prediction.

Meanwhile, *uncertainty* is another problem that has to do with the access to data. In Rescher's judgment, it is usual that not *all* the relevant information is available, so he considers uncertainty as the main epistemological obstacle to prediction: "In view of the inevitable incompleteness of our information, we cannot eliminate the risk of error in prediction; even the best of predictions can in principle go awry" (Rescher 1998a, p. 59). Unlike ignorance, uncertainty does not involve the complete lack of information, but it has to do with the *limited* character of the available information, so it can affect the reliability of the prediction. Thus, Rescher links uncertainty with fallibilism.

Generally, we can deal with uncertainty to the extent that the available information about past and present phenomena is increased or improved. In this way, it is possible to reduce the uncertainty that affects some predictive issue. In turn, this feature has repercussions on the reliability of the prediction, since, in principle, the better we know the phenomena that we want to predict, the more reliable the prediction about those phenomena will be.

However, this is not a general rule. As Rescher notes, "the access of further information can sometimes make the future less predictable" (1998a, p. 58). If someone we know is going to take a trip, we can predict quite securely that he will use his or her car. But, if we later know that his or her car is in the garage, not only does the initial prediction turn out to be false, but also it is not possible now to make a reliable prediction about his or her means of transport. By this and other examples from ordinary life, Rescher illustrates how sometimes the access to additional information increases the uncertainty regarding the future (1998a, pp. 58–59).

Therefore, a prediction can turn out to be false due to the presence of errors in the initial data; but also due to incomplete data (Rescher 1998a, p. 59). In that case, the access to further information can have two main consequences: first, it may allow us to predict certain phenomenon in a more reliable way; and, second, it might increase the uncertainty regarding what is going to happen in the future. Despite this, in my judgment access to further information always has a positive character for our knowledge about the future, either because it increases the security and reliability of the prediction, or because it might highlight that certain prediction is incorrect.

Besides the problem of access to data, there is the question about the *quality* of those data: what characteristics should they have in order to make the prediction possible. Rescher identifies three features that, for him, characterize the data required for prediction: (1) they should be obtained timely; that is, prior to the occurrence of the phenomenon or event that we want to predict; (2) they should be accurate; and (3) they should be reliable (Rescher 1998a, p. 86).[34]

The first of these characteristics has to do with the temporal factor of prediction. Thus, the data required for prediction must be about the past and present development of the phenomena that we want to predict; i.e., the information, from a chronological perspective, deals with a reality that is previous to the referent of the prediction. This temporal feature of the information is connected with Rescher's proposal according to which a scientific prediction is a statement oriented towards

---

[34] It should be noted that Rescher does not consider "relevance," as such, as a criterion regarding the data that is required in order to predict. However, he offers a series of criteria that the relevant data for prediction should meet, such as accuracy or reliability.

the future. Then, in order to anticipate the future, information is necessarily about previous events or phenomena: "for to evidentiate our predictive claims about the future, we have no alternative but to look to the past-&-present" (Rescher 1998a, p. 86).

Besides the temporal factor, from an axiological perspective, Rescher considers accuracy and reliability as two major values that must characterize the information for the prediction. Accuracy deals with the correctness of the data. It is a relevant value, since error in the information usually leads to incorrect predictions. Meanwhile, data reliability has to do with the security that can be attributed to those data; and, therefore, involves a level of certainty.

Within Rescher's approach, where credibility is more important than correctness, reliability appears as an especially important value.[35] Regarding the relevant data for the prediction, reliability can be evaluated according to issues such as the *source* from which the data have been obtained or the *means* used to obtain the data. Since it is usual that the process of prediction is performed in a context of limited information, the more accurate and reliable the data are, the higher will be, in principle, the probability of attaining true predictions.

5.4.2 Pattern Discernability

The second precondition for rational prediction has to do with pattern discernability. Once the data have been obtained, the researcher should first discern the patterns followed by the phenomena; and, secondly, he should anticipate their future behavior. On the one hand, this question leads to an ontological dimension; and, on the other, it is related to an epistemic feature. In effect, it is assumed that phenomena that we want to predict should follow some patters; and, then, it is accepted that the subject who makes the prediction has to be able to discern those patterns from the available data.

Here, the problem of complexity (both structural and dynamic) can be crucial. In effect, when the system we want to predict is a complex system, this feature can make the task of establishing the patterns of the relevant variables in the system difficult. In this regard, the main contribution of Rescher is in the identification of the different modes of complexity, which in his judgment belong to two dimensions: the epistemological and the ontological (see, in this regard, Chap. 7, Sect. 7.4 herein). Thus, complexity can have repercussions on the concrete obstacles (epistemological and ontological), which makes pattern discernability difficult.

---

[35] According to Rescher, we are not really interested in prediction as such; but our interest is in reliable predictions. In his judgment, the reliability of a prediction might rest partly on the evidence and partly on the kind of phenomenon at issue (for example, if it is stable or volatile). In this case, the ontological aspect involves the attention to the context, which may influence the behavior of the phenomena. Rescher, *Personal communication*, 2.6.2015.

On this basis, in order to study the problem of the patter discernability, two steps should be followed. First, the ontological and epistemological aspects of this issue can be considered (the existence of regularities or patterns and our capacity to grasp those patterns). And, second, the incidence of complexity—in its epistemological and ontological modes—on scientific prediction can be analyzed, because complexity can be an obstacle when we try to grasp the regularities of a system.

Regarding pattern discernability, the initial matter to take into account is of an ontological character. Thus, when we want to predict certain phenomenon or process, there are two initial possibilities: that it is a regular phenomenon or that, on the contrary, it is what Rescher calls an "anarchic" phenomenon (Rescher 2009b, p. 101); that is, a phenomenon that has followed no patterns. It is usual to consider pattern discernability as a necessary condition for predictability, which in principle excludes the possibility of predicting phenomena characterized by being completely irregular (i.e., anarchic phenomena).[36]

David F. Hendry has also insisted on the existence of regularities as a necessary condition for scientific prediction. His approach is with regard to economic forecasts. In his judgment, the success of economic forecasts is only possible if a series of requirements are fulfilled: "(a) there are regularities to be captured, (b) the regularities are informative about the future, (c) the proposed method captures those regularities, and yet (d) it excludes non-regularities" (Hendry 2003, p. 24).

It can be maintained that this proposal coincides, in general lines, with Rescher's conception about the preconditions for rational prediction; although Rescher's approach is, I believe, more complete. This is because, besides the ontological feature and the strictly methodological component, he takes into account what has to do with the information needed for the prediction—data availability and the characteristics of the data—and he also adds an epistemological aspect: that there are regularities is not good enough, but the human capabilities to discern those regularities should be taken into account.

Despite these differences, both Rescher—from the perspective of the requirements needed for scientific prediction, in general—and Hendry—in the concrete realm of the economic forecasts—consider that the existence of regularities is needed in order to obtain reliable predictions. According to Rescher, "all modes of rational prediction call for scanning the data at hand in order to seek out established temporal patterns, and then set about projecting such patterns into the future in the most efficient way possible" (1998a, p. 86).

Meanwhile, it is possible to think that the complete absence of regularities can make prediction impossible. In this regard, Rescher considers anarchy as one of the main ontological obstacles to predictability. He defines anarchy as "lawlessness—the absence of lawful regularities to serve as linking mechanisms" (2009b, p. 101). Insofar as anarchy encompasses the absence of patterns in phenomena, it implies that the second precondition for rational prediction—pattern discernability—cannot be fulfilled. Therefore, it might make the task of predicting impossible.

---

[36] On the implications of anarchy for predictability, see Sect. 7.3.1 of this monograph.

Rescher highlights this feature when he points out that "irregularity of process—the eccentricity of modus operandi at issue in anarchy—precludes rational prediction. A world without a stable order—even if only a probabilistic one—must inevitably fail to be predictively tractable" (2009b, p. 104). Thus, anarchic phenomena might be unpredictable, instead of being merely non-predictable. However, when prediction deals with an anarchic system, there are several possibilities:

First, it might be the case that the system is not really anarchic. Then, the apparent absence of patterns or regularities may be due to the incapacity of the agents to discern those patterns. Therefore, the obstacle to prediction is not an ontological limit, but an obstacle of an epistemological character. This happens more frequently when the system at issue is a complex system.

Second, it is not probable that the system changes in the short or middle run; that is, anarchy—as an ontological feature—will be a characteristic of the system in the future (in the short and middle term). However, it is possible that its dynamics changes in the long run, so something that is now unpredictable can be predictable in the future. In that case, the phenomenon or process is just non-predictable, instead of being genuinely unpredictable.

Third, Rescher takes into account another possibility that, in principle, excludes unpredictability in the strong sense. Thus, in his judgment, "we can safely predict that they [the phenomena at issue] will keep on being anarchic, since no order-engendering processes are (by hypothesis) at work" (Rescher 1998a, p. 136). In this way, he contemplates non-predictability regarding anarchic phenomena, but it is a non-predictability that only affects the possibility of obtaining specific predictions. Meanwhile, generic prediction is possible with regard to anarchic systems: we can securely predict that they will be anarchic in the future, at least, in the short and middle run.

Besides the ontological problem—the lack of regularity—Rescher takes into account an epistemic difficulty. Thus, even when phenomena are regular, the second precondition for rational prediction may not be satisfied. This might happen when the agents have difficulties in discerning the patterns. In this regard, "inferential incapacity" should be taken into account, which consists in "the infeasibility of carrying out the needed reasoning (inferences / calculation)—even where we may have the requisite data and know the operative inferential linkages" (Rescher 2009b, p. 102).

An example can be found in some of the predictive models, where there is the possibility of ignoring some minor effects—although they can be potentially important—in order to carry through the required inferential processes (Rescher 1998a, p. 153). Seen in these terms, inferential incapacity can be an especially important epistemological limit to the prediction of complex systems (for example, the economic system). This issue has been highlighted by the conception of bounded rationality. Thus, when a system consists of a large number of variables with complex interactions among them (so the whole is more than the sum of the parts) [Simon 1996, p. 184], our capacities can be inadequate in order to encompass the system.

In effect, in the case of economics, it is usual to think of complexity as one of the features that raise more problems for prediction in this realm: "Very frequently—

both from an epistemological viewpoint and from a methodological perspective—complexity is among the main reasons for maintaining that, in principle, the prediction of economic phenomena—its possibility and reliability—is more difficult than the prediction about natural happenings (including the weather forecast or, even, the prediction about the climate change)" [Gonzalez 2012b, p. 145].

It seems to me that Rescher's proposal regarding complexity (above all, epistemological and ontological) can shed light on this problem. His framework of complexity, where different modes of complexity are identified and analyzed, highlights that there are a large number of factors—mainly epistemological and ontological—that are at stake when the aim is the prediction of a complex system.[37] For this reason, inferential incapacity, which can affect pattern discernability, has more weight when the prediction is made in a context of complexity. In a complex system, the number of variables and the interactions among them can exceed the human capacities to encompass the system and grasp in an adequate way the regularities that are important for the prediction.

Related to this problem, Rescher analyzes the difficulties for scientific prediction that have to do with factor exfoliation. In his judgment, "when effective prediction requires the resolution of various subordinated issues, we may have a situation where the chain is no stronger than its weakest link. For if any one of the subordinate factors is predictively intractable, the whole problem remains unresolved. Where the overall issue is systematically holistic, malfunction in a single component may well engender an overall breakdown" (Rescher 1998a, pp. 153–154).

So he considers that "issues of this factor-exfoliating sort can readily prove to be predictively intractable because the outcome becomes veiled in the fog of a complexity into which we have—and can obtain—little or no secure insight" (1998a, p. 155). In my judgment, parsimonious factors should be taken into account when a prediction is about a system of this kind (modulated by complexity). According to Simon, parsimony should not be equated to simplicity; since parsimony does not seek what is merely simple, but the necessary and sufficient factors in order to make the system manageable (Simon 2001).

When the systems are complex, the methodological role of the parsimonious factors conception should be emphasized, as a way of dealing with the prediction of those complex systems. In effect, "due to the high number of factors at stake, prediction should start with something 'tractable' or feasible: it should search for those factors that, in principle, are more relevant in order to encompass all the field of interest" (Gonzalez 2012b, p. 153). Thus, it seems clear that the methodological conception of parsimony can provide solutions when the difficulty of prediction is related to pattern discernability within a complex system. This involves both the structural realm and the dynamic component (Gonzalez 2013b).

Therefore, when the whole system is not manageable, a solution might be to focus in the patterns of the necessary and sufficient factors in order to encompass the system. But, besides the structural complexity that usually centers Rescher's attention, the dynamic dimension should be also considered, which has to do with

---

[37] Rescher's proposal regarding complexity is analyzed in Chap. 7, Sect. 7.4.

the changes over time. This issue leads to the third precondition for rational prediction: pattern stability. In this regard, historicity can be crucial in order to predict future phenomena (above all, in the social and artificial realms).

5.4.3 Pattern Stability

Once data have been obtained and patterns have been detected, the success of prediction requires that these patterns are stable towards the future (either in the short, middle, or long run). In this regard, Rescher identifies two obstacles to scientific prediction that are related to *pattern stability*: (1) *Volatility* or absence of nomic stability; and, therefore, the lack of manageable laws from a cognitive perspective; and (2) *haphazard*, which implies that the linking mechanisms do not permit the secure inference of particular conclusions. In turn, haphazard can be due to the presence of three factors: (a) chance and chaos; (b) arbitrary choice; and (c) change and innovation (Rescher 1998a, p. 134).[38]

From the perspective of the need for stability in the patterns, the importance of the dynamic viewpoint is clear when the preconditions for rational prediction are considered. Thus, besides the structural dimension, there is a dynamic component of complexity, which is related to change over time. That change can involve novelties (for example, when human creativity intervenes) that are difficult to predict. Furthermore, as Rescher notices, *"the more important the innovation, the less predictable it is, because its very unpredictability is a key component of importance"* (2012a, p. 152).

However, the structural aspect has primacy over the dynamic dimension of complexity in Rescher's account. Moreover, when he addresses the dynamic complexity, he uses the notion of "process." In his judgment, the metaphysics of processes has advantages over a substantialist approach in analyzing the future from an ontological perspective. In this regard, he maintains that "the processual nature of the real means that the present constitution of things always projects beyond itself into one as yet unrealized future" (Rescher 1996, p. 54).

In my judgment, the notion of "process" is not good enough by itself to address the study of the complex dynamics (both internal and external) and its repercussions on scientific prediction. As Wenceslao J. Gonzalez has noted, in order to make advancements in the study of the changes in the phenomena and systems, historicity must be taken into account, insofar as it is a broader notion than the concept of "process" (Gonzalez 2012a). In rigor, *historicity* deals with three successive levels of analysis: science, agents, and the reality itself researched (above all, in the social and artificial realms) [Gonzalez 2011a, p. 43]. Thus, the notion of "historicity" is required—in my judgment—when the problem of the absence of pattern stability is analyzed.

---

[38] The epistemological and ontological limits to predictability are analyzed in more detail in Chap. 4, Sect. 4.4 and Chap. 7, Sect. 7.3.

But Rescher mainly focuses on the problem of predictability regarding the natural sciences, where historicity has less weight than in the social sciences or in the sciences of the artificial. Thus, in a science such as economics, which is a dual science (i.e., social and artificial), *historicity* can be considered as a factor that poses more difficulties for economic prediction than for prediction in the natural sciences (Gonzalez 2012a, p. 95). In effect, the regularity of the phenomena (and, therefore, the stability of the patterns) is generally higher in natural sciences than in the social sciences and the sciences of the artificial, where the component of *historicity* leads to a dynamic dimension that involves changes over time.[39]

Despite these difficulties, prediction is still possible if the stable elements are emphasized (or those elements that have some kind of regularity) in the different processes. Volatility and haphazard (in its diverse forms: chance and chaos, arbitrary choice, change and innovation) make the task of predicting difficult, and they can have negative repercussions on the accuracy and reliability of the predictions. Pattern stability of the processes is crucial for scientific prediction, so predictive procedures and methods must take into account the operations of the phenomena in order to predict them.

In this regard, Rescher's proposal offers, in my judgment an adequate synthesis of the preconditions required for predictive success, which encompass different factors that are important from a methodological perspective: (i) data availability; (ii) pattern discernability; and (iii) pattern stability. With this approach he highlights the requirements that are necessary for predictability. Thus, by using the relevant information about the phenomena's operations in the past and in the present, prediction requires discerning the stable patterns followed by those phenomena, so it is possible to infer their future behavior.

Furthermore, it should be highlighted Rescher's interest in making explicit the obstacles—above all, epistemological and ontological—that can affect each one of these three successive levels of the rational process of prediction. His perspective is preferentially focused on natural science, so he barely pays attention to the social sciences and the sciences of the artificial. For this reason, the notion of historicity is not conveniently emphasized. This feature makes his approach, in my judgment, unsatisfactory for addressing the problem of change (either ontological or epistemological) and its repercussions on scientific prediction.

References

Ayyub, B.M. 2001. *Elicitation of expert opinions for uncertainty and risks*. Boca Raton: CRC Press.

Bacon, M. 2012. *Pragmatism: An introduction*. Cambridge/Malden: Polity Press.

Batitsky, V., and Z. Domotor. 2007. When good theories make bad predictions. *Synthese* 157 (1): 79–103.

---

[39] On the dynamic dimension of complexity and the role of historicity, especially in the social sciences and the sciences of the artificial, see Chap. 7, Sect. 7.4.3 of this monograph.

Bell, W. 2003. *Foundations of futures studies. History, purposes, and knowledge, human science for a new era*, vol. 1. Piscataway: Transaction Publishers (5th rep. 2009; 1st ed. 1997).

Bereijo, A. 2011. The category of "Applied Science." An analysis of its justification from 'Information Science' as design science. In *Scientific realism and democratic society: The philosophy of Philip Kitcher, Poznan studies in the philosophy of the sciences and humanities*, vol. 101, ed. W.J. Gonzalez: 329–353. Amsterdam/New York: Rodopi.

Burke, F.Th. 2013. *What pragmatism was*. Bloomington: Indiana University Press.

Caamaño, M. 2013. Pragmatic norm in science: Making them explicit. *Synthese* 190: 3227–3246.

Dasgupta, S. 1994. *Creativity in invention and design*. Cambridge: Cambridge University Press.

Friedman, M. 1953. The methodology of positive economics. In *Essays in positive economics*, M. Friedman, 3–43. Chicago: The University of Chicago Press (6th reprint, 1969).

Gonzalez, W.J. 1997. Progreso científico e innovación tecnológica: La 'Tecnociencia' y el problema de las relaciones entre Filosofía de la Ciencia y Filosofía de la Tecnología. *Arbor* 157 (620): 261–283.

———. 1999. Ciencia y valores éticos: De la posibilidad de la Ética de la Ciencia al problema de la valoración ética de la Ciencia Básica. *Arbor* 162 (638): 139–171.

———. 2005. The philosophical approach to science, technology and society. In *Science, technology and society: A philosophical perspective*, ed. W.J. Gonzalez, 3–49. A Coruña: Netbiblo.

———. 2007. Análisis de las Ciencias de Diseño desde la racionalidad limitada, la predicción y la prescripción. In *Las Ciencias de Diseño: Racionalidad limitada, predicción y prescripción*, ed. W.J. Gonzalez, 3–38. A Coruña: Netbiblo.

———. 2008. Rationality and prediction in the sciences of the artificial. In *Reasoning, rationality, and probability*, ed. M.C. Galavotti, R. Scazzieri, and P. Suppes, 165–186. Stanford: CSLI Publications.

———. 2010. *La predicción científica. Concepciones filosófico-metodológicas desde H. Reichenbach a N. Rescher*. Barcelona, Montesinos.

———. 2011a. Conceptual changes and scientific diversity: The role of historicity. In *Conceptual revolutions: From cognitive science to medicine*, ed. W.J. Gonzalez, 39–62. A Coruña: Netbiblo.

———. 2011b. Complexity in economics and prediction: The role of parsimonious factors. In *Explanation, prediction, and confirmation*, ed. D. Dieks, W.J. Gonzalez, S. Hartman, Th. Uebel, and M. Weber, 319–330. Dordrecht: Springer.

———., ed. 2011c. *Scientific realism and democratic society: The philosophy of Philip Kitcher*, Poznan studies in the philosophy of the sciences and the humanities. Amsterdam: Rodopi.

———. 2012a. La vertiente dinámica de las Ciencias de la Complejidad. Repercusión de la historicidad para la predicción científica en las Ciencias Diseño. In *Las Ciencias de la Complejidad: Vertiente dinámica de las Ciencias de Diseño y sobriedad de factores*, ed. W.J. Gonzalez, 73–106. A Coruña: Netbiblo.

———. 2012b. Complejidad estructural en Ciencias de Diseño y su incidencia en la predicción científica: El papel de la sobriedad de factores (*parsimonious factors*). In *Las Ciencias de la Complejidad: Vertiente dinámica de las Ciencias de Diseño y sobriedad de factores*, ed. W.J. Gonzalez, 143–167. A Coruña: Netbiblo.

———. 2012c. Methodological universalism in science and its limits. Imperialism versus complexity. In *Thinking about provincialism in thinking*, Poznan studies in the philosophy of the sciences and the humanities, vol. 100, eds. K. Brzechczyn and K. Paprzycka, 155–175. Amsterdam/New York: Rodopi.

———. 2012d. New reflections on an old problem: Freedom and determinism in the scientific context. In *Freedom and determinism: Social sciences and natural sciences*, monographic volume of *Peruvian Journal of Epistemology*, vol. 1, ed. W.J. Gonzalez, 3–20. Cambre: Imprenta Mundo.

———. 2013a. The roles of scientific creativity and technological innovation in the context of complexity of science. In *Creativity, innovation, and complexity in science*, ed. W.J. Gonzalez, 11–40. A Coruña: Netbiblo.

———. 2013b. The sciences of design as sciences of complexity: The dynamic trait. In *New challenges to philosophy of science*, ed. H. Andersen, D. Dieks, W.J. Gonzalez, Th. Uebel, and G. Wheeler, 299–311. Dordrecht: Springer.

———. 2014. The evolution of Lakatos's repercussion on the methodology of economics. *HOPOS: The Journal of the International Society for the History of Philosophy of Science* 4 (1): 1–25.

———. 2015. *Philosophico-methodological analysis of prediction and its role in economics.* Dordrecht: Springer.

Guillán, A. 2013. Analysis of creativity in the sciences of design. In *Creativity, Innovation, and Complexity in Science*, ed. W.J. Gonzalez, 125–139. A Coruña: Netbiblo.

Hendry, D.F. 2003. How economists forecast. In *Understanding economic forecasts*, ed. D.F. Hendry and N.R. Ericsson, 15–41. Cambridge, MA: The MIT Press.

Howson, C. and Urbach, P. 1989. *Scientific reasoning: The Bayesian approach.* La Salle: Open Court (reprinted in 1990; 2nd edition in 1993).

Margolis, J. 2002. *Reinventing pragmatism. American philosophy at the end of the twentieth century.* Ithaca: Cornell University Press.

Martínez Solano, J. F. 2012. La complejidad en la Ciencia de la Economía: De F. A. Hayek a H. A. Simon. In *Las Ciencias de la Complejidad: Vertiente dinámica de las Ciencias de Diseño y sobriedad de factores*, ed. W.J. Gonzalez, 233–266. A Coruña: Netbiblo.

Mellor, D.H. 1975. The possibility of prediction. *Proceedings of the British Academy* 65: 207–223.

Niiniluoto, I. 1980. Scientific progress. *Synthese* 45: 427–462.

———. 1984. *Is science progressive?* Dordrecht: Reidel.

———. 1993. The aim and structure of applied research. *Erkenntnis* 38 (1): 1–21.

———. 1995. Approximation in applied science. *Poznan Studies in the Philosophy of Sciences and the Humanities* 42: 127–139.

———. 1997. Límites de la Tecnología. *Arbor* 157 (620): 391–410.

Rescher, N. 1973. *The Primacy of Practice.* Oxford: Basil Blackwell.

———. 1977. *Methodological pragmatism. A systems-theoretic approach to the theory of knowledge.* Oxford: Blackwell.

———. 1990. Our science as o-u-r science. In *A useful inheritance. Evolutionary aspects of the theory of knowledge*, N. Rescher, 77–104. Savage: Rowman and Littlefield.

———. 1993. *Pluralism.* Oxford: Oxford University Press.

———. 1997. Pragmatism in crisis. In *Profitable speculations. essays on current philosophical themes*, N. Rescher, 27–48. Lanham: Rowman and Littlefield.

———. 1995. Predictive incapacity and rational decision. *European Review* 3 (4): 327–332. Compiled in Rescher, N. 2003. *Sensible decisions. Issues of rational decision in personal choice and public policy*, 39–47. Lanham: Rowman and Littlefield..

———. 1996. *Process metaphysics. An introduction to process philosophy.* New York: State University of New York Press.

———. 1998a. *Predicting the future. An introduction to the theory of forecasting.* New York: State University of New York Press.

———. 1998b. *Complexity: A philosophical overview.* N. Brunswick: Transaction Publishers.

———. 1998c. Collective responsibility. *Journal of Social Philosophy* 29: 44–58. Compiled in Rescher, N. 2003. *Sensible decisions. Issues of rational decision in personal choice and public policy*, 125–138. Lanham: Rowman and Littlefield.

———. 1999a. *The limits of science*, revised ed. Pittsburgh: University of Pittsburgh Press.

———. 1999b. Risking democracy (some reflections on contemporary problems of political decision). *Public Affairs Quarterly* 12: 297–308.

———. 1999c. *Razón y valores en la Era científico-tecnológica.* Barcelona: Paidós.

———. 2000. *Realistic pragmatism.* Albany: State University of New York Press.

———. 2003. Technology, complexity, and social decision. In *Sensible decisions. Issues of rational decision in personal choice and public policy.* N. Rescher, 81–98. Lanham: Rowman and Littlefield.

———. 2004. Pragmatism and practical rationality. *Contemporary Pragmatism* 1 (1): 43–60.

———. 2009a. *Free will: A philosophical reappraisal.* New Brunswick: Transaction Publishers.

———. 2009b. *Ignorance. On the wider implications of deficient knowledge.* Pittsburgh: University of Pittsburgh Press.

———. 2012a. The problem of future knowledge. *Mind and Society* 11 (2): 149–163.

———. 2012b. Pragmatism at the crossroads. *Pragmatism. The Restoration of its Scientific Roots,* N. Rescher, 1–19. New Brunswick: Transaction Publishers. (Originally published as Rescher, N. 2005. Pragmatism at the crossroads. *Transactions of the C. S. Peirce Society* 41 (2): 355–365.

———. 2012c. Political pragmatism. In *Pragmatism. The restoration of its scientific roots,* N. Rescher, 205–215. New Brunswick: Transaction Publishers.

Rorty, R. 1982. *Consequences of pragmatism.* Minneapolis: University of Minnesota Press.

Simon, H.A. 1989. The state of economic science. In *The state of economic science. Views of six Nobel laureates,* ed. W. Sichel, 97–110. Kalamazoo: W. E. Upjohn Institute for Employment Research.

———. 1990. Prediction and prescription in systems modeling. *Operations Research* 38: 7–14. Compiled in *Models of bounded rationality. Vol. 3: empirically grounded economic reason,* ed. H.A. Simon, 115–128. Cambridge, MA: The MIT Press.

———. 1995. Problem forming, problem finding, and problem solving in design. In *Design and systems: General applications of methodology,* ed. A. Collen and W.W. Gasparski, vol. 3, 245–257. New Brunswick: Transaction Publishers.

———. 1996. *The sciences of the artificial.* 3rd ed. Cambridge, MA: The MIT Press. (1st ed., 1969; 2nd ed., 1981).

———. 2001. Science seeks parsimony, not simplicity: Searching for pattern in phenomena. In *Simplicity, inference and modeling. Keeping it sophisticatedly simple,* ed. A. Zellner, H.A. Keuzenkamp, and M. McAleer, 32–72. Cambridge: Cambridge University Press.

———. 2002. Forecasting the future or shaping it? *Industrial and Corporate Change* 11 (3): 601–605.

White, R. 2003. The epistemic advantage of prediction over accommodation. *Mind* 112 (448): 653–683.

Chapter 6
Methods of Prediction and Their Scientific Relevance

Abstract Specific methodological characters of scientific prediction are made explicit in this chapter. Its specific trait deals with the different predictive procedures and methods and their scientific import. Following the framework offered by Rescher, the processes of prediction are researched. This leads to distinguishing between estimative procedures of prediction and discursive or formalized processes of prediction, which can be either elementary processes or scientific methods. In turn, this pathway goes more deeply into the reliability and characteristics of the predictive procedures and methods, within an approach that assumes *de facto* a methodological pluralism regarding prediction.

Keywords Predictive methods • Predictive procedures • Delphi procedure • Predictive models

When Rescher analyzes the preconditions for rational prediction—data availability, pattern discernability, and pattern stability—he maintains that "rational prediction pivots on the existence of some sort of appropriate *linkage* that connects our predictive claims with the input data that provide for their justification" (1998a, p. 87). Thus, he considers that there are a variety of predictive methods and procedures, according to the kind of process used to establish that linkage between the available information and the anticipation of the possible future.

Rescher classifies the predictive processes into two main types: (i) *estimative or judgmental procedures*, and (ii) *formal or discursive methods* (1998a, ch. 6, pp. 85–112; especially, pp. 86–88). The former are developed on the basis of the personal estimations of the experts, while the latter follow a series of well-articulated rules or inferential principles. In turn, formal methods can be of two kinds: (a) *elemental* discursive processes, such as trend extrapolation or the use of analogies; and (b) *scientific* discursive methods, which are mainly the inference from laws and the predictive models.

This chapter, which complements the methodology of prediction addressed in Chap. 5, analyzes the different predictive processes. This means taking into account problems like the reliability of those processes and their adequacy in order to address

© Springer International Publishing AG 2017
A. Guillán, *Pragmatic Idealism and Scientific Prediction*, European Studies in Philosophy of Science 8, DOI 10.1007/978-3-319-63043-4_6

different predictive issues. To do this, first, the estimative procedures of prediction are studied. Among them the Delphi procedure is highlighted, because Rescher himself contributed to its creation.[1] Secondly, the elemental discursive processes are considered, which include trend extrapolation and the use of analogies. Thirdly, the scientific discursive methods are evaluated, which, in principle, are the most reliable processes. Finally, an assessment of the different predictive processes is offered.

6.1 Estimative Procedures of Prediction

By "estimative" or "judgmental procedures" Rescher recognizes those predictive procedures that are developed on the basis of personal estimations of individuals, so they always have an important subjective component. This procedure does not necessarily involve the absence of rules or inferential principles; but those rules, if they are used, are not explicitly shown. In this way, our reliance on prediction's correctness is directly based in the authority of the predictor. Usually, predictors should be experts in the issue that prediction is about, so that they can use their knowledge to discern the patterns in phenomena.

Estimative procedures are not scientific methods of prediction. This is emphasized by Rescher when he notes that, in this kind of procedure, "no substantive rationale for the claim itself [the prediction] need be discernible: our knowledge *that* can outrun our knowledge *why*" (1998a, p. 89). For this reason, an estimative prediction is not, in rigor, a scientific prediction. In a scientific context, it is not good enough to state that something will happen (that is, to make a prediction), because scientific prediction must be supported by reasons. Thus, besides predictive knowledge, there should be some explicative knowledge,[2] and this knowledge is not accessible in the case of estimative predictions.

Within the estimative procedures, two initial possibilities might be distinguished: (I) the individually made prediction (that is, a prediction made by one expert), or (II) the combination of predictions of different experts, that can be made by using different techniques, such as aggregation processes (averaging) or the Delphi procedure (that Rescher himself contributed to creating) [Rescher 1998a, pp. 91–96]. Both cases alike require a *metaprediction*; that is, a prediction about the reliability of the estimative prediction of the expert or experts,[3] which adds an *evaluative* component that should be taken into account.

---

[1] On the characteristics of Delphi procedure, see Ayyub (2001, pp. 99–105) and Bell (2003, pp. 261–272).

[2] This feature leads to Rescher's "harmony thesis" between explanation and prediction. See, in this regard, Chap. 3, Sect. 3.4.2.

[3] On "metaprediction" see Gonzalez (2010, pp. 84–85, 251, 259 and 267).

6.1.1 The Role of the Experts

When prediction is the result of the estimation of one individual, the main methodological problem has to do with the reliability of the result. Since the *modus operandi* of the processes that lead to the prediction is unknown, "we certainly cannot provide any sort of cogent account for why the predicted result will indeed obtain" (Rescher 1998a, p. 89). For this reason, faced with an estimative prediction, the receiver or receivers of the prediction have a prominent role, insofar as they should evaluate the reliability of the prediction without knowing the reasons that can lead to the predictive statement.

Because the reasons are unknown, the only available criterion for the evaluation of the prediction is the expert's authority, so some objective indicator should be sought in order to evaluate that authority. In this regard, Rescher notices that "we will have to be in a position to see that track record of successful past predictive performance as providing a cogent (inductive) ground for expecting the predictor's analogous future predictions to come true" (1998a, p. 89). Then, if a predictor has been successful in the past in predicting issues related to certain field, this feature provides a sufficient reason to trust that the expert's predictions will succeed in the future.

In this way, there is an initial limitation that affects this kind of predictions and that has to do with their subject matter. In effect, "the fact is that predictive expertise has to be established with reference to a particular subject-matter area and limited issue domain" (Rescher 1998a, p. 90). Thus, in order to assess the credibility of an estimative prediction, the most reliable indicator is the record of the successes that, in the past, the predictor in question has achieved with regard to predictive questions within a concrete subject matter.

A difficulty might arise that has to do with the possibility of conflicting predictions (Rescher 1998a, p. 90). This difficulty is more frequent in estimative predictions than when more elaborated predictive processes are used. It is a problem that makes the evaluation of the prediction difficult, above all when two experts whose predictions have been successful in the past now present conflicting predictions about the same topic. If this happens, there is not, in principle, any rational basis that allows us to give more credibility to either prediction.

Undoubtedly, although there were reasons to consider an estimative prediction as credible, there is always a possibility of error, which is higher than in the case of formal methods of predictions. This is because, in comparison with the formal or scientific methods, in estimative predictions there is a clear prominence of subjective elements. In this regard, there are many well-known examples of incorrect predictions, above all, in the realm of technology. For instance, in 1943 the founder of IBM, Thomas Watson, said that there was a world market for about five computers (Rescher 1998a, p. 90); and in the mid-nineties several experts in computer sciences predicted the end of the company Apple (Pogue 2006).

This kind of error may be due to a series of cognitive biases of the experts, which affect the correctness of the estimative prediction. Among those biases, the following can be highlighted: illusory correlations (i.e., false beliefs about the existence of a correlation between two variables), selective perception (that is, discounting information because it is inconsistent with the predictor's beliefs or expectations), underestimating uncertainty, optimism, overconfidence, etc. (Önkal-Atay et al. 2002, p. 137).

Rescher considers these types of biases as cognitive or psychological obstacles to prediction. Among them, he highlights (1998a, pp. 218–222): (1) Imminence and scale exaggeration, which is the tendency to consider that an event or happening will occur "earlier in timing and larger in extent than the actual course of events will bear out in due course" (1998a, p. 219). (2) Conservatism, which is the exaggeration of the stability and durability of current patterns. (3) Wishful (or fearful) thinking, which makes the predictor predict that things will happen because he or she thinks that they ought to. 4) Probability misjudgment, which can be due either to incorrect evaluations or to erroneous combinations.

Rescher considers that the mentioned factors can affect the predictive task in different circumstances; that is, independently of what concrete methods or procedures have been used, to the extent that predicting is a *human activity*. Thus, he does not circumscribe these problems to the specific realm of the estimative procedures of prediction. But it seems clear to me that these obstacles can affect to a greater extent estimative predictions which have an important subjective component. For this reason, in my judgment, the acknowledgment and analysis of this kind of psychological or cognitive biases is an especially relevant issue, which can contribute to elaborating the basis to overcome the problems of accuracy and precision of the estimative predictions.

6.1.2   Procedures for Combining Predictions

One way to overcome the difficulties posed by the predictions made by individual experts is to use procedures that allow the combination of individual predictions. Rescher divides these procedures for amalgamating expert predictions into two main groups (1998a, p. 91): (i) non-interactive mechanical procedures; and (ii) consensus-formation processes such as the Delphi procedure (1998a, pp. 92–96). The former are simpler than the latter, which require further elaboration.

Within the first group—non-interactive mechanical procedures—Rescher takes into account two possibilities (1998a, p. 91): (a) selecting the majority option; and (b) averaging the answers of different predictors. The first procedure is more appropriate for binary issues; for example, predictive questions whose answer can be "yes" or "no." Meanwhile, the second procedure can be used in numerical predictions, so the final prediction would be the average answer of the different predictors.

The adoption of the principle of the majority rule—the first option—is the simpler procedure. It is used to achieve a single prediction from the estimations of several experts. Certainly, the application of this procedure is limited, since it is habitually used to answer binary predictive issues. Furthermore, the risk that is taken when this procedure is used seems clear, since it involves the assumption that the majority answer is also the correct answer, and this is not always the case (Rescher 1998a, p. 91).

The second procedure mentioned consists in finding the "average" answer. This can be done when different experts are questioned about a predictive answer that requires a numerical answer (for example, 'what will life expectancy be in Spain in 2030?'). In these cases, instead of taking the answer of one of the experts, the average answer can be used, so one single prediction is obtained, which is the result of the combination of different predictions.

By employing this second procedure, the predictive answer obtained will be certainly closer to the right answer than the worst of the individual predictions of the experts. But there is nothing else to say in favor of this predictive procedure. In this case, Rescher highlights the lack of reliability of this kind of prediction, when he notes that the average is not always better than most individual predictions (1998a, p. 91).[4]

Therefore, non-interactive mechanical procedures of combination of predictions have, generally, little reliability. On the one hand, the majority option can be an incorrect option; and, on the other, the average answer can be further from truth than most of the individual answers. However, they are, in my judgment, procedures that can be useful when the only alternative is the suspension of judgment (for example, when there are conflicting predictions and there are no reliable indicators to choose one of the predictions, or when it is not possible to apply any other more sophisticated combination process).

6.1.3 The Delphi Procedure

An alternative to these mechanical procedures is the application of other predictive procedures that seek the consensus of the experts. The Delphi procedure is a predictive process of this kind. Since Rescher himself contributed to the creation of this predictive procedure, its analysis is especially interesting when his approach to the methodology of scientific prediction is considered.

Olaf Helmer, Norman Dalkey, and Nicholas Rescher conceived the Delphi procedure in the fifties, when they worked at the RAND Corporation (in Santa Monica, California), which was an institution that offered research support to the United States armed forces. The creation of this procedure of prediction should be related to the socio-political context of that time. Thus, in an international scene marked by the Cold War, the Delphi procedure was developed with the aim of predicting the impact that technological innovation would have on the conflict between the United States and the Soviet Union.

With the Delphi procedure, the aim is to achieve one prediction from the individual predictions of a group of experts, so avoiding any direct interaction among them. Thus, an "aggregate prediction" is sought from the consensus of the different

---

[4] However, in his judgment, there are some contexts where the averaging process has been useful and outperformed most of the individual predictions of the experts. This happens, for example, in the case of the economic forecasts about the production and employment rates. Cf. Rescher (1998a, p. 92).

predictors (Rescher 1998a, p. 92). In order to achieve this aim, the experts should answer anonymously a series of questionnaires in several successive rounds. Besides the experts, there should be one or more person monitoring the process. Their role is to elaborate the questionnaires and to collect the answers of the experts in statistical terms.[5] After each round of questionnaires, the experts are given the results of the group along with a new series of questionnaires. In this way, the predictors have the chance to review their own initial answers.

If the procedure works as planned, the answers of the experts will progressively converge after each successive round of questionnaires. Thus, in the final round, the opinions of the experts will be closer than in the first round, so finally a prediction is achieved with the consensus of opinion of the experts. This final result is achieved by giving the predictors the answers of the group and allowing them to reconsider their own answers in the light of the results obtained. Moreover, there is the possibility of favoring this convergence of the different opinions by using other means, for example, ruling out extreme answers (Rescher 1998a, p. 93).

There are two possible versions of the Delphi procedure: the pen and paper version, called "Delphi exercise," and the computer version, called "Delphi conference" (Linstone and Turoff 1975). In the Delphi exercise there is a facilitator or a small group of facilitators that direct the process, make the questionnaires, collect the answers, put them in statistical terms and make the necessary adjustments. In the "Delphi conference" the computer replaces to a large extent the figure of the facilitator. A computer that has been programmed to process the results of the questionnaires means the process is faster.

Although the Delphi procedure has different versions, these always have the following characteristics: (i) anonymity of the experts; (ii) structuring of information flow; (iii) feedback; and (iv) presence of one or more facilitators.[6] These characteristics make the Delphi procedure different from other interactive procedures that seek the consensus of the experts.

Together, the main characteristics of the Delphi procedure—anonymity, structuring of information flow, feedback, and the presence of facilitators—minimize the effects that social pressure can have (either when it is exerted by one individual or by the majority) on the individual opinions. Moreover, to the extent that it is oriented towards the predictor's consensus, it offers—in principle—better results than the mechanical procedures of aggregation of predictions (such as the majority principle or the average of different numerical predictions). Finally, since the aim is the achievement of one collectively-made prediction, the individual psychological factors have less weight than when the prediction is made by only one expert.

---

[5] In its "classic" form, the first round of questionnaires of the Delphi procedure serves to identify the relevant questions for the prediction. Instead of giving the experts an elaborated questionnaire, first they are asked to identify the problems that should be considered in relation to the predictive issue. Then, the facilitator or facilitators of the process select the key questions, and they elaborate a structured questionnaire that the experts should answer. Cf. Rowe and Wright (2001, p. 126).

[6] Cf. "Delphi Method," *Wikipedia*. Available in: http://en.wikipedia.org/wiki/Delphi_method (access on 3.7.2013).

When Delphi was created, it was mainly used to obtain predictions about the development of technology.[7] Later, it was also used to make predictions about the social realm; for example, in economics. Within the realm of economics, it seems clear to me that the Delphi procedure has some advantages over other estimative procedures of prediction. In effect, "Delphi assumes an especially critical role when geographically separated experts with diverse knowledge bases need to interact under conditions of scarce data—an archetypical economic forecasting scenario given the globalization process" (Önkal-Atay et al. 2002, p. 141).

The application of Delphi to planning issues (especially, in the design of public policy) is related to its use in order to obtain forecasts (Ziglio 1996 and Rescher 1969).This has been done since the 1970s and has, moreover, involved the introduction of a series of methodological contributions, among which the following can be highlighted:[8]

1. Besides the strictly predictive issues, some aspects that have to do with the problems, the aims, and the possible options should be considered. This leads to the introduction of different scales of evaluation, which are not used when the Delphi procedure—in the standard version—is oriented towards prediction. Thus, parameters are included such as a proposal's desirability and feasibility (in both a technical and political sense), importance, and probability (Turoff 1970).
2. When Delphi is oriented towards planning, the arguments provided by the experts to defend their opinions are more important. This is because of the complexity of the problems that they want to solve. For this reason, the experts are asked to prepare lists with the pros and cons of each specific option and, moreover, there is always the possibility of adding new relevant questions for planning.
3. The complexity of the problems also means a higher complexity of the process, so the processes of measuring frequently require more sophisticated methods, such as multi-dimensional scaling. It should be noted that measures should take into account complex reality, which has two dimensions: structural and dynamic.[9]

In hindsight, it can be appreciated that the realms of application of the Delphi procedure have been broadened and diversified since its creation in the 1950s. This has also involved the incorporation of variations regarding its initial configuration. However, its use is generally limited to those situations in which information about the future is required and it is not possible to apply another method that provides more

---

[7]An example of this use of the Delphi procedure can be found in the report made by T. J. Gordon and O. Helmer. The report presents the results of a study whose aim was to forecast the long run scientific and technological development in several realms: main scientific discoveries, population growth, automation, space development, prevention of war, and future weapon systems. Cf. Gordon and Helmer (1964).

[8]Cf. "Delphi Method," in *Wikipedia*. Available in: *http://en.wikipedia.org/wiki/Delphi_method*

[9]Both prediction and prescription are related with structural and dynamic complexity. This is addressed in Gonzalez (2012a, b, c).

guarantees.[10] Thus, Delphi can be adequate to obtain predictions about some complex phenomena—above all, in the social realm—and when there is uncertainty.

But this limitation has repercussions on the reliability of the Delphi procedure as a process of prediction. In effect, insofar as it is used in order to predict matters when uncertainty makes it impossible to use other more reliable predictive processes, errors frequently might appear in the results. This has a double reading: on the one hand, it is not a completely reliable method—in fact, it is strictly a "procedure"[11]—but, on the other, it is a procedure that has utility, since it allows us to obtain forecasts in those cases where it is not possible to use other predictive processes.

Another issue to be considered when the reliability of this procedure is assessed has to do with the subject-matter of the prediction (economic, technological, sociopolitical, etc.). In this regard, as Rescher admits, the success achieved by estimative predictions varies, in general, according to their thematic realm. Thus, he notes that its performance is "good in engineering, fair in medicine, shaky in economics, and distinctly poor in sociopolitical affairs" (Rescher 1998a, p. 90).

It might be thought that the same happens with the use of the Delphi predictive procedure, insofar as the reliability of the achieved predictions varies according to the thematic field of the forecast: technological issues or social matters (Rescher 1998a, pp. 94–96). Thus, when the aim is to predict the future development of technology, Delphi has been successful in the forecast in the short run; while it is less reliable when it is oriented towards prediction in middle and in the long run.[12]

However, according to Rescher, "it would seem that if we are looking for a forecasting method with a proven track record in the area of 'technological forecasting' over the nearer term then we can do no better than to employ the Delphi technique" (1998a, p. 95). Meanwhile, regarding social prediction, Rescher thinks that "all the indications are that on such matters one would not be well advised to put much reliance on Delphi methods" (1998a, p. 96). Therefore, there is a clear acknowledgement of the existence of limits—at least methodological—that imply epistemological limitations regarding prediction.

To sum up, with regard to the reliability of the Delphi procedure, several methodological conclusions with epistemological incidence can be drawn: (i) The results are generally more reliable than in the case of the predictions made by only one expert, since by seeking an "aggregate prediction" the biases that affect the predictors at the individual level are minimized. (ii) It gives more guarantees than the mechanical procedures of combination of predictions (such as averaging or choosing the majority answer), since the prediction is the result of an active search for consensus.

---

[10] This is acknowledged by Rescher (1998a, p. 96).

[11] On the distinction between predictive "procedures" and "methods," see Gonzalez (2015, pp. 255–257).

[12] This can be clearly seen in the report by Gordon and Helmer, published in 1964. The forecasts for the year 1984 achieved generally a higher degree of accuracy than the predictions for the year 2000. Cf. Gordon and Helmer (1964).

Moreover, it can be pointed out that the procedure has some methodological advantages and undeniable epistemological limitations. (iii) The characteristics of the procedure—anonymity, structuring of information flow, feedback, and the presence of facilitators—prevent the social pressure and other biases commonly present in group interactions from directly affecting the result. (iv) Delphi allows forecasts about complex problems and under conditions of uncertainty, so certain level of error in the results would be expected. (v) The reliability of the procedure varies according to the thematic field of the prediction and its temporal projection. In practice, it is more effective for making technological predictions—above all, in the short run—than in social prediction.

6.2 Elementary Formalized Processes

Although Rescher contributed to the creation of the Delphi procedure, which is one of the better-known procedures for obtaining judgmental predictions and also one of the most used processes, he acknowledges that "it is fortunate that the use of experts is not our only predictive resource" (Rescher 1998a, p. 97). Thus, besides the estimative or judgmental predictive procedures, based on expertise knowledge, there are formalized or discursive processes, which are based on well-defined rules or principles.

Rescher divides the formalized procedures into two categories: *elementary discursive processes* and *scientific discursive processes*. The former are procedures mainly oriented to establishing patterns followed by the phenomena in the past in order to project those patterns into the future; while the latter use scientific laws or predictive models. Within the first group, he considers two main processes: (1) trend extrapolation, which can be divided into three options: (a) linear extrapolation, (b) non-linear extrapolation, and (c) cyclical analysis; and (2) the use of analogies (Rescher 1998a, pp. 97–102).

6.2.1 Trend Extrapolation

Trend extrapolation is a frequently used predictive process. In its simplest version, it consists in the linear projection of patterns or trends that have been observed in the phenomenon at issue. This involves the assumption that the phenomenon is characterized by being regular (that is, it is highly stable). In that case, its future development can be anticipated through the projection into the future of its past behavior (for this reason, it is mainly used for forecasts in the short run) [Armstrong 2001].

On this basis, the reliability of the trend extrapolation completely depends on the above-mentioned feature: that the predicted processes are characterized by being stable. Consequently, this makes their results very sensitive to changes in the prevailing conditions (Rescher 1998a, p. 98). So, if a change occurs that alters the prior patterns, prediction will not be successful (at least, in terms of accuracy and precision).[13] Moreover, the changes are especially difficult to anticipate, to the extent that the prior patterns do not allow us to predict them.

The assumption that the previous patterns will project themselves into the future without alterations is usually problematic. An example of this problem is provided by the predictions based on the assumption that growth rates of the past will continue into the future. According to Rescher, in most phenomena and processes, both natural and social, exponential growth does not continue indefinitely, but there comes a saturation point: "most growth phenomena eventually conform to an S-shaped saturation curve" (Rescher 1998a, p. 99).

In these cases, non-linear extrapolations are used instead of linear trend extrapolation. In this way, "it can assume the shape of a curve of some nonlinear sort (exponential, sinusoidal, S-shaped, etc.)" [Rescher 1998a, p. 99]. Thus, although the initial assumption is that the phenomenon is stable in its development, non-linear extrapolation deals better with the possibility of alterations (for example, anticipating the saturation point when there has been an exponential growth in the phenomenon predicted).

When this method is used to make a prediction, then the analysis of the available data is crucial in order to achieve predictive success. In this case, the aim is to find a function to accommodate data, at least in a general way (Rescher 1998a, p. 100). Statistics is especially useful for this task, since it provides the required mechanisms for detecting patterns from big data (Rescher 1998a, p. 100). But there is also a series of values that should accompany the strictly mathematical or statistical considerations, such as simplicity, smoothness, and symmetry (1998a, p. 100).

However, it is usual to consider to what extent simplicity in the processes (statistical, mathematical, etc.) used for trend or pattern extrapolation is better than complexity. One option might be the principle according to which simple representations of trends should be used when there are no good reasons to do another thing (Armstrong 2001, pp. 227–228). Rescher supports this option when he notices that, when a function is sought in order to represent a process, "the pivotal task is to find one that fits reasonably well, is plausible considering the overall situation, and is no more complex than need be" (1998a, p. 99).

Several studies provide support for the thesis that it is not usual for the complexity of an extrapolation to lead to an increase in the accuracy of the prediction obtained (see, for example, Mahmoud 1984, Schnaars 1984, and Meade and Islam 2001). However, as Rescher maintains, the crucial point is that there is no more complexity than need be. J. Scott Armstrong goes in the same direction when he defends that simplicity should be valued in relation with the need for realism. In that case, "complexity should only be used if it is well-supported. Simplicity is especially important when few historical data exist or when the historical data are unreliable or unstable" (Armstrong 2001, p. 227).

Within the procedures of trend extrapolation, cyclical analysis can also be considered, although it has a series of specific characteristics (Armstrong 2001, p. 233). It is a process of prediction that seeks to detect patterns in the data that repeat over time (for example, the business cycle, where there are periods of economic growing

---

[13] On these notions, see Gonzalez (2007).

followed by periods of contraction). In this case, prediction is based on the fact that the phenomenon in question follows a cyclical development. In this way, it is possible to anticipate the evolution of that phenomenon once the current stage of the cycle is known.

It is a procedure commonly used in the social sciences, in general, and in economics, in particular. However, its scientific character can be questioned, since it is only based on pattern detection. Prediction needs *reasons* (Gonzalez 1996), so the simple detection of a cyclical development is not good enough for making a scientific prediction. For this reason, although it is used frequently, it is also habitual to consider its use as problematic (Rescher 1998a, pp. 100–101 and Armstrong 2001, p. 23).

According to Rescher, locating the current moment within a cycle is one of the main problems that this procedure poses. In his judgment, "the unfortunate reality is that this matter of positioning [in a cycle] is something that can all too often be done only with the wisdom of hindsight" (1998a, p. 101). Obviously, this feature makes prediction difficult, since the anticipation of the future stages of a cycle depends to a large extent on the correct knowledge about the current stage of that cycle.

One problem that generally affects extrapolation (both linear and non-linear) is establishing in which situations it is adequate to use it as a predictive procedure. According to J. S. Armstrong, there are a series of conditions that justify the use of this procedure. These conditions are: (a) when it is not possible to ascertain the costs of other predictive processes; (b) when the forecaster has little knowledge about the problem in question; (c) when the phenomenon to be predicted is stable; d) when alternative methods would be subject to biases; and (e) when it is used a support for another predictive process (Armstrong 2001, pp. 236–237).

1. For Armstrong, the use of extrapolations can be justified when it is not possible to ascertain the costs (in terms of effort, time, resources, etc.) that might involve the use of other more sophisticated methods of prediction (for example, when many forecasts are required) [Armstrong 2001, p. 236]. However, the success of the predictions or forecasts depends on phenomena themselves. For this reason, in my judgment, the selection of trend extrapolation as predictive procedure is questionable when it is based solely on economic criteria or values.

2. Armstrong thinks that, when the predictor has little knowledge about the problem, it is reasonable to assume that the processes will follow the same patterns in the future as those observed in the past (2001, p. 236). Rescher takes into account this possibility when he notes that "too often, trend projection is simply a matter of the crude extrapolation of everyday experience" (Rescher 1998a, p. 98). In this case, the result of the extrapolation is not a scientific prediction, to the extent that there is not an understanding of the mechanisms that connect past facts with future developments.[14] Furthermore, the absence of reasons has negative repercussions on the reliability of the procedure: "where we lack an understanding of underlying processes, we become particularly vulnerable to defeat by the unexpected" (Rescher 1998a, p. 98).

---

[14]As it is noticed in different places of this book, predictivist approaches have background epistemological problems.

3. If the predictive issue is stable, trend extrapolation is frequently the best method available. In effect, it allows us to reliably predict phenomena that are characterized by being stable and it also has the advantage of being a simple, fast, and low-cost procedure. But, even in this case, there are limitations regarding the use of this predictive procedure. Usually, trend extrapolation works better when prediction is in the short run than when it is in the middle or long run (Gardner and McKenzie 2011). This is because in the long run, changes are more difficult to predict, so making a prediction in the long run based on the assumption that the patterns will continue into the future is always risky.

4. If the alternative processes can be subject to the predictor's biases (psychological, cognitive, etc.), the use of trend projection is—according to Armstrong—appropriate (2001, p. 236). This implies the assumption that trend extrapolation, as a formal process of prediction, is always better than any estimative or judgmental procedure. But this is not always the case. If the predictive issue is unstable, estimative procedures can be the best option when it is not possible to use other formal methods.

5. Finally, Armstrong considers that trend extrapolation can be used as a benchmark in assessing the effects of policy changes (2001, p. 236). When an expert wants to predict an issue of this kind, he can use the results obtained from trend extrapolation. But, in this case, trend extrapolation is not used in rigor as a predictive procedure; but it might serve as a support for an estimative or judgmental prediction.

After considering the cases that Armstrong maintains favor the use of trend extrapolation, it seems clear to me that conditions 1 and 4, by themselves, do not justify the use of this predictive procedure. In turn, in cases 2 and 5 trend extrapolation is not a formal method of prediction. In effect, in condition 2, it is not a scientific method, since there is no knowledge about the working of phenomena; while in case 5 it is only used as a support for an estimative procedure. Thus, only when condition 3 is the case (i.e., when the patterns are stable), can the use of extrapolation as predictive processes be justified.

Therefore, trend extrapolation is a reliable method, but its realm of application is certainly limited. It should be used only for the prediction of stable processes and phenomena, since its success depends on the premise that the past patterns will continue into the future. For this reason, when there are changes or discontinuities, it is very likely that the use of trend projection as predictive procedure will result in false predictions (Armstrong 2001, p. 237). Thus, it also has little interest from a practical viewpoint. As D. F. Hendry notes, "forecasts are most useful when they predict changes in tendencies, and extrapolative methods can never do so" (Hendry 2003, p. 25).

6.2.2   The Use of Analogies

Rescher also considers the use of analogies as an elementary predictive process (1998a, pp. 101–102). Like trend projection or cyclical analysis, analogy is also a procedure based on extrapolation. But, in this case, the patterns discerned in a certain phenomenon are used in order to predict other phenomena. For this reason, this predictive process is frequently used when there are no available data about the predictive issue (for example, when the prediction is about the sales of a new product, it is possible to resort to data about the sales of other similar products in the past).

The first issue to consider regarding this procedure is how the analogy can be established (i.e., on what basis is the decision made that the use of the patterns from one realm can be used to predict in another realm). In this regard, Rescher considers that an analogy can rest on several types of similitudes: (i) descriptive similarities; (ii) shared structures; and (iii) common processes (1998a, p. 101). Obviously, the more similar the phenomena are—either in descriptive, structural, or procedural terms—the higher the probability of predictive success will be.

Consequently, to a large extent, the use of analogies as a predictive process will be reliable if the analogies are well-grounded. Thus, Armstrong maintains that "information from analogies can reduce the effects of potential biases because analogies provide objective evidence" (2001, p. 379) is, in my judgment, questionable. In effect, although there are many similitudes between two phenomena, these are never identical, so there is always a certain risk of error: "The situations we analogize are never exactly identical, so that all analogies are of limited elasticity and have an eventual breaking point" (Rescher 1998a, p. 102).

Furthermore, sometimes the prediction might be false because the predictor has the false belief that two processes or events are analogous. Thus, the truthful information about the working and characteristics of phenomena is especially important. Certainly, it is a useful procedure when there are no data about a certain phenomenon, but we have enough information to establish an analogy with a different phenomenon. However, taking a wary attitude towards the forecasts obtained though this procedure seems to be advisable. The reason is clear: there is always a certain risk of error when, in order to predict a phenomenon, data about a different phenomenon are used, even if there are many similarities between them.

6.3 Formalized Scientific Methods

Unlike estimative procedures, formalized processes are based on a series of well-articulated rules or inferential principles. Within them, there are "scientific" methods. Formalized scientific methods are the most reliable predictive processes. The crucial factor is an epistemological one: they are articulated on the basis of a rigorous (scientific) knowledge of the predicted phenomena. Among them the following can be highlighted: (a) the use of correlations as predictive indicators; (b) the inference from scientific laws; and (c) the use of predictive models (Rescher 1998a, pp. 106–110).

6.3.1 Correlation as a Predictive Indicator

A method that is frequently used to predict the possible future consists in "employing *predictive indicators* that are based on an empirical finding that two (usually quantitative) factors are closely correlated in such a way that the behavior of the one foreshadows the behavior of the other (either invariably or with statistical preponderance)" [Rescher 1998a, pp. 102–103]. From this perspective, correlations have predictive value: they can serve as indicators that make the prediction possible.

"Correlation" is a linear association (positive or negative) between two variables. The correlation coefficient (r) covers the interval [-1, 1]. If $0 < r < 1$, then there is a positive correlation between two variables. If $r = 1$, there is a positive perfect correlation; that is, a total dependence between the variables. Thus, if the value of one of them increases, the values of the other one will increase in constant proportion. But the correlation can also be negative. This happens when $-1 < r < 0$. The negative correlation is perfect if $r = -1$. In that case, if the values of one the variables increase, the values of the other decrease in constant proportion. Meanwhile, there is no linear correlation between the variables when $r = 0$.

When the existence of a correlation between two variables is known, this knowledge can be used to predict changes in those variables. In fact, it is possible to anticipate either an increase in the value of one of the variables, when the correlation is positive, or a decrease in its value, when the correlation is negative, if it has been observed previously that the value of the other variable has increased. Obviously, the closer the correlation coefficient (r) is to the extreme values -1 and 1, the more reliable the prediction will be.

However, correlation does not necessarily imply a causal relation[15] (even if it is a perfect correlation). In this regard, the notion of "spurious correlation" is used for those correlations that do not imply causality (Simon 1954). But spurious correlations can be the result of a causal structure, for example, when the two events or processes that form the correlation are related with a third variable, which is the cause of them (Ward 2013, p. 701).

Certainly, the prediction made on the basis of a spurious correlation can be problematic. In effect, sometimes there is no explicative knowledge about the correlation: we do not know why two variables are correlated. Prediction, in this case, is possible; but its scientific character can be easily questioned. The Babylonians, for example, could predict eclipses with accuracy on the basis of correlations, but this predictive success is not supported by a scientific knowledge of astronomical phenomena (Rescher 1998a, p. 103). A scientific prediction cannot be limited to claiming that something will happen, there should also be a theoretical framework that provides an account about why the predicted phenomenon or process will happen (Rescher 1958, p. 286, 1998a, pp. 165–169).

---

[15] A discussed issue is what "causal relation" is. In recent decades, there have been many proposals, among them the one offered by Woodward (2003). See in this regard Campaner (2011).

Although in the case that we cannot provide reasons why two variables are correlated, correlation has anyway predictive value. In Rescher's judgment, "people's reliance on such unexplained predictive indicators is itself reasonable and explicable, being validated by the very fact of that established correlation" (1998a, p. 103). So he considers that, for example, "if suicide rates are coordinated with phases of the moon, that by itself is good enough; the question of 'the reason why' is in this context something secondary" (1998a, p. 103).

Despite this practical value of spurious correlations, it is preferable to use as predictive indicator the correlations that do involve a causal relation between the two variables, or to use them when the causal structure of the correlation is known (for example, when there is a third variable that is the cause of both the correlated variables). Rescher admits this when he holds that "unless predictive indicators rest on a secure explanatory foundation, those who rely on them build their houses on sand" (1998a, p. 104).

Sometimes, the only thing that can be done is to rely on predictive indicators that we cannot explain. But, in this case, a wary attitude should be maintained about the prediction finally obtained. In effect, the better we know the predicted phenomena, the more reliable it will be—in principle—the predictions achieved. Meanwhile, a prediction that is not supported by reasons is always more vulnerable to error, due to the presence of obstacles such as chance, change, or chaos.

A field where the use of correlations as predictive indicators is more usual is the realm of the applied sciences. In this regard, Rescher provides an example from economics, where a correlation has been established between an increase in travel between two regions and a rise in commercial activity between them (Rescher 1998a, p. 103). Thus, when an increase in travel between two countries is observed, this feature can be used as a predictive indicator in order to anticipate that there will be a rise in commercial activity between them. In this case, correlation does not involve a causal relation (the increase in the travel is not the cause of the rise in commercial activity), but the prediction is, in principle, reliable.

Besides the problem of the reliability of these predictions, which are not based on knowledge oriented towards causes, in the realm of applied science the use of correlations poses other problems that have to do with prescription. Thus, as applied science is oriented to the solution of concrete problems, the use of prediction as a previous step to prescription is usual in this context.[16] However, when Rescher analyzes the use of correlations as predictive indicators, he does not take into account this issue. This is because his attention is mainly focused on prediction, instead of on prescription.

However, this is an especially important issue when the role of correlations in applied sciences is addressed, both in order to obtain reliable predictions and to prescribe in an effective way. Moreover, it is relevant for the application of science, where prediction can be the basis of agents' decision-making. In this regard, the practical value of correlations has been considered; that is, the value of correlations

---

[16]On the role of prediction and prescription in applied sciences, Cf. Simon (1990), and Gonzalez (1998, 2008).

in order to design strategies oriented towards the achievement of a concrete aim (Leuridan et al. 2008 and Ward 2013). In a paper published in 1979, entitled "Causal Laws and Effective Strategies," Nancy Cartwright maintained that an effective strategy—that is, an effective and efficient way of achieving a goal sought—should be supported by knowledge oriented towards the causes (1979, pp. 419–437).

In this regard, Cartwright distinguished between "laws of association" and "causal laws." Laws of association are correlations that do not involve causality: they "tell how often two qualities co-occur; but they provide no account of what makes things happen" (Cartwright 1979, p. 419). Meanwhile, causal laws do provide account of what makes things happen (for example, "force causes change in motion"). In her judgment, a spurious correlation—or "law of association"—is not an adequate basis for action (Cartwright 1979, p. 430). If there is a correlation between two events E_1 and E_2, which are effects of a third variable that is the common cause of both, then a strategy designed to produce a change in E_2 by modifying the values of E_1, and vice versa cannot be justified.

Instead of this, effective strategies should be supported by causal relations, insofar as "the objectivity of the strategies requires the objectivity of causal laws" (Cartwright 1979, p. 436). Thus, although it could be possible to use spurious correlations as predictive indicators, the prediction obtained in this way is not a good basis for prescription. Knowledge about the causes that explain the correlation is also required. For this reason, "knowledge of spurious correlations, along with knowledge of the underlying causal structure and its stability, is sufficient for identifying effective strategies of change" (Ward 2013, p. 710). Therefore, the practical value of correlations is certainly limited.

In my judgment, the existence of these limits that affect the predictive value of correlations, which have also repercussions on their practical value as a basis of prescription, calls into question the following Rescher's claim: "for purposes of rational planning in economic policy matters actual prediction is not necessary; it suffices to have information about the ways of *influencing the probabilities of outcomes*" (1998a, p. 198). However, when the use of correlations as predictive indicators is analyzed, it seems clear that rational planning (as well as prescription) cannot be only supported by knowledge of probabilities. Instead, a prediction or, at least, a specific forecast (the less secure kind of prediction) is required.

In order to describe a prediction as *scientific*, there should be rational bases that support the prediction. In this way, prediction "should tell us not only *what* to expect but also *why* to expect it" (Rescher 1998a, p. 164). This why-knowledge is also required in order to prescribe (and also for planning). If there is a correlation between two events E_1 and E_2, but there is no a causal relation between them so E_1 is the cause of E_2, the knowledge of this correlation is not a good basis in order to prescribe. In this case, promoting a modification of the values of E_1 will not lead necessarily to a change in the values of E_2, and vice versa.

Therefore, both prediction and prescription need to be supported by rigorous knowledge. On the one hand, reliability of prediction depends, to a large extent, on the reliability of the knowledge about the phenomena in order to exert an effective control on those phenomena. And, on the other hand, prediction needs to be supported by knowledge—predictive or prescriptive—which is as reliable as possible.

6.3.2 *Inference from Laws*

Another scientific method of prediction consists in inferring a statement about the future from scientific laws, plus a series of initial conditions. In principle, it is the most reliable method of prediction, in that the patterns that have been observed in phenomena can be expressed in the form of a scientific law (either in deterministic or probabilistic terms), instead of being simple trends or regularities.

As Rescher puts it, "while all forms of pattern detection call for screening the empirical/phenomenological data in the endeavor to detect structure in this domain, the laws of nature (or what we accept as such) are the most important and fundamental sorts of patterns there are. Accordingly, our most sophisticated predictive method is that of *inference from formalized laws* (generally in mathematical form), which govern the functioning of a system" (Rescher 1998a, p. 106).

When there are scientific laws, reliability is higher in the different methodological levels, which have to do with data availability, pattern discernability, and pattern stability. The presence of laws means that there are nomic regularities that can serve as linking mechanisms between the events of the past and the future phenomena. Moreover, nomic stability allows the secure inference of future statements, from universal laws and initial conditions.

Hans Reichenbach especially insisted on the use of scientific laws—in the probabilistic sense—as a means to make predictions. In his judgment, "scientific propositions make assertions about the *future*. Indeed, there is no scientific law which does not involve a prediction about the occurrence of future events; for it is of the very essence of a scientific law to assure us that under certain given conditions, certain phenomena will occur" (Reichenbach 1936, p. 152). Thus, according to him, a scientific law includes statements that are oriented towards the future.[17] This position is in tune with Rescher's proposal about the predictive utility of scientific laws, which are the basis of "our most sophisticated predictive method" (Rescher 1998a, p. 106).

In effect, he considers that the predictive power of the natural sciences is, to a large extent, related to the availability of laws regarding natural phenomena: "nature is predictable insofar as its phenomena exhibit discernible patterns that reveal rulish lawfulness in its operations" (Rescher 1998a, p. 176). Therefore, the scientific laws have *predictive import*, insofar as they involve regularities that allow us to infer concrete conclusions about future occurrences or events (for example, the prediction of an eclipse).

Within this framework, predictability is closely related with the availability of laws. Moreover, Rescher considers that anarchy—the absence of laws in the strict sense—can involve unpredictability (or, at least, non-predictability) [2009, p. 104]. Thus, a question arises as to what extent predictability of phenomena depends on the availability of laws regarding the behavior of those phenomena. In my judgment, the possible answer to this question should take into account, at least, two different methodological levels: (i) the general methodology of science, and (ii) the

---

[17] This linkage with the future of the scientific laws and propositions leads Reichenbach to reject the logical positivism of the Vienna Circle. Cf. Gonzalez (2010, pp. 23–26).

special methodology of science (Gonzalez 2000). In turn, within the special methodology, it should take into account: (a) the methodology of the natural sciences, and (b) the methodology of the social sciences.

From the perspective of the general methodology of science, the inference from laws has a high value as a method of prediction. The presence of lawful regularities and nomic stability have a positive repercussion on the *reliability* of the process, insofar as the preconditions for rational prediction are fulfilled: (1) there are enough data about the working of phenomena; (2) those data exhibit the presence of patterns; and (3) the patterns are stable into the future.

However, the availability of laws is not a necessary condition for predictability. There are other methods that can be used to predict: estimative procedures, trend extrapolation, use of analogies, predictive models, etc. They can be used when there is no nomic stability, either due to epistemic failures (the "ignorance of the law") or due to the very character of phenomena (when there is chance, chaos, change, etc.).

Within the broad scope of the methodology of science, which has to do with the methodology of the natural sciences and the methodology of the human and social sciences, there is, in my judgment, a duality regarding the relation between predictability and law availability. In this regard, Rescher considers that, to a large extent, predictability in natural sciences is due to the presence of patterns in phenomena that can be expressed by laws (1998a, p. 176). Meanwhile, human and social phenomena are less predictable, because they are generally more complex (Gonzalez 2011, pp. 319–330; especially pp. 319–321, and 2013, pp. 299–311), so it is not so usual to have laws regarding their working.

According to Rescher, "the difficulties that underlie the mixed-to-poor predictive performance of economic models are easy to see. In the case of modeling in physics, we understand the background phenomenology and its laws pretty well, and since they are stable we can apply this well-established theoretical information in making our predictive models. (…). But in economic modeling, it is the model itself that is supposed to provide for our understanding of the phenomenology—we have no prior, independent, well-confirmed laws on which to base its operations" (1998a, p. 197). There is, then, a background problem, which affects the methodological approaches to prediction in social sciences: is it possible to have social laws? And, if so, what might their characteristics be?

Merrilee Salmon has made important contributions to the study of scientific prediction in the social sciences (Salmon 2005). Regarding the existence and characteristics of the social laws, she maintains that "if it turns out that the only laws possible for social science are statistical, this is no great defect. Laws are an indispensable part of science, but genuine scientific laws may be either statistical or universal" (Salmon 1992, p. 416). Thus, she considers that, unlike in the natural sciences, "laws of social science tell us what *usually*, *typically*, or *rarely* happens rather than what always, without exception, or never happens" (1992, p. 416). From this perspective, prediction from laws is also possible in the social sciences, but there is a background instability that affects the reliability of the prediction.

Rescher, to some extent, is in tune with this approach. He considers that there can be laws with a mathematical expression, but they are usually of a statistical character. However, he admits that these kinds of laws are also used in natural sciences (which include bio-medicine) or in the sciences of the artificial (for instance, in pharmacology).[18] Nevertheless, he thinks that it is usual that the laws in social sciences express qualitative relations instead of quantitative relations.[19]

In Rescher's judgment, also the qualitative laws have predictive utility: "the utility of those predictive principles is not negated by their lack of mathematical form; qualitative or phenomenological relationships can meet the needs of the situation" (1998a, p. 107). Thus, it is possible to predict on the basis of social laws. In this case, they are usually statistical laws or statements that express qualitative relations among several phenomena or processes.

From the approaches of M. H. Salmon and Rescher, it seems that, in principle, two main differences between laws in the natural sciences and laws about social phenomena can be noted: (1) while in the natural sciences there can be quantitative laws of a universal character, quantitative laws in social sciences are usually of a statistical character; and (2) in social sciences the use of qualitative laws for prediction is frequent, while these kinds of laws are not as usual in the natural sciences.

The first difference between the two groups of sciences has to do with the *indeterminism*, which habitually characterizes human matters, and also with the diversity of variables, both endogenous and exogenous, which are open to the future (Gonzalez 2010, p. 273). Meanwhile, the second feature mentioned—qualitative laws—deals with the complexity of selecting the social factors that can be quantified and the study of the representative interrelations in quantitative terms, instead of being merely qualitative (Gonzalez 2010, p. 221).

These issues have incidence on the complexity and reliability of social predictions. Thus, the indeterminism of the human matters complicates having social laws of a universal character. Furthermore, the problems related to the quantification of the social variables means that frequently the notion of "social law" refers to the expression of a qualitative relation between two variables. However, both the "statistical laws" and the "qualitative laws" can be used in order to infer statements about the future.

Nevertheless, there are a series of obstacles that might affect the inference of a statement about the future from laws, both in the natural sciences and, to a larger extent, in the social sciences (Rescher 1998a, p. 107). On the one hand, *chance* and

---

[18] Cf. Rescher, *Personal communication*, 17.6.2014. In this regard, Rescher insists on the importance of the value of reliability. In order to consider that a prediction is reliable, it can be good enough that it is based on some understanding of regularities. On the one hand, this feature involves that there is not, in his judgment, a radical discontinuity between scientific prediction and other kinds of rational predictions; and, on the other hand, it involves that this radical discontinuity neither exists between prediction in natural sciences and prediction in social sciences. Thus, obstacles such as chance, chaos, etc. might intervene in both realms (natural and social), so regarding certain predictive issues, predictive accuracy can be only achieved at the statistical, aggregate level. Cf. Rescher *Personal communication*, 2.6.2015.

[19] A defense of the existence of general laws in the social sciences, which can be used as a basis for social explanations, in general, and historical explanations, in particular, is in Joynt and Rescher (1959).

chaos can intervene, so the relevant variables follow random or stochastic trajectories that break the previous paths and make prediction difficult. And, on the other, *change* can play a role, which means that the discernible patterns are continuously broken.

Those obstacles add complexity to the process of prediction. Rescher admits this feature when he maintains that "in domains whose phenomenology is so complex as to put the operative laws outside the range of our cognitive vision, we obviously cannot accomplish our predictive work by their means" (1998a, p. 107). Frequently, this is what happens in social sciences, in general, and in economics, in particular. Thus, the level of complexity of social phenomena—among them, economic phenomena—is generally comparatively higher than the complexity of the natural phenomena (Gonzalez 2012b, pp. 73–106; especially, pp. 94–96).

In philosophy and methodology of economics, the problem of complexity and its repercussions on prediction has been widely discussed over the last decades (see, for example, Rosser 1999). In this regard, it is usual to compare prediction in economics with prediction in the natural sciences; and, more concretely, with predictions in physics.[20] Through this comparison, it is possible to see that there is a difference between the two kinds of predictions—economic and physical—that has to do with the level of regularity of the events (Rescher 1998a, p. 198). Thus, while natural phenomena have in general a higher level of regularity, "economic regularities cannot be usually derived from 'economic laws' (whose universality and, even, existence is questioned)" (Gonzalez 2012b, p. 95).

Thus, in economics it is usually accepted that "more generally, economists must resort to statistical inference" (Hahn 1993, p. 78). Moreover, some economists think that economics is a complex system. This is, in general, what Rescher actually thinks. In fact, in his judgment, "economic systems (and social systems in general) are in large measure chaotic" (Rescher 1998a, p. 197). Furthermore, the *creativity* of the agents adds other feature of complexity to economic predictions, to the extent that it increases the possibility of novelty in phenomena (Gonzalez 2012c, p. 147). In this way, the obstacles to predictive inference from laws (chance, chaos, and change) have an important weight. For this reason, other predictive processes are frequently used, such as time-series or econometric models.

6.3.3 Predictive Models

Within the "scientific" or "sophisticated" methods of prediction, Rescher considers the use of predictive models as one of the most reliable supports for methods of prediction. When the use of models is oriented towards scientific prediction, it consists in "exploiting for predictive purposes an artificially structured collection of processes that parallels the system's operations" (Rescher 1998a, p. 107). Thus, the predictive models should select the important variables—endogenous and

---

[20] "Economists cannot and will never be able to predict with the very high degrees of probability and accuracy which many natural scientists can achieve for their predictions" (Hutchison 1977, p. 10).

exogenous—that intervene in the system and their interactions, so the model would be *representative* of the system at issue.

Rescher distinguishes two main types of models: *physical models* and *symbolic models* (1998a, p. 107). Physical models are oriented towards obtaining predictions by simulating the properties and physical operations of a system (for example, a scale plane model that uses aerodynamic tunnels in order to test its stability under different atmospheric conditions). Meanwhile, symbolic models use symbolic representations, habitually of a numeric kind (for example, an econometric model).[21]

According to Rescher, the success of a model—both physical and symbolic—depends on an epistemological factor: to what extent the model corresponds with the system at issue. For this reason, he considers that predictive models should have realist assumptions. Thus, he explicitly criticizes Milton Friedman's methodological conception (Rescher 1998a, p. 109), who maintained that "the only relevant test of the validity of a hypothesis is comparison of its predictions with experience" (Friedman 1953, pp. 8–9). In fact, for Friedman, an economic model cannot be evaluated on the basis of the realism of its assumptions, but only through its predictive capacity according to the result.

In contrast to the methodological instrumentalism defended by Friedman, Rescher maintains that the realism of the assumptions is crucial in order to have predictive success. In his judgment, "a model's faithfulness to its original and its predictive efficacy are bound to go hand in hand, and a model's options will generally not manage to prefigure a reality whose operations it does not reflect" (1998a, p. 109). For this reason, a model that does not reflect reality in an adequate way will not be, in principle, predictively successful.

It seems to me that Rescher is right when he defends the thesis that the realism of the assumptions is crucial regarding the reliability of the predictive models. Realism of the assumptions is, in my judgment, the main criterion of demarcation between scientific prediction and non-scientific prediction. Thus, the issue is not only will a model that does not reflect reality in an adequate way have problems predicting with success, but that, even if it did achieve predictive success, there are no theoretical bases that allow us to justify the scientific character of the model.

It is usual to think of predictive models as the best option to obtain predictions about a complex system (natural, social, or artificial). In fact, some authors think that the main aim of the so-called "sciences of complexity" is to overcome the simplifications that can lead to the design of non-realist models (Chu et al. 2003, p. 19). However, there is no widely accepted definition of "complex systems" or "complexity." In order to address this problem, stressing the structural dimension of complexity can be an option. This is what H. A. Simon did. He considered a complex system as "one made up of a large number of parts that have many interactions" (Simon 1996, pp. 183–184).

Instead, Rescher considers that Simon's conception of complexity is too generic. For this reason, besides the compositional complexity (the number and variety of

---

[21] However, not every symbolic model should have quantitative variables: "Computers can process symbols of all kinds, including symbols that represent natural language or diagrams," Simon (1990, p. 13).

the elements that form a system) and the structural complexity (that takes into account the way in which the different subsystems are structured within the complex system), Rescher thinks that the complexity of a system can also be functional (Rescher 1998b, p. 9), which opens the door to complexity in dynamical terms. Therefore, in order to model a complex system, different ontological features of complexity should be taken into account. These features have to do with the composition, the structure, and the function of the system. But, as Rescher notes, complexity can also be epistemic, which adds difficulty to the task of designing a model. [22]

In my judgment, besides the ontological dimension and the epistemological aspect, the methodological component of complexity should also be considered, which has to do with the difficulty of the *processes* (Gonzalez 2012c, p. 152). In the case of the scientific models and, concretely, predictive models, methodological complexity seems clear. In effect, in order to deal with complex systems, "we must separate what is essential from what is dispensable in order to capture in our models a simplified picture of reality which, nevertheless, will allow us to make the inferences that are important to our goals" (Simon 1990, p. 7).

Certainly, this task is difficult, so complications can appear that have to do with the process of designing a model that has predictive success. This feature has been highlighted in the case of economics, where it is usual to consider economic systems as more complex than physical systems. Thus, according to Donald Saari, "even the simple models from introductory economics can exhibit dynamical behavior far more complex than anything found in classical physics or biology" (1995, p. 222). This is because the difficulties in calculating the interactions of the agents when they make decisions add an aspect of complexity to economics that does not exist in other disciplines (for example, in physics) [Rosser 1999, p. 171].

In effect, when a model about a complex system (for example, an economic system) is developed with the aim of predicting, difficulties can arise that mainly have to do with the selection of the variables and the kind of interactions among those variables. In this regard, it can be considered to what extent a high degree of sophistication and complexity in the models can contribute to the accuracy and precision of the predictions. In principle, it seems that, the more elaborated a model is, the higher its capacity to adequately grasp the reality of a system will be. However, a high degree of sophistication will lead invariably to a series of problems of practical kind (difficulties for information accessing and processing, an increase of the possibility of error, complications in the evaluation of the model, etc.) [Rescher 1998a, p. 109].

Regarding this problem, Rescher's proposal does not seek to offer solutions. Instead, he tries to show how a high level of sophistication in a model is a desirable characteristic, but one that is difficult to achieve in practice. For him, "the difficulty is that most realistic forecasting models are based on a vast host of interrelated assumptions on which the outcomes predicted hinge in a highly sensitive way. And in the study of complex phenomena (…) it is somewhere between difficult and impossible to establish the tenability of these assumptions sufficiently firmly to engender ready confidence in the resultant predictions" (1998a, p. 110).

---

[22] Rescher's approach to complexity is further analyzed in Chap. 7, Sect. 7.4.

In these epistemological-methodological terms, it seems that the choice is between more sophistication and complexity in the models to make them more realist, or the simplification of the model in order to make it more manageable form a cognitive viewpoint. But, in my judgment, parsimony of factors can be crucial in dealing with complexity, instead of seeking mere simplicity. Simon has highlighted this aspect (Simon 2001). Unlike simplicity, the methodological conception of parsimony, understood as austerity in the selection of the essential elements, leads to focusing on the necessary and sufficient elements in order to encompass the complex system.

Methodologically, regarding parsimonious factors there are two main problems: (a) the identification of the necessary and sufficient factors to grasp the complex system; and (b) the consideration of what their contribution to the system is, so they can be the basis of the elaboration of a predictive or prescriptive model (Gonzalez 2012c, p. 154).

Regarding the first problem (which elements can configure the "parsimonious factors" and what are the relevant relations among them), W. J. Gonzalez considers four forms of analysis that can lead to the identification of the parsimonious factors of a complex system (natural, social, or artificial):

(1) The holological analysis, which deals with the relation between the whole and its parts, so allowing us to see the nexus among the different components of the system. (2) The etiological analysis, which seeks the causes in the system and studies the relations with the effects. (3) The teleological analysis, which considers the dynamics of the system regarding the relations from means to ends, either from a synchronic perspective or form a diachronic viewpoint. (4) The logical analysis, which studies the relations in relation to conceptual contents (Gonzalez 2012c, pp. 154–157).

From these four forms of analysis, and once the factors have been identified, it is possible to address the second problem, which has to do with the contribution of the parsimonious factors to the system analyzed (Gonzalez 2012c, pp. 157–160). Thus, once there is an understanding of the key elements of the system and their relations, a model oriented towards the future can be designed on a realist basis and, furthermore, it can be manageable from a cognitive viewpoint. In fact, it seems that an inadequate selection of factors explains many of the problems of reliability that some economic models have.[23]

Therefore, regarding this problem, the methodological conception of parsimonious factors can be crucial in order to overcome the complexity of predictions, in general, and economic predictions, in particular. Concretely, it can increase the accuracy and precisions of the predictive models (Gonzalez 2012c, p. 148). Thus, by focusing on the necessary and sufficient elements to encompass the system, the realism of the assumptions is possible, as is the overcoming of practical problems

---

[23] "Actual econometric models incorrectly omit important linkages and include irrelevant ones. This class of problems is called model mis-specification, and it adds to the forecasters' difficulties. In particular, model mis-specification complicates calculating the likely magnitudes of forecast errors," Hendry and Ericsson (2001, p. 6).

that affect the models that either try to grasp all the factors at stake or do not select adequately the relevant factors.

6.4   Assessment of the Predictive Procedures and Methods

The methodological analysis of Rescher about prediction highlights that there is a *diversity of procedures and methods* oriented towards predicting the future. This methodological plurality reflects different degrees of rigor in predictive processes. First, there are *estimative* or *judgmental* procedures that are developed on the basis of the personal judgment of individual agents (the experts on the topic), which follow non-formalized or even "inarticulable" processes of reasoning. Secondly, predictive methods can be *formalized or inferential*, which follow explicitly articulated modes of reasoning. In turn, formalized methods can be divided into two groups: (a) *elementary* processes of prediction (such as trend extrapolation or the use of analogies); and (b) *scientific* or sophisticated methods (such as the inference from laws or the use of predictive models).

6.4.1   Reliability and Characteristics of Predictive Procedures and Methods

An assessment of the methods of prediction and their scientific import should emphasize the reliability and characteristics of the different predictive procedures and methods. Certainly, there are different levels of rigor and sophistication in the processes, which vary from the lowest level—the estimation of the experts—to the maximum possible rigor, when the methods are properly scientific (as in the case of the predictive models or the inference from scientific laws). Moreover, the rigor of the processes has repercussions on the reliability of the predictions. Thus, it is possible to maintain that "the level of reliability of the former (judgmental or intuitive in proceeding and personally mediated) is clearly lower than the latter (discursive in proceeding and methodologically oriented), because this can have a higher level of objectivity" (Gonzalez 2015, p. 258).

Furthermore, the *estimative procedures* of predictions are not scientific methods; since they are based on the judgment of the experts, instead of being based on modes of reasoning that follow clearly articulated rules or principles. For this reason, there is a clear prominence of subjective elements, which can affect both the *credibility* and the *reliability* of the predictions. In this regard, the main contribution of Rescher is, in my judgment, the identification of some of the most important obstacles to the estimative prediction: (i) imminence and scale exaggeration; (ii) conservatism; (iii) wishful or fearful thinking; and (iv) probability misjudgment (Rescher 1998a, pp. 218–222).

The identification of these cognitive or psychological obstacles has two main methodological roles: on the one hand, they allow us to understand many of the errors of the estimative predictions; and, on the other, it can contribute to elaborating the basis in order to increase the accuracy and precision of this kind of predictions (a task that can be made through the development of mechanisms in order to introduce corrections in the forecasts) [Harvey 2001].

Generally, the cognitive or psychological biases of predictors have more weight when the prediction is made by only one expert. For this reason, one way of overcoming those obstacles can be the use of procedures to combine predictions (Stewart 2001, pp. 81–106; especially, pp. 95–96). Rescher divides these procedures into two main groups: (1) non-interactive mechanical procedures (such as averaging or the majority option); and (2) interactive procedures (as in the case of the Delphi procedure). The former are less reliable, while the latter are more sophisticated procedures that seek a consensus among the experts.

But consensus offers no guarantees regarding the success of predictions.[24] Rescher admits this feature, although he personally contributed to the creation of the Delphi procedure. In effect, he notes that "the history of science amply illustrates that consensus in error is no less common that consensus in truth" (1998a, p. 96). Thus, it can be concluded that formal methods (elementary or scientific) are preferable to estimative procedures. But the possibility of using formal methods depends on the available resources and on the character itself of the phenomena in question. For this reason, when more sophisticated methods cannot be implemented, estimative procedures are a good option for obtaining predictive statements about the future.

Also the *elementary discursive procedures* have clear limits regarding their application. Thus, the use of trend extrapolation seems only appropriate for the prediction of phenomena and processes that are characterized by being stable, since their success depends on the principle of continuity: i.e., that the patterns of the past will be maintained in the future. For this reason, if discontinuities or changes occur, trend extrapolation will lead to false predictions. In addition, the reliability of the analogies as a predictive procedure requires a real similitude between the phenomena. But, although this condition is fulfilled, an analogy is never indefinitely valid; and, certainly, this procedure does not allow us to anticipate the moment when the analogy will disappear.

Generally, elementary discursive procedures do not lead to anticipating changes in phenomena, so they do not grasp, in principle, the historicity that accompanies human activity.[25] On the one hand, this feature means that predictions obtained through these processes have little interest, from both a theoretical and a practical perspective. And, on the other hand, the possibility of error in these predictions is

---

[24] On consensus, there are interesting remarks in Rescher (1993).

[25] The analysis of historicity is not usually considered in the predictive analysis in economics. It is not just that phenomena have a past and a potential future, but that there is variability in phenomena according to decisions made in the framework of the "economic activity" and "economics as activity." See Gonzalez (2013).

always present. In effect, as Rescher notes, we can safely predict that changes will happen in trends (1998a, p. 102),[26] above all in the long run. However, the problems of reliability are not only limited to the use of elementary processes. In this regard, it should be noted that "no predictive methodology (...) can ever immunize forecasting against the prospect of failure" (Rescher 1998a, p. 102); although there are methods that, in principle, are more reliable.

This is the case of the formalized or *discursive scientific methods*, insofar as they have the highest level possible of rigor. Rescher addresses some of these scientific methods of prediction (a) the use of correlations as predictive indicators; (b) the inference from scientific laws; and (c) the use of models oriented towards prediction (1998a, pp. 88, 102–104 and 106–110). On these methods, his proposal offers some of the key issues that are required in order to analyze them from the philosophy and methodology of science. But he does not encompass all the issues at stake, because his approach is mainly focused on natural sciences, with special attention to basic research.

On the use of correlations as predictive indicators, Rescher is right when he notes that "unless predictive indicators rest on a secure explanatory foundation, those who rely on them build their houses on sand" (1998a, p. 104). The success of a prediction made on the basis of a correlation does not depend on our knowledge about the causes that explain the correlation. Thus, a spurious correlation can have predictive value.[27] But, in this case, there are no reasons for seeing the predicted result as credible. In fact, if there is no knowledge about the background structure of the correlation, prediction is more vulnerable to error, due to the presence of obstacles such as chance, chaos, or change.

These reflections can be directed to the realm of prescription, when correlations are used to design strategies oriented towards the achievement of concrete goals. This problem does not receive Rescher's attention, because he is mainly focused on prediction. In this regard, a question arises about to what extent knowledge of the causes (instead of simple correlations) is required in order to prescribe.

N. Cartwright's and A. Ward's reflections on this problematic point to the need to know the causal relations for it to be possible to exert an effective influence over the effects (Cartwright 1979 and Ward 2013). Undoubtedly, knowledge about the causes is preferable in order to make effective prescriptions, rather than knowledge about probabilities or simple trends. Thus, once the causes are known, the effects can in principle be anticipated, and this anticipation of the possible future (prediction) is the previous step to prescription.

Due to the problems posed by the use of correlations, the inference from scientific laws is undoubtedly a more reliable method of prediction. In effect, when there are scientific laws about the working of phenomena it is possible to give reasons

---

[26] He also considers that, in human matters, the capacity to anticipate changes is especially important. For this reason, the anticipation of when and how conditions will change is especially useful. Cf. Rescher, *Personal communication*, 17.6.2014.

[27] The logic of conditionals shows that from something false something true can follow, since from false things can follow anything, even something true. So the eventual result by itself is not good enough.

about why it is expected that something will occur. Regarding the use of laws as a basis for predicting, Rescher proposal is mainly focused on the natural sciences. In his judgment, predictability of natural phenomena is related to the availability of laws (Rescher 1998a, p. 176). For this reason, prediction in social sciences is more problematic, since a background issue here is if there can be genuine "social laws" (Hausman 2000).

Regarding these problems, M. H. Salmon's and Rescher's proposals accept the possibility of having social laws (Salmon 1992 and Rescher 1998a, p. 107). These laws can be based on knowledge about the causes. For this reason, admitting causality in human matters can be crucial for prediction (Salmon 2002),[28] although a series of problems have to be taken into account that make prediction difficult, especially in this realm: (1) the presence of *chance* and *chaos* can mean that the relevant variables follow random or stochastic paths that break the previous patterns; and (2) *change* is also an obstacle to prediction, since the previous patterns are broken.

From this perspective, when prediction is about a complex phenomenon or system, it is not possible to predict only from laws (Rescher 1998a, p. 107). Usually, predictive models are the best option in order to predict complex systems (natural, social, or artificial). Both in the methodology of prediction, in general, and in the predictive models, in particular, *the realism of the assumptions* is crucial. In this regard, Rescher's proposal is, in my judgment, right, insofar as he considers that demarcation between scientific and non-scientific prediction begins with the realism of the assumptions.

6.4.2 Parsimonious Factors

But Rescher does not take into account the methodological conception of parsimonious factors, which can be the key in overcoming the problems that can lead to designing non-realist models. In my judgment, parsimonious factors—understood as austerity in the selection of the essential elements—can increase the accuracy and precision of the predictive models, to the extent that they allow us to focus on the necessary and sufficient factors to encompass a system. Moreover, this is especially important when the system studied, such as some biological systems or certain social systems, is very complex.[29]

Parsimonious factors can be useful for prediction in the context of structural complexity and can also be important in the sphere of dynamic complexity (see Gonzalez 2013). The reason is clear: when we need to deal with a complex system (natural, social, or artificial), having a number of factors—a few of the total avail-

---

[28] An analysis of this problem is in Gonzalez (2010, pp. 212–215).

[29] However, as Rescher notes, simple systems can also be difficult to predict (and even unpredictable), for example systems that exhibit Brownian motion. And conversely, some complex systems might be highly predictable (a modern airplane, for instance). Cf. Rescher, *Personal communication*, 2.6.2015.

able—that give the necessary and sufficient conditions to embrace, in principle, the whole system is particularly useful for addressing the predictive problems. Parsimonious factors contribute to establishing which variables should concentrate the attention of the researcher.

Although there is a clear progression in the rigor of predictive processes, from the less rigorous procedures (the estimative procedures) to the more sophisticated methods (the scientific discursive methods), Rescher does not maintain that there is a methodological superiority of some processes over others regarding the success of the results. Thus, for him, "determining the effectiveness of any particular predictive method with respect to a given predictive issue is therefore something that must always await the lessons of experience. Only by trial and error can one eventually assess (…) the efficacy of a particular predictive method with respect to a particular sort of predictive issue" (1998a, pp. 111–112).

Through this pragmatic criterion, the methodological superiority of a process over other ones cannot be established *a priori* and in a general way. The capacity of a concrete process to provide accurate and precise predictions is a question that should be assessed according to the characteristics of the predictive issue (both from an epistemological perspective and an ontological viewpoint). For this reason, although the discursive or inferential methods are more rigorous than the estimative procedures, this feature by itself does not provide more guarantees regarding the success of the predictions (for example, when a phenomenon is highly stable, an estimative procedure such as Delphi can be a better option than a discursive process, such as the use of analogies). In this regard, it would be desirable to have more studies about the comparative efficacy of the different predictive processes.

References

Armstrong, J.S. 2001. Extrapolation for time-series and cross-sectional data. In *Principles of forecasting: A handbook for researchers and practitioners*, ed. J.S. Armstrong, 217–243. Boston: Kluwer.

Ayyub, B.M. 2001. *Elicitation of expert opinions for uncertainty and risks*. Boca Raton: CRC Press.

Bell, W. 2003. *Foundations of futures studies. History, purposes, and knowledge. Human science for a new era*, vol. 1. Piscataway: Transaction Publishers (5th rep. 2009; 1st ed. 1997).

Campaner, R. 2011. Causality and explanation: Issues from epidemiology. In *Explanation, prediction, and confirmation*, ed. D. Dieks, W.J. Gonzalez, S. Hartman, Th. Uebel, and M. Weber, 125–135. Dordrecht: Springer.

Cartwright, N. 1979. Causal laws and effective strategies. *Nous* 13 (4): 419–437.

Chu, D., R. Strand, and R. Fjelland. 2003. Theories of complexity. Common denominators of complex systems. *Complexity* 8 (3): 19–30.

Delphi Method. *Wikipedia*. Available in: http://en.wikipedia.org/wiki/Delphi_method. Accessed on 3 July 2013).

Friedman, M. 1953. The methodology of positive economics. In *Essays in positive economics*, M. Friedman, 3–43. Chicago: The University of Chicago Press (6th reprint, 1969).

Gardner, E.S., Jr., and E. McKenzie. 2011. Forecasting trends in time series. In *Forecasting. Vol. 1: Traditional time-series and computer-intensive methods*, ed. R. Fildes and P.G. Allen, 45–57. London: Sage.

Gonzalez, W.J. 1996. On the theoretical basis of prediction in economics. *Journal of Social Philosophy* 27 (3): 201–228.

———. 1998. Prediction and prescription in economics: A philosophical and methodological approach. *Theoria* 13 (2): 321–345.

———. 2000. Marco teórico, trayectoria y situación actual de la Filosofía y Metodología de la Economía. *Argumentos de Razón Técnica* 3: 13–59.

———. 2007. The role of experiments in the social sciences: The case of economics. In *General philosophy of science: Focal issues*, ed. Th. Kuipers, 299–325. Amsterdam: Elsevier.

———. 2008. Rationality and prediction in the sciences of the artificial. In *Reasoning, rationality and probability*, ed. M.C. Galavotti, R. Scazzieri, and P. Suppes, 165–186. Stanford: CSLI Publications.

———. 2010. *La predicción científica. Concepciones filosófico-metodológicas desde H. Reichenbach a N. Rescher*. Barcelona: Montesinos.

———. 2011. Complexity in economics and prediction: The role of parsimonious factors. In *Explanation, prediction, and confirmation*, ed. D. Dieks, W.J. Gonzalez, S. Hartman, Th. Uebel, and M. Weber, 319–330. Dordrecht: Springer.

———. 2012a. Las Ciencias de Diseño en cuanto Ciencias de la Complejidad: Análisis de la Economía, Documentación y Comunicación. In *Las Ciencias de la Complejidad: Vertiente dinámica de las Ciencias de Diseño y sobriedad de factores*, ed. W.J. Gonzalez, 7–30. A Coruña: Netbiblo.

———. 2012b. La vertiente dinámica de las Ciencias de la Complejidad. Repercusión de la historicidad para la predicción científica en las Ciencias Diseño. In *Las Ciencias de la Complejidad: Vertiente dinámica de las Ciencias de Diseño y sobriedad de factores*, ed. W.J. Gonzalez, 73–106. A Coruña: Netbiblo.

———. 2012c. Complejidad estructural en Ciencias de Diseño y su incidencia en la predicción científica: El papel de la sobriedad de factores (*parsimonious factors*). In *Las Ciencias de la Complejidad: Vertiente dinámica de las Ciencias de Diseño y sobriedad de factores*, ed. W.J. Gonzalez, 143–167. A Coruña: Netbiblo.

———. 2013. The sciences of design as sciences of complexity: The dynamic trait. In *New Challenges to Philosophy of Science*, ed. H. Andersen, D. Dieks, W.J. Gonzalez, Th. Uebel, and G. Wheeler, 299–311. Dordrecht: Springer.

———. 2015. *Philosophico-methodological analysis of prediction and its role in economics*. Dordrecht: Springer.

Gordon, T. J., and Helmer, O. 1964. Report on a long-range forecasting study. *RAND Corporation Research Paper P-2982*. Available in: http://www.rand.org/content/dam/rand/pubs/papers/2005/P2982.pdf. Accessed on 12 July 2013.

Hahn, E. 1993. Predicting the economy. In *Predicting the future*, ed. L. Howe and A. Wain, 77–95. Cambridge: Cambridge University Press.

Harvey, N. 2001. Improving judgment in forecasting. In *Principles of forecasting: A handbook for researchers and practitioners*, ed. J.S. Armstrong, 59–80. Boston: Kluwer.

Hausman, D.M. 2000. ¿Necesita leyes la Economía? *Argumentos de Razón Técnica* 3: 115–137.

Hendry, D.F. 2003. How economists forecast. In *Understanding Economic Forecasts*, ed. D.F. Hendry and N.R. Ericsson, 15–41. Cambridge, MA: The MIT Press.

Hendry, D.F., and N.R. Ericsson. 2001. Editors's introduction. In *Understanding economic forecasts*, ed. D.F. Hendry and N.R. Ericsson, 1–14. Cambridge, MA: The MIT Press.

Hutchison, T. W. 1977. On prediction and economic knowledge. In *Knowledge and ignorance in economics*, ed. T.W. Hutchison, 8–33 and 145–151. Oxford: Blackwell.

Joynt, C.B., and N. Rescher. 1959. On explanation in history. *Mind* 68 (271): 383–388.

Leuridan, B., E. Weber, and M. Van Dyck. 2008. The practical value of spurious correlations: Selective versus manipulative policy. *Analysis* 68 (4): 298–303.

Linstone, H. A., and Turoff, M. 1975. *The Delphi Method. Techniques and applications.* Reading: Addison-Wesley.

Mahmoud, E. 1984. Accuracy in forecasting: A survey. *Journal of Forecasting* 3: 139–159.

Meade, N., and T. Islam. 2001. Forecasting the diffusion of innovations: Implications for time-series extrapolation. In *Principles of forecasting: A handbook for researchers and practitioners*, ed. J.S. Armstrong, 577–595. Boston: Kluwer.

Önkal-Atay, D., M.E. Thomson, and A.C. Pollock. 2002. Judgmental forecasting. In *A companion to economic forecasting*, ed. M.P. Clements and D.F. Hendry, 133–151. Oxford: Blackwell.

Pogue, D. 2006. When Apple hit bottom. *The New York Times*, 20.9.2006. Available in http://pogue.blogs.nytimes.com/2006/09/20/21pogue-email/?_r=0. Accessed on 3 July 2013.

Reichenbach, H. 1936. Logistic empiricism in Germany and the present state of its problems. *Journal of Philosophy* 33 (6): 141–160.

Rescher, N. 1958. On prediction and explanation. *British Journal for the Philosophy of Science* 8 (32): 281–290.

———. 1969. Delphi and values. *RAND Corporation Research Paper P-4182.* Available in http://www.rand.org/content/dam/rand/pubs/papers/2008/P4182.pdf. Accessed on 12 July 2013.

———. 1993. *Pluralism. Against the demand for consensus.* Oxford: Clarendon Press.

———. 1998a. *Predicting the future. An introduction to the theory of forecasting.* New York: State University of New York Press.

———. 1998b. *Complexity: A philosophical overview. N.* Brunswick: Transaction Publishers.

———. 2009. *Ignorance. On the wider implications of deficient knowledge.* Pittsburgh: University of Pittsburgh Press.

Rosser, J. B., Jr. 1999. On the complexities of complex economic dynamics. *Journal of Economic Perspectives* 13 (4): 169–192. Compiled in Rosser Jr., J.B. (ed.). 2004. *Complexity in economics. Vol. 1: Methodology, interacting agents and microeconomic models,* 74–97. Cheltenham: E. Elgar.

Rowe, G., and G. Wright. 2001. Expert opinions in forecasting: The role of the Delphi technique. In *Principles of forecasting: A handbook for researchers and practitioners*, ed. J.S. Armstrong, 125–144. Boston: Kluwer.

Saari, D.G. 1995. Mathematical complexity of simple economics. *Notices of the American Mathematical Society* 42 (2): 222–230.

Salmon, M.H. 1992. Philosophy of the social sciences. In *Introduction to the philosophy of science*, ed. M.H. Salmon et al., 404–425. Englewood Cliffs: Prentice Hall.

———. 2002. La explicación causal en Ciencias Sociales. In *Diversidad de la explicación científica*, ed. W.J. Gonzalez, 161–180. Barcelona: Ariel.

———. 2005. Predicción en las Ciencias Sociales. *Enrahonar* 37: 169–179.

Schnaars, S.P. 1984. Situational factors affecting forecast accuracy. *Journal of Marketing Research* 21: 290–297.

Simon, H.A. 1954. Spurious correlation: A causal interpretation. *Journal of the American Statistical Association* 49 (267): 467–479.

———. 1990. Prediction and prescription in systems modeling. *Operations Research* 38: 7–14. Compiled Simon, H.A. 1997. *Models of bounded rationality. Vol. 3: Empirically Grounded Economic Reason,* 115–128. Cambridge, MA: The MIT Press.

———. 1996. *The sciences of the artificial,* 3rd ed. Cambridge, MA: The MIT Press (1st ed., 1969; 2nd ed., 1981).

———. 2001. Science seeks parsimony, not simplicity: Searching for pattern in phenomena. In *Simplicity, inference and modeling. Keeping it sophisticatedly simple*, eds. A. Zellner, H. A. Keuzenkamp, and M. McAleer, 32–72. Cambridge: Cambridge University Press.

Stewart, Th.S. 2001. Improving reliability of judgmental forecasts. In *Principles of forecasting: A handbook for researchers and practitioners*, ed. J.S. Armstrong, 81–106. Boston: Kluwer.

Turoff, M. 1970. The design of a policy Delphi. *Technological Forecasting and Social Change* 2 (2): 149–171.

Ward, A. 2013. Spurious correlations and causal inferences. *Erkenntnis* 78: 699–712.

Woodward, J. 2003. *Making things happen: A theory of causal explanation*. New York: Oxford University Press.

Ziglio, E. 1996. The Delphi method and its contribution to decision-making. In *Gazing into the oracle. The Delphi Method and its application to social policy and public health*, ed. M. Alder and E. Ziglio, 3–33. London: Kingsley Publishers.

Part III
From Reality to Values: Ontological Features, Axiological Elements, and Ethical Aspects of Scientific Prediction

Chapter 7
Ontology of Scientific Prediction

Abstract Ontologically, the problem of prediction connects with the reality of the phenomena. This feature leads to the specific characteristics that phenomena of different realms of the reality (natural, social, or artificial) might have. From this perspective, the attention goes first to the repercussions (above all, epistemological and methodological) of the reality of phenomena on scientific prediction. Secondly, the problem of the characterization of future phenomena is considered, which is connected with the time horizon of prediction and the possibility of control over phenomena. Thirdly, the ontological limits to scientific prediction are analyzed, because there are obstacles to predictability that are rooted in the reality of phenomena. Finally, the ontology of scientific prediction is analyzed from the angle of complexity (structural and dynamic), which means emphasizing the notion of *historicity*. Thus, the reflection on complexity is developed from the perspective of historicity, which is especially important for the social sciences and the sciences of the artificial.

Keywords Ontology of scientific prediction • Natural sciences • Social sciences • Sciences of the artificial • Prediction and control • Ontological limits • Complexity • Historicity

Although Nicholas Rescher's philosophy is a system of pragmatic idealism, his ontology of science has basically a realistic character. Idealism is open to realistic contributions in his approach, because idealism is characterized mainly in conceptual terms whereas realism is commonly focused on ontological features. These realistic contributions are accepted insofar as they are compatible with pragmatism in the style of Charles S. Peirce (Gonzalez 2010, p. 256). Moreover, ontological realism—especially, his idea of objectivity together with the notion of "fact"—is a key aspect of Rescher's thought.[1] In effect, ontological realism is the support for the

---

[1] Among the books that best reflect the ontology of Rescher can be highlighted Rescher (1987, 1997, 2005a, b, 2010). On the ontological realism, within the varieties of realist approaches, cf. Gonzalez (1993, pp. 11–58; especially, pp. 50–55).

© Springer International Publishing AG 2017   219
A. Guillán, *Pragmatic Idealism and Scientific Prediction*, European Studies in
Philosophy of Science 8, DOI 10.1007/978-3-319-63043-4_7

epistemological and methodological realms of science, in general, and scientific prediction, in particular.

From this viewpoint, the analysis of the ontological features of scientific prediction is an especially relevant issue. In order to carry out this analysis, a series of steps are followed in this chapter. In the first place, the research is oriented towards the characteristics of the reality that is predicted, which can be natural, social or artificial. In the second place, the future is analyzed from an entitative perspective. This perspective involves taking into account several issues: (i) the reality of future phenomena; (ii) the entitative incidence of the prediction's temporal horizon; and (iii) the relations between prediction and control. In the third place, the attention goes to the ontological limits to scientific prediction, because there are obstacles to predictability that are rooted in the reality of phenomena.

Among the possible obstacles, Rescher highlights the following: (a) anarchy and volatility; (b) chance, chaos and arbitrary choice; and (c) creativity (1998a, pp. 133–156). Thereafter, the ontology of scientific prediction is analyzed from the angle of complexity (structural and dynamic).[2] In my judgment, this perspective can shed light on the reality of future phenomena and the difficulties in predicting them. In this regard, historicity can be seen as a key notion when investigating the complexity of the real things (above all, in the social and artificial realms) and its impact on prediction.

7.1 Scientific Prediction and the Reality of Phenomena

Even though Rescher's thought is a system of pragmatic idealism, where the idea of a *system*—with an insistence on the conceptual realm—is combined with the primacy of *practice*, his approach to ontology of science is basically of a realistic kind (see, in this regard, Sect 1.2.). When he describes his own philosophical approach, he characterizes it as a "middle-of-the-road idealism that makes significant concessions to realism" (Rescher 1992a, p. xiv). Furthermore, he thinks that "the sensible move is to opt for the middle ground and to combine a plausible version of realism with a plausible version of idealism" (1992b, p. 324).

On the one hand, Rescher defends the existence of a reality that is independent of the mind of the knowing subject, and, on the other, he accepts that reality has its own properties that are accessible to the knowing subjects.[3] Nevertheless, there are idealistic aspects of his thought, in particular in his insistence on theory as mediated by mind (1992b, p. 324). From this viewpoint, he maintains that science is the result of the interaction between the researcher and the reality that is researched (above all, the natural environment) [Rescher 1990]. So human categories and, in general,

---

[2] The research on the ontological limits of predictability from the perspective of complexity is in Guillán (2016b).

[3] These two features are crucial for the objectivity of knowledge that goes with a realist approach, cf. Gonzalez (1986, p. 37). See also Niiniluoto (1984, ch. 1, pp. 1–9).

concepts allow us to articulate the reality, so that there is a distinction between reality *as such* and reality *as it presents itself to us* (1990, p. 77). Nonetheless, he thinks that there is objectivity in scientific knowledge, since reality—which has its own properties—is open to the mind that investigates it (Rescher 1998c).

Within this framework of pragmatic idealism, which is clearly open to ontological realism, ontology of science has a fundamental role, because, for Rescher, ontology is the support for epistemology and methodology of science. In the more specific case of scientific prediction, this means that the kind of predictions that we can make (with regard to their reliability, accuracy, precision, etc.) depends, to a large extent, on the reality that is predicted. In this regard, a relevant distinction is that related to the differentiation between natural reality, social realm, and artificial field.

7.1.1 Natural Phenomena

Rescher is usually mainly interested in the natural sciences.[4] In effect, when he deals with scientific prediction, he basically pays attention to natural phenomena. Ontologically, he admits that there are natural phenomena that are not homogeneous, so there are relevant differences between the phenomena that are studied by the different natural sciences. These ontological differences have an impact on the epistemological differences and affect prediction. Thus, there are some disciplines where it is possible to predict accurately (for instance, astronomy). Meanwhile, in other sciences prediction is generally based on probabilities, so that it has less accuracy and precision (as can happen in the realm of biomedicine).[5]

Therefore, he considers that the ontological side is crucial for science, because only "through the learning on the characteristics of phenomena we can know if it is possible to predict and to what extent it is possible."[6] To a certain extent, his conception links the possibility of predicting in a rigorous way in natural sciences with the availability of laws. This is because scientific prediction needs to be supported by the regularities that phenomena have shown in the past and in the present; and "the laws of nature (or what we accept as such) are the most important and fundamental sorts of patterns there are" (Rescher 1998a, p. 106).[7]

On the basis of the existence of this property of the real things (that is, to the extent that natural phenomena are regular), these regularities can be expressed in the form of scientific laws. Afterwards, these laws can be used as a support for the scientific procedures oriented towards prediction (Rescher 1998a, pp. 106–110).[8]

---

[4] This can be seen in books like Rescher (1978, 1999a, b, 2009c).

[5] Cf. Rescher, *Personal communication*, 15.7.2014.

[6] Rescher, *Personal communication*, 15.7.2014.

[7] This issue is developed from a methodological point of view in Sect. 5.3.1.

[8] Cf. Rescher, N., *Predicting the Future*, pp. 106–110. On the inference from laws as a method for predicting, see Sect. 6.3.2 herein.

Nevertheless, there are ontological limits to predictability with regard to the natural phenomena. These limits arise when the future that we want to predict is open to the development of unexpected patterns due to the contingencies of chance, volatility or chaos.[9] In these cases, the patterns that the phenomena have followed until now are broken, so predictions that can be reached are generally less reliable (for instance, in certain meteorological phenomena).

When possibilities such as chance, volatility, chaos, etc. appear in natural phenomena, then one basic issue has to do with the temporal horizon of prediction. Thus, the ontological obstacles such as chance or chaos do not intrinsically prevent prediction, but they do affect the type of prediction that can be achieved and its reliability. So predictions are generally more reliable—as can be seen in the phenomena of nature—in the short run than in the long run (Rescher 1998a, p. 77); especially when the predicted phenomena are volatile (for instance, in a prediction about the weather) or when they become chaotic.

Nevertheless, Rescher maintains that natural phenomena are generally more stable than social phenomena.[10] For this reason, the unreliability of prediction is a more frequent problem in social sciences than in natural sciences. In addition, he thinks that, when changes occur in the natural sciences (especially, in the phenomena of physics), these changes are usually evolutionary changes. In principle, this makes these phenomena predictable, because "evolution involves some kind of continuity."[11] Meanwhile, in the social realm, changes are more variable and discontinuous. From this perspective, there are more methodological difficulties to prediction in the social sciences than in the natural sciences, in accordance with the kind of issues addressed by the social sciences.

7.1.2 The Social Realm

Although Rescher focuses on the natural sciences, he also pays attention to prediction in the social sciences. In his judgment, the main problems to prediction in these sciences are of an ontological rather than epistemological kind. This is because "what impedes prediction in this domain is not a mere lack of information, a mere failure to develop the discipline as far as one can. Rather, the root cause is something very different, something that lies in the very nature of the operative realities at issue. It is the nature of the phenomenology of the domain—its volatility, instability, and susceptibility to chance and chaos—that is responsible for our predictive

---

[9] On the epistemological and ontological limits to predictability, see Rescher (2009b, pp. 91–122).

[10] Cf. Rescher, *Personal communication*, 29.7.2014.

[11] Rescher, *Personal Communication*, 29.7.2014. Here the concept of "evolution" that Rescher uses is the current idea of evolution as change over time that follows a certain line towards the future, instead of being the ramified idea of Darwinian evolution (as can be seen in the graph that Darwin made in chapter IV of *On the Origin of Species*, in the first edition of 1859).

incapabilities here, rather than our imperfections as investigators" (Rescher 1998a, p. 202).

Comparatively, the possibility of *novelty* in the social realm is greater than in the natural phenomena, where, in principle, regularity prevails to a larger extent[12] and is the main element in achieving reliable predictions. Novelty in the phenomena studied is an issue that, in my judgment, should be considered with regard to the problem of complexity. This is because when prediction is about the social realm, it has to deal with the complexity of the social systems. In this regard, it is usual to consider that complexity in the social realm is greater than the complexity of the phenomena of nature (Gonzalez 2011b, pp. 319–330; especially, pp. 319–321, 2013b).

This is something that Rescher acknowledges expressly. In his judgment, the limits of social prediction "lie in the intractability of the issues, so that there is little reason to think that the relatively modest record of the past will be substantially improved upon in the future" (Rescher 1998a, p. 202). Nevertheless, his analysis does not contemplate *historicity* as an ontological feature of the social reality. In this regard, it seems to me that this is a lacuna, since historicity gives the appropriate framework to address the complexity of the social realm. Historicity affects both structural complexity and dynamic complexity, and this has clear repercussions on the possibility of predicting the social systems and on the kind of predictions achievable (with regard to their reliability, accuracy, precision, etc.).[13]

However, for Rescher, there are two main reasons that shed light on the comparatively higher difficulty of social prediction and its frequent unreliability. On the one hand, social prediction is often about the actions and choices of the agents. These actions and choices are difficult to predict, because they are made on the basis of what agents think about reality (their beliefs, wishes, expectations, etc.), instead of being based on reality itself.[14] Therefore, social prediction should deal with psychological factors that are subjective, and this involves a greater difficulty in predicting.

On the other hand, Rescher acknowledges that social systems are not independent of the environment (social, political, ecological, etc.). Thus, although he hardly pays attention to the external factors, he admits that there is an interrelation of social systems and the environment where they are placed (which is of a changing character). In his judgment, this is a feature that limits the predictive capability of social sciences. In effect, he thinks that "insofar as developments in social affairs reflect human choices, and insofar as these are influenced by circumstances (such as technical innovation and fashion) that develop in a chaotic way, the course of development in human affairs can only be predicted to a small extent" (1998a, p. 202).

---

[12] The possibility of "new facts" in the realm of the social sciences (for example, in economics) is comparatively higher than in the case of the natural sciences. There are a series of contextual elements that have repercussions on economics ("Economics as activity") together with the components of variability of the economic activity itself ("economic activity"), Cf. Gonzalez (1994).

[13] This feature has been highlighted by Wenceslao J. Gonzalez with regard to the sciences of design. Cf. Gonzalez (2012b).

[14] Cf. Rescher, *Personal communication*, 10.6.2014.

But these two features that Rescher points out—psychological factors and the influence of the external factors in social systems—are better addressed, in my judgment, within the framework offered by the notion of historicity. (1) Historicity is a feature of *science*, in general, and of each science, in particular. It is a characteristic that can be seen in the structure of science (in the configuration of its constitutive elements) and in its dynamics. It can be seen both from an internal perspective, which takes into account aims, processes, and results, and from the external angle, which deals with the relations with the environment (Gonzalez 2012a, pp. 13–14). (2) Historicity is a feature that configures the *agents* who develop the scientific research, since they are human beings in a concrete historical context. Historicity affects both the approaches to address the research topics and the circumstances that surround the research of the agents. (3) Historicity modulates the *reality* itself that is researched, above all, when it is a social or artificial reality (Gonzalez 2011a, p. 43). In effect, social reality is historical in itself, so that historicity is a key ontological feature of social systems. From this perspective, social reality has in itself a component of variability that adds complexity to prediction.

Nevertheless, it is important to point out that *historicity*—which includes the change in the real things—is compatible with the objectivity of scientific knowledge (Gonzalez 2008b, pp. 68–71). In this way, historicity should not be seen just as an obstacle to prediction in social sciences, but rather as a feature of the social reality, which is a complex reality (both from the point of view of its structure and from the perspective of its dynamics). Thus, when the prediction is made within the social realm, it often has to deal with a wide variety of factors, where changes are frequent. Frequently, this aspect makes social prediction more difficult and less reliable than prediction in the natural sciences.[15]

7.1.3 The Artificial Field

Frequently, when Rescher analyzes science, in general, and scientific prediction, in particular, he does not pay attention to the artificial field. However, prediction is a main aim in the sciences of the artificial, which are applied sciences.[16] Within these sciences, prediction is usually the previous step to prescription. Thus, the anticipation of the possible future (prediction) is required before we can give indications about what should be done (prescription) when we want to solve a concrete problem (Gonzalez 2008a, pp. 165–186; especially, pp. 179–183).

A characterization of the artificial field from an ontological perspective was offered by Herbert A. Simon. According to him, the sciences of the artificial are disciplines that have to do with the reality of the human-made (Simon 1996, see

---

[15] The roles of complexity and historicity in social sciences are further developed in Sect. 7.4.3 of this chapter.

[16] An especially influential proposal on the different kinds of scientific activity—basic, applied, and of application—is in Niiniluoto (1993, 1995).

Sect. 5.3.2 herein). Thus, their sphere is the field of what is made by humans, where the ontological aspect of historicity is a key feature: "insofar as it is a realm of specific human elaboration that seeks to enlarge the existing potentialities or to achieve new and more ambitious aims, there is always a contextual component that is modulated by historicity" (Gonzalez 2012a, p. 14). However, Simon did not manage to grasp this relevant feature of historicity, which is in line with the creativity of the designs.

Certainly, the sciences of the artificial—especially as sciences of design—are disciplines that are not limited to the knowledge or the reality that is researched, but which actively search for new possibilities to enlarge that reality. Thus, the reality of the artificial realm is changeable. In effect, it is a field that is open to novelty, which can appear through the creativity of the agents.[17] This creativity involves a clear obstacle to prediction when it deals with artificial things, because "human creativity and inventiveness defies predictive foresight" (Rescher 1998a, p. 149).

So historicity and creativity are two main features of the artificial field. Through these features, there is a great possibility of novelty in the artificial phenomena. This adds complexity to the task of making predictions in the sciences of the artificial. In this way, it can be seen that the ontological basis has direct epistemological and methodological repercussions. On the one hand, changes and the continuous presence of novelty affect the kind of achievable knowledge about the future (with regard to its reliability, accuracy, precision, etc.); and, on the other, predictive models must address the nature of the things, where creativity and the interactions with the environment have an important role.

7.2 Characterization of the Future from an Entitative Perspective

An important feature of scientific prediction, as Rescher sees it, is that prediction is a statement with a content oriented towards a potential future, independently of the specific reality it is about (i.e., natural, social or artificial). He considers that every "authentic" prediction is oriented towards an open future, which is inaccessible to observation at the moment the prediction is made (Rescher 1998a, p. 59). Thus, in principle, prediction foretells that something may occur or can happen when there are some given circumstances. Consequently, a genuine "retrodiction" or "prediction about the past" can be rejected based on ontological reasons (see Sect. 2.3.3. herein. See also Gonzalez 2014).

This approach to scientific prediction leads to the problem of the characterization of the future, which is studied here from an entitative perspective. In this regard, three issues are considered: (i) how to characterize the reality of future phenomena, since the future is something that does not yet exist; (ii) what can be the entitative

---

[17] On the role of creativity in the sciences of design, see Guillán (2013).

incidence of the time horizon of prediction, which can be, in principle, in the short, middle or long run; and (iii) the relation between scientific prediction and the effective control of phenomena (natural, social or artificial).

7.2.1 The Reality of Future Phenomena

For Rescher, scientific prediction is a statement about the future. From an ontological point of view, he considers that the future has two main characteristics: (a) it does not yet exist, by definition; and (b) it is something that unavoidably will come, in one way or another (Rescher 1998a, p. 2).[18] Another feature of a different kind—an epistemological one—comes from these characteristics: the information we have about the future is usually incomplete; and, consequently, the control we can exert on future phenomena is commonly very limited (1998a, p. 2). Therefore, there are limits to scientific predictability due to the reality itself of the future phenomena, which make these phenomena difficult for our knowledge and which it is not easy to have control over.

In effect, "there are ontological limits in the starting point: the *future* is the ontological axis of prediction; but, strictly speaking, the future is something that *does not yet exist*, since its being depends on what will happen" (Gonzalez 2010, p. 268). This is the first feature that Rescher points out in his ontological characterization of the future. Thus, in his judgment, "the future is nonexistent" (Rescher 1998a, p. 70). In this way, the future has not facticity (it does not yet exist) and, consequently, it is not accessible to us from an ontological viewpoint.

But, although the future *is not* available ontologically yet, it can be considered if the future can have effectiveness on the facts of the present. In this regard, the possibility of a genuine backward causation has been discussed. This involves that the future can have a causal effect on the present, a possibility considered by M. Dummett (1954, 1964), among others.[19] In backward causation, causes do not necessarily precede the effects and effects might go before the causes. In this way, the future can have a genuine causal influence on the present.

Rescher explicitly rejects the idea of backward causation. In this judgment, the future cannot have a causal influence on the present, since the future, strictly speaking, has not happened yet (Rescher 1998a, p. 70). However, he admits that the decisions we make in the present are influenced by our ideas about what is to happen in the future.[20] Furthermore, he thinks that "to act, to plan, to survive, we must anticipate the future" (1998a, p. 65). Therefore, what can have influence on the decision making of the human beings is the cognitive anticipation of the possible future—the prediction—instead of being the future itself what has an influence (causal or

---

[18] Rescher does not defend an ontological determinism. Rather, he makes it clear that there is a reality of the future as something that will show up.

[19] Although Dummett considers this possibility, he actually does not support it.

[20] Cf. Rescher, *Personal communication*, 8.7.2014.

non-causal) on the facts of the present. Possible examples are predictions of natural disasters, such as hurricanes or tsunamis. If predictions have a high degree of probability and are stated several days before the disaster, they can move authorities to take actions.

In a scientific context, this is what usually happens in the applied sciences, where prediction is the previous step to prescription. In this case, prediction provides the knowledge about the future and this knowledge is used to establish what should be done to solve a concrete problem (Gonzalez 2008a, pp. 179–183). Thus, the knowledge offered by prediction, which deals with what can be expected to happen in the future, allow us to direct the action towards the sought aims. In this case, it is not the future as such that influences the decision making, but that the decisions are made on the basis of prediction, which offers knowledge about the possible future.

According to Rescher, the second main feature that characterizes the future is that it is something that unavoidably will be (in one way or another) [1998a, p. 2]. Thus, he thinks that the future will happen or show up even though we do not want it to. But Rescher does not support a fatalistic proposal or an ontological determinism, according to which the future is something fixed beforehand.[21] In effect, he thinks that, to some extent, it is possible for us to shape the future, although he acknowledges that there are important limitations to this task (Rescher 1998a, pp. 232–238). In addition, he defends that human beings have free will, so human actions do not happen necessarily (Rescher 2009a).[22]

These ontological features that characterize the future have an epistemological incidence, since the character itself of the future phenomena leads to the difficulties in obtaining information about them. Thus, Rescher considers that the future remains open, so it is possible to think that it has, in principle, many different possibilities. Additionally, the future is something that does not yet exist, so we cannot have access to it through observation (1998a, p. 59). Consequently, the information we can obtain about the reality of the future phenomena is always limited. This, in turn, has negative repercussions on our ability to control those future phenomena, since the effective control requires, in principle, that the information needed becomes available (Rescher 1998a, p. 2).

7.2.2   The Entitative Incidence of the Time Horizon

The time horizon has to do with the temporal distance between the anticipated phenomenon and the present moment. This concerns prediction insofar as it pays attention to a possible future. Regarding this issue, Rescher distinguishes between the short-range forecasts and the long-range forecasts, depending on the temporal distance of the predicted phenomenon with regard to the moment when the prediction

---

[21] On the topic of determinism both in social sciences and in the sciences of nature, see the set of papers published in Gonzalez (2012d).

[22] In this volume, Rescher offers renewed approaches in favor of human freedom.

is stated (1998a, pp. 76–78). However, the diverse sciences usually classify predictions in three different types in accordance with the length of time: the short, the middle, and the long run.

Even though there is no clear-cut division between them, commonly (1) prediction is *in the short run* when the scientific statement is up to a near future. (2) When the prediction deals with a distant future (a number of years), then it is a prediction *in the long run*. (3) When the predictive statement is about a future that is in an intermediate point between the near future and the more distant future, then the prediction is *in the middle run* (Gonzalez 2015a, pp. 65–67). It is, in my judgment, a more complete classification than Rescher's distinction, since scientific prediction certainly can be about a future that is between the short and the long run.

An element that makes it difficult to specify in detail this classification consists in establishing where are situated the short, middle, and long run. In this regard, the ontological aspect has a key role, since where the short, middle, and long run are located is an issue that depends, to a large extent, on the kind of reality researched. For Rescher, the volatility of phenomena that we want to predict is one of the key factors we must take into account. Thus, when the phenomena are characterized by being volatile, the long run encompasses a shorter temporal distance than when the phenomena are more stable.[23]

In effect, the short, middle, and long run in meteorological predictions is, for instance, different from the short, middle, and long run in the astronomical predictions; because the reality meteorological predictions are about has different characteristics from the reality studied by astronomy. But there are other factors that should be taken into account, such as the kind of variables used or the type of data available (Gonzalez 2015a, p. 66). These features lead to there being differences within a scientific discipline. An example in this regard is provided by economics, where "the short, middle, or long run in unemployment could be different from the case of business cycles" (Gonzalez 2015a, p. 66).

Thus, even when the classification of scientific prediction in three types in accordance with the temporal distance (in the short, middle or long run) could be generally valid, achieving a highest level of sophistication to establish where we can place each one of these types is a task that must be done with regard to each discipline. In economics, for example, a distinction has been proposed which typifies five types of predictions, where the time horizon is specified: the immediate horizon (between 0 and 3 months), the short run (between 3 and 12 months), the middle run (between 1 and 3 years), the long run (between 3 and 10 years) and the very long run (beyond 10 years) [Firth 1977. On this classification see Gonzalez 2015a, p. 66].

Concerning this topic, philosophers commonly accept that the time horizon of the prediction has epistemological relevance. Thus, the kind of prediction that we can obtain (with regard to its reliability) is something that varies in accordance with the temporal distance to the reality that is predicted. In Rescher's judgment, "in

---

[23] "A weather forecast is long range if it looks ahead for more than a month, an economic forecast is long range if it looks ahead for more than a year, a population forecast would have to look several generations ahead to qualify as long run," Rescher (1998a, p. 78).

general and by and large it is more difficult to predict matters in the more distant future" (1998a, p. 77). Regarding this issue, he quotes a well-known statement of John Maynard Keynes: "in the long run we are all dead"; and he adds: "and in the *very* long run the universe will presumably be so as well" (1998a, p. 77).

On this ontological basis, in principle, the larger the temporal distance is, the higher the possibility of error will be. However, it is possible to find phenomena that are more predictable in the long run than in the short run. Rescher gives an example related to the meteorological prediction that can illustrate this possibility: "It is March, a very changeable month for our region. And so we cannot predict the weather for next week—though we can be pretty sure that three months hence, in July, the weather will be sunny and warm" (1998a, p. 78). But these are exceptions, since those things that are in the near future are generally more attainable. Thus, it should be acknowledged that "the variables are, in principle, more knowable when they are close to the researcher (and, thus, within the field of experience and expertise to the researcher)" [Gonzalez 2015a, p. 219].

7.2.3 Scientific Prediction and the Control of Phenomena

A central issue in Rescher's approach to scientific prediction is its relation to the control of phenomena, which is one of the features that accompany his characterization of science as *our* science. In his approach, control "consists in bending the course of events to our will, of attaining our ends within nature" (Rescher 1998a, p. 233). So his analysis of control is commonly focused on the realm of the natural sciences, although this problem can certainly be posed in relation to the social and artificial phenomena and systems.

Control is a notion that Rescher relates to the concept of "intervention."[24] Because control requires, in his judgment, having "*the capacity to intervene in the course of events so as to be able both to make something happen and to preclude it from happening, this result being produced in a way that is not only foreseen but intended or planned.* Control thus calls for the possibility of causal participation ('intervention') in the course of event ('to make something happen or preclude it') with a power that can be exercised both positively ('to make happen') and negatively ('to preclude from happening')" (Rescher 1998a, p. 235).

Both if the control is negative (preclude something from happening) and if the control is positive (make something happen), the relation with prediction is clear, since "the limits of predictability set limits to control as well" (Rescher 1998a, p. 237). From this perspective, the question at issue is to exercise control over the future phenomena, and to do this prediction is required. But control may also be considered with regard to the variables used to predict the future in a scientific way. In this respect, the difficulties to control the variables that are relevant to the

---

[24] It should be noted that the concept "intervention" and its differences with "representation" received especial attention after the publication of the book by Hacking (1983).

prediction—both with regard to the knowledge of these variables and in relation to the possibility of measure them—set limits to the ability to predict the phenomena.[25]

Thus, the relation between the prediction and the control of the phenomena can be contemplated in two ways: (a) prediction as a *necessary condition* to the control of future phenomena (in the positive meaning and in the negative sense); and (b) the control of the variables as an *element required* to prediction. Usually, Rescher attention goes to the first kind of relation, since he thinks that "a future we cannot foresee is *a fortiori* a future we cannot control" (1998a, p. 236). Implicit here is the methodological path frequently followed in applied science, where prediction is needed in order to prescribe.

Within the applied sciences, to the extent that "they do not describe reality, but rather tell what we ought to do in order to realize our goals" (Niiniluoto 1995, p. 128), there is a clear interest in exercising a control over the systems (natural, social, and artificial). Rescher considers that prediction is required in order to make that control possible, together with an understanding of how the phenomena at issue work.[26] However, there are phenomena that, due to their very nature, we cannot control, even when there is an understanding of their operation and we can predict them.

Certainly, as Simon points out—with a pragmatic vision—predictive models in applied sciences are usually oriented towards predicting matters that are impossible to control, so the main aim is the prescription: to suggest patterns favoring the best possible adaptation to the anticipated problems (for example, in the case of the weather or business cycles).[27] Thus, it seems clear that the very nature of the phenomena sets limits to our ability to exercise an effective control over the future occurrences, even when it is possible to predict them.

If we think of the relation between scientific prediction and the control of phenomena in the second sense that has been pointed out—the control of the variables as an element required to predicting—then the notion of "control" in question is more epistemological and methodological than ontological. The reason is clear: it has to do with the knowledge of the variables relevant to the prediction and with the possibility of measuring them. In this regard, there are important differences between the degree of control that is achievable in the sciences of nature (for example, in physics) and the degree of control that is possible in the social sciences (for instance, in sociology) and the sciences of the artificial (for example, in economics).

On this issue, it should be pointed out that "physical predictions, in spite of the problems generated by the theory of chaos, work often with a range of events that, in principle, could be under our control, insofar as they are repeatable" (Gonzalez

---

[25] Furthermore, it is possible to distinguish four types of scientific prediction (foresight, prediction, forecasting, and planning) according to the degree of control of the variables. Cf. Gonzalez (2015a, pp. 68–72). See also in this regard Sect. 2.4.2 herein.

[26] Cf. Rescher, *Personal communication*, 10.6.2014.

[27] See Simon (1990, pp. 10–11). On this matter, see also Simon (2002).

2015a, p. 194). Meanwhile, in other sciences, such as economics, control is more difficult. This leads to the difficulties to predict. Thus, (1) the repeatability of the events might be questioned (are they actually the "same" events?); and (2) there are more factors that should be considered and, in addition, they could be really varied and show complex interactions (Gonzalez 2015a, p. 194).

These considerations show the relevance of analyzing the diversity of the reality (natural, social, and artificial) that is researched by science, which can be complex (both with regard to its structure and in respect of its dynamics). Within this thematic framework, a crucial issue is the study of the ontological limits to scientific prediction, which are those difficulties to prediction that derivate from the very status of the phenomena. They are features that certainly involve a source of complexity for prediction and also have repercussions on our ability to control (for example, when a system is chaotic or when it is open to changes through human creativity).

7.3 Ontological Obstacles to Scientific Prediction

Certainly, the knowledge of the extramental reality is an indispensable aspect of prediction, since predicting is a cognitive anticipation of a possible future. Thus, prediction "is a cognitive venture whose successful pursuit is inseparably bound up with factual matters regarding the nature of the world's *modus operandi*" (Rescher 1998a, p. 81). In this case, the ontology of science is in the service of the epistemology and the methodology of science. Obviously, it is an important field within the study of scientific prediction.

Due to this unbreakable nexus between the reality of the phenomena and the success in the prediction about those phenomena, it seems clear that the reliability and credibility that must characterize a successful prediction can only be achieved under certain conditions. In this regard, one of the main contributions of Rescher to the research about scientific prediction is, in my judgment, the identification and analysis of the obstacles to predictability (above all, epistemological and ontological). Some ontological obstacles can be highlighted: (i) anarchy and volatility, (ii) chance, chaos, and arbitrary choice, and (iii) creativity.[28] Finally, the obstacles on predictors are also considered.

7.3.1 Anarchy and Volatility

Ontological limits that affect prediction, above all because prediction is about phenomena which have not happened yet and, therefore, are still open. This is highlighted by Rescher when he considers the obstacles to prediction. In his judgment,

---

[28] The analysis of these limits to predictability is also in Guillán (2016b).

"ontological limits exist insofar as the future of the domain at issue is *developmentally open*—causally undetermined or underdetermined by the existing realities of the present and open to the development of wholly unprecedented patterns owing to the contingencies of choice, chance, and chaos" (1998a, p. 134).

The very nature of the phenomena may make it difficult or even impossible to meet the preconditions for rational prediction: the availability of the relevant data, pattern discernability, and the stability of the patterns followed by the phenomena (Rescher 1998a, pp. 86–87). This leads to admitting the existence of ontological obstacles to predictability.[29] In this way, he brings again to the forefront the ontological level, as a support for the epistemology and methodology of science. The complexity of the reality conditions the complexity of knowledge and the consequent method in this regard. This is because we do not want just to predict, but to achieve reliable, accurate, and precise predictions.

From this ontological perspective, we should take into account the distinction between unpredictability (the *complete* impossibility to predict) and non-predictability (the *current*) impossibility to achieve a prediction).[30] The former is a methodological-ontological issue (the very reality of the phenomena and events makes the prediction impossible); while the latter involves a methodological-epistemological feature (future knowledge might allow us to overcome the current difficulties) [Gonzalez 2015a, pp. 54–56].

Unpredictability can appear as a result of the presence of phenomena of anarchic character (Rescher 1998a, pp. 136–138). Because, in this case, the second precondition of the rational process of prediction cannot be met (i.e., pattern stability): there are no regularities that allow the inference oriented towards the future. However, it seems that, in principle, it can be too drastic to claim that there is a clearly unpredictable phenomenon or event, because the dynamic character of scientific knowledge involves historicity. From this perspective, it is problematic to give examples of unpredictable issues without taking into account the historical dimension of knowledge (Gonzalez 2015a, p. 56).

For Rescher, only science can inform us of its own limits,[31] so it is not the society what sets the limits. This position implies that the limits of science always have to do with a historical moment of scientific development. Therefore, even when he admits that there are anarchic phenomena, which can be intractable from a predictive viewpoint, he takes a wary attitude with regard to the existence of "unpredictability" in the strict sense. This is because, for him, we can only claim that a phenomenon is anarchic on the basis of the available knowledge about that phenomenon, which can change in the future.[32]

To be "not-predictable" has to do with the *current* impossibility to obtain a prediction. In Rescher's account of the ontological obstacles to scientific prediction, to be "not-predictable" is related to the volatility of the phenomena or events (1998a,

---

[29] On the preconditions for rational prediction, see Sect. 5.4.

[30] For further details on this distinction, see Sect. 2.5.

[31] Cf. Rescher, *Personal communication*, 17.6.2014.

[32] Cf. Rescher, *Personal communication*, 1.7.2014.

p. 79). Usually, the current impossibility to predict is due to the instability of phenomena, so that the discernible patterns from the past can vary in the future (1998a, pp. 79–82).

When volatility is considered as an ontological limit to predictability, the time dimension is crucial, because the processes are frequently more stable in the short run than in the long run. When prediction is in the long run, it is more likely that errors appear; while predictions in the short run are generally more reliable. Here the information used is another factor to consider, since the available information to predict in the short run is generally more relevant than when prediction is in the long run (Granger 1989, p. 7).

In contrast to "anarchy", volatility of phenomena or events is a matter of degree. In this regard, Rescher suggests classifying the volatile processes into two groups: (a) moderately volatile processes (for example, the weather in the temperate regions); and (b) very volatile processes, which are characterized as processes which exhibit sudden and fortuitous changes (for example, wind velocity in the English Channel) [Rescher 1998a, p. 79]. Thus, a prediction in the long run that is about a moderately volatile process can be more reliable than a prediction in the short run about a very volatile process.

7.3.2 Chance and Chaos

In Rescher's judgment, the main ontological obstacles to predictive success are chance, chaos, and arbitrary choice (1998a, p. 134). He describes chaos and chance as stochastic or random processes that make the laws at issue probabilistic laws (1998a, p. 134). Furthermore, arbitrary choices can be considered as another important obstacle to prediction. In effect, in his approach, predictability of human actions depends, to some extent, on their rational character (Rescher 1995). Thus, actions and choices that are arbitrary (and consequently non-rational) may be, in principle, not-predictable.

The presence of chance is an obstacle to predictability. This is the case because, "where chance is at work, the world can exhibit a fixed past and nevertheless confront us with distinct but altogether feasible futures—situations that are descriptively indistinguishable may unfold in different ways and issue in totally different results" (Rescher 1998a, p. 138). From this perspective, chance does not make it impossible to predict, but it affects the possibility to achieve predictions with certainty (Rescher 2009b, p. 108).

Chance does not involve unpredictability in a strict sense, but it may involve a phenomenon that is not-predictable. However, the degree in which a phenomenon or event characterized by chance is not-predictable is an issue that depends, in turn, on the kind of predictive question posed about that phenomenon or event. As Rescher points out, when there is chance, it is usually possible to achieve generic predictions and predictions on a large scale: "Individual chance events are indeed unpredictable,[33] but the very randomness of chance fluctuation means that large-scale phenomenology will be predictable via the laws of chance codified in probability theory" (Rescher 2009b, p. 107).

This leads one to acknowledge the interrelation between the epistemological and the ontological realms, which is especially clear in the case of the scientific prediction. This is because predictability of phenomena or events does not depend only on the nature of those phenomena (the presence of chance, chaos, arbitrary choices, etc.), but also on the questions posed and, therefore, on the type of prediction that we expect to obtain (generic or specific, conditional or categorical, qualitative or quantitative, etc.).

When chance is considered, what is actually an epistemological limitation might get mixed up with an ontological obstacle. This is what Rescher calls "fallacy of misattributed chance" (1998a, p. 141). This consists of attributing erroneously to chance something that belongs to another factor. This is because, occasionally, we think that our difficulties in predicting are due to the presence of chaos, when actually they are due to ignorance (or, potentially, to another factor). This can happen when the phenomena at issue are complex and we do not have an appropriate understanding of their working.

Together with chance, chaos is, for Rescher, another ontological factor that makes it difficult to predict to a greater extent. He associates chaos with extreme volatility, so in a chaotic system small differences in its initial state can cause big differences in the result. He considers that "a process is chaotic if *observationally indistinguishable* initial conditions can eventuate in different results irrespective of how sophisticatedly we make our observations" (1998a, pp. 144–145).

M. Mitchell points out that chaos, in this sense of sensitive dependence of initial conditions, has been observed "in cardiac disorders, turbulence in fluids, electronic circuits, dripping faucets, and many other seemingly unrelated phenomena. These days, the existence of chaotic systems is an accepted fact of science" (Mitchell 2009, p. 20). In effect, chaos can intervene in a great variety of phenomena (natural, social, and artificial). This affects the very possibility of predicting, as well as the level of accuracy and precision that can be achieved by predictions.

Chaos theory in natural sciences has pointed out that even a deterministic system can give rise to unpredictable results, since it might be sensitive to variations in the initial conditions (Gonzalez 2005, pp. 36–38). Thus, the difficulties to predict are due to chaos, instead of being dependent on chance or randomness (Rescher 1998a, p. 143). However, prediction in social sciences is usually more difficult than prediction in the natural sciences. In this regard, the ontological aspect is crucial, since "it seems clear that the differences in the kind of subject matter have incidence in the role of prediction" (Gonzalez 2015a, p. 62).

Chaos can intervene in social systems, just like in natural phenomena. According to Rescher "in any system whose workings are subject to a very large number of intricately interacting factors, there is going to be a great sensitivity to parameter

---

[33] Note that in Rescher's terminology, "unpredictability" is used instead of "not-predictability." See Sect. 2.5 of this monograph.

determination, so that even a small variation on input values will amplify into substantial variations in output values. This of course is the characteristic situation of chaos (…). Economic systems (and social systems in general) are in large measure chaotic in exactly this sense" (1998a, p. 197). Thus, some of the difficulties to social prediction can be due to the presence of chaos.

Although chaos can appear in social systems, there are other obstacles, such as arbitrary choice, that have to do to a greater extent with social phenomena (see Sen 1986). For Rescher, rationality is an important predictive instrument when human activities are considered (Rescher 2009b, p. 113). Thus, when the behavior of the agents is characterized by being rational, this involves, to some extent, the possibility of anticipating their choices. Meanwhile, irrational behavior makes it difficult to anticipate agents' choices in a reliable way (see Sect. 4.2.2. of this book. See also Guillán 2016b, pp. 188–190).

7.3.3 Creativity as an Obstacle to Prediction

When creativity is considered as an obstacle to scientific prediction there are, in principle, two successive levels of analysis: (i) creativity as an element that *configures the object studied* by science (the economic, social, cultural, or politic reality, among others), where agents' creativity can generate changes that involve novelty; and (ii) creativity as a factor that *is part of the very scientific activity* of research (basic or applied) that is relevant both from an structural point of view and from a dynamic perspective (Guillán 2013).

In the first case—when there is creativity in phenomena or events that are the object of study of science—agents' creativity can be a clear obstacle to scientific prediction (above all, in social sciences and in the sciences of the artificial). In this regard, Rescher relates creativity to unpredictability. Fundamentally, he does this with regard to contexts such as arts, where the presence of creativity allows us to account the difficulties to prediction and might even result, in his judgment, in the impossibility of achieving a prediction.[34]

When human creativity intervenes, the pattern followed by a system in the past does not allow us to infer its development in the future (Rescher 2012b, p. 210). For this reason, Rescher considers that "human creativity and inventiveness defies predictive foresight" (1998a, p. 149).[35] Even more, one might consider that creativity is a source of complexity. This has been highlighted, above all, in economics, where it is usual to see the creativity of the agents as an obstacle to economic predictions.

Regarding this issue, Wenceslao J. Gonzalez points out that economic agents "frequently develop *creativity* when they perform some actions (e.g., in the design

---

[34] "The future of American poetry is impredictable: we simply have no grip on any laws or regularities that provide for rational prediction," Rescher (2009b, p. 103).

[35] Rescher suggests this issue with regard to humanistic realms, like literature, so he does not consider how it can affect the scientific disciplines related to designs.

of new financial products). This creativity of the agents adds another element of complexity while making economic predictions, because it increases the possibility of novelty in the phenomena instead of [having] the prevalence of the regularity, which is the main ingredient for promoting the reliability of the predictions" (2012c, p. 147). From this perspective, agents' creativity is another element that gives additional complexity to the economic predictions.

But, together with human creativity *in general*, there is another level of analysis—the second mentioned above—that has to do with *scientific* creativity, in particular. This is because creativity can also be a feature of scientific activity; both form a structural point of view and from a dynamic perspective. Nevertheless, it is difficult to characterize what *creativity* consists in, as "the term 'creativity' as commonly used covers too much ground. It refers to very different entities, thus causing a great deal of confusion" (Csikszentmihalyi 1996, p. 25).

This difficulty has been acknowledged by several authors, such as Subrata Dasgupta or Margaret A. Boden. For Dasgupta, there are "many interpretations—some differing in only subtle ways—that may be placed on the very *idea* of creativity" (1994, P. 16). Meanwhile, Boden has noted that, in the study of creativity, problems may arise "because of conceptual difficulties in saying what creativity *is*, what *counts* as creative" (2004, p. 13). In spite of these difficulties, it is usual to accept that the notion of "creativity" involves, in principle, something positive, which is related to something both valuable and novel or original.[36]

In this regard, for the sake of a more rigorous characterization of the notion of "scientific creativity," this concept should be distinguished from the term "innovation." Thus, "creativity" is usually related to a scientific context; while the notion of "innovation" is frequently connected with the realm of technology, where there are often modifications of existing realities rather than something completely new or original (Neira 2012, p. 217). However, "creativity" and "innovation" are interrelated notions, since there are relevant connections between scientific creativity and technological innovation (for instance, in the case of the sciences of design and the information and communication technologies) [Gonzalez 2013a, pp. 11–14].

When scientific creativity is considered in relation to prediction, then creativity can have, in my judgment, a double role: (a) it might be, in principle, an obstacle for predicting the future science; and (b) it can be a key factor for problem-solving, so it can contribute to overcoming the current unpredictability of the phenomena and events. Usually, Rescher's attention is focused in the first case: creativity as an impediment to the prediction about the development of the future science (Rescher 2012a). But he offers a thematic framework that makes the second option also possible; i.e., to analyze creativity as a medium to deal with the obstacles to predictability (above all, in the epistemological and methodological levels).

Regarding the first role—creativity as an obstacle to the prediction of the future science—Rescher focuses his attention on one of the aspects of scientific creativity: conceptual creativity (see Guillán 2016a, pp. 144–146). This is because, within a framework of pragmatic idealism, concepts are crucial for the articulation of

---

[36] On the notion of creativity, see Gonzalez (2013a).

knowledge. In this way, his thought gives priority to the epistemological realm, as a path to address the problems posed. This also happens when he considers scientific creativity. Thus, he thinks that science is open to conceptual creativity, so changes occur regarding scientific concepts.[37]

It is a key feature in addressing scientific progress and the problem of the limits of science, since research is always carried out from a conceptual framework open to changes. So it happens that taking a concrete conceptual framework sets, in practice, limits to scientific research, insofar as "the major discoveries of later stages are ones which the workers of a substantially earlier period (however clever) not only have failed to make but which they could not even have understood, because the requisite concepts were simply not available to them" (Rescher 2012a, p. 151). Thus, conceptual creativity affects the predictive ability of science, since certainly we cannot predict those things that we cannot conceive (Rescher 2012a).[38]

However, conceptual creativity is, in my judgment, one aspect *among others* of scientific creativity. In effect, from a structural point of view, creativity can be considered in relation to the constitutive elements of science: language, structure, knowledge, method, activity, ends, and values.[39] Therefore, there might be creativity in scientific language (through the introduction of new terms or through changes of meaning in the existing terms), in the structure of scientific theories (where there are alternatives to the dominant theory), in scientific knowledge (through new concepts), in methods (where there are changes in the procedures), with regard to scientific activity (that involves novelty regarding the aims, processes, and results of the research) and in scientific values (internal and external) that modulate the selection of ends (Gonzalez 2013a, pp. 15–17).

From this perspective, scientific creativity might involve the possible "not predictable" situation or even the "unpredictability" about the development of future science in the diverse elements pointed out (language, structure, knowledge, method, activity, ends, and values), instead of being something due only to changes regarding concepts. At the same time, it might be noted that the problems to predict future science, which are mainly the result of the presence of creativity, depend to a large extent on the time horizon of the scientific prediction. Thus, it seems possible to anticipate with enough reliability the aims, processes, and results of science in the short run; while it could be certainly impossible to predict them in the long run (Gonzalez 2016).

It happens that scientific creativity, besides being an important obstacle to the predictability of future science, can also be a medium to overcome the limits of science, in general, and the limits to scientific prediction, in particular. This can happen both in a context of basic science, where creativity connects above all with explanation and prediction, and in the realm of applied science, where creativity is mainly related to prediction and prescription. In addition, creativity can be considered in

---

[37] On conceptual change, see Gonzalez (2011a).

[38] From this point of view, it is a mistake to think of the experiments as prior to the concepts related to them or the technological doing as prior to the conceptual support for its use in the research.

[39] On the constitutive elements of science, see Gonzalez (2005, pp. 3–49; especially, pp. 10–11).

relation to the application of science, which has to do with the use of scientific knowledge by agents, so creativity can lead to new applications of the available knowledge (Gonzalez 2013a, pp. 17–18).

Therefore, while human creativity, in general, is a clear obstacle to scientific prediction (for example, in economics, insofar as it favors a high degree of novelty), scientific creativity can be dual in this regard. On the one hand, creativity is certainly an obstacle that makes it difficult to predict the future science, above all when the prediction is in the long run. But, on the other hand, it can contribute to overcoming the limits to predictability, because it allows us to open new routes, regarding explanation and prediction in basic science as well as with regard to prediction and prescription in applied science.

7.3.4 Obstacles on Predictors

Clearly, the limits to the task of predicting—among them, the ontological limits—make it impossible to achieve perfection in the predictive domain. From the point of view of predictors, the existence of a perfect predictor is also unfeasible, both from a theoretical point of view and from a cognitive or epistemic perspective. In the first place, there are what Rescher calls "inherent limits of prediction," that are mainly related to the presence of unsolvable "self-insight" problems (i.e., a kind of self-prediction).[40] In the second place, there are cognitive limitations on predictors that make it difficult to achieve accurate predictions (Rescher 1998a, pp. 218–222).

Regarding the first problem—the *inherent* limits to prediction—Rescher thinks that the existence of a perfect predictive engine is something impossible; that is, he considers unfeasible an artifact designed to predict everything that, in principle, is considered as predictable. This is because problems of inconsistency inevitably appear when the predictive engine is asked about its own working and, more concretely, about its future predictions. Thus, for example, if a predictor (either a human subject or an engine designed to predict) is asked "will you answer this very question by 'no'?", three answers are possible: "yes," "no" or "I can't say." If the predictor answers "yes" or "no", then the prediction about its own answer is false; and, in the case of answering "I can't say", then the predictor cannot achieve an actual prediction (Rescher 1998a, p. 213).

In view of this kind of problem, we can claim that it is possible to establish the impossibility of perfection in the predictive domain. In effect, there are reflexive questions about the future predictions of an agent that cannot be answered in an adequate way. In this regard, Gregor Betz maintains that such *a priori* limits are not relevant (2006, pp. 3–4). However, I consider that they are relevant when the limits of science are considered. This is because, on the basis of the inherent limits of

---

[40] Cf. Rescher (1998b, pp. 159–161). What Rescher means by "self-insight" problems is in fact a kind of "meta-prediction" about the limits (e.g., of a universal problem-solver).

prediction, scientific prediction can be clearly seen as a limited venture, so it is impossible for science to achieve perfection in the predictive domain.

Besides these theoretical limits that affect predictors, Rescher also takes into account other limitations that are cognitive or epistemic. He highlights the following: (i) *immanency and scale exaggeration*, that is, the tendency to exaggerate the magnitude and promptness of the predicted developments; (ii) *conservatism*, which affects prediction when the stability and durability of the existing order of things is exaggerated; (iii) *wishful or fearful thinking*, which affects prediction when it is influenced by those thing we wish or fear; and (iv) *probability misjudgment*, which are due to errors in evaluations or combinations (Rescher 1998a, pp. 218–222).[41]

Taken together, the ontological limitations to predictability involve the impossibility of achieving what Rescher calls "predictive completeness," i.e., that science could accurately predict those things that are in principle recognized as predictable by science itself (Rescher 1999a, pp. 137–148). So, due to the limits that affect prediction, the predictability of phenomena is not something that can be established beforehand. Thus, "only the course of experience can inform us about the extent to which the phenomena of a particular domain are predictable. And with predictability in general, just as with specific issues of prediction, one must simply wait and see" (Rescher 1998a, pp. 81–82).

7.4 Ontology of Prediction from the Perspective of Complexity

The perspective of complexity is relevant when scientific prediction is analyzed from an ontological viewpoint.[42] Prediction might be about a reality (natural, social or artificial) that can be characterized as complex. Complexity has repercussions that go in two directions: on the one hand, it can affect the very possibility of predicting; and, on the other, it modulates the configuration of the predictive processes and the type of prediction that can be achieved (its reliability, accuracy, precision, etc.).[43] In this regard, it should be taken into account that complexity is a twofold notion: it has a structural dimension and a dynamic trait (Gonzalez 2013b).

The structural dimension of complexity has to do with the parts that compound the complex system and the interactions between those parts. Meanwhile, the dynamic trait of complexity has to do with the changes that occur in the system over time, both from an *internal* point of view (usually in terms of aims, processes, and results) and from an *external* perspective (that deals with the relations with the environment). Usually, Rescher pays more attention to the structural dimension of

---

[41] These limitations have received attention in Sect. 6.1.1 of this monograph.

[42] This analysis of complexity is also in Guillán 2016b.

[43] It has been noted that complexity sets limits to science: "the study of the characteristics of complex dynamic systems are showing us exactly why limited knowledge is unavoidable—or, to be more precise, why knowledge *has to be* limited," Cilliers (2007, p. 82).

complexity than to the dynamic trait. A broader approach to complexity should also pay attention, in my judgment, to the complex dynamics ("internal" and "external"), where historicity can be a key aspect, above all when the reality researched is social or artificial.

7.4.1 Varieties and Relevant Modes of Complexity

When the complexity of a system (natural, social or artificial) is considered, two different dimensions of complexity should be taken into account in principle: (I) the structural dimension (that has to do with the parts that compound the system and their interactions); and (II) the dynamic trait (that deals with the change over the time in the system).[44] Frequently, the studies of complexity are focused on the structural perspective. Furthermore, there are characterizations of complexity, in general, that only address the complex structure, without taking into account, in fact, the dynamic complexity.[45]

An interesting and especially influential approach to complexity is that offered by Simon, centered on the sciences of the artificial, in general, and the sciences of design, in particular. His focus is on "organized complexity," instead of paying attention to a kind of complexity that is disorganized (for example, a chaotic structure). From this perspective, Simon considers that designs consist of a series of elements that are coordinated, and they may have a hierarchic configuration, such that complex systems are nearly-decomposable. Thus, their components cannot be completely separated, since they are eventually interdependent (Simon 1996, ch. 8, pp. 183–216).

In this way, Simon considers that, in scientific research, *analysis* is the complement of the task of synthesis carried out by design sciences (Simon 1995, p. 246). From the structural perspective, his conception of the analysis is commonly *holological*, and it is oriented towards near-decomposability. So, when the aim is to clarify the structural complexity of design sciences, he thinks that a holological analysis is required. This is so because "the complexity of a whole should be decomposed (…), where there are numerous parts that interact. When this operation is carried out, insofar as the subsystems that compound the complex system are interdependent, it is not possible that the subsystems can split up completely ones from others" (Gonzalez 2012a, p. 17).

Simon's view of a complex system is based mainly on a holological perspective: the relations between the parts and the whole are emphasized, although he avoids

---

[44] A recent study of both aspects is in Gonzalez (2015b).

[45] "Complexity depends on the number of elements of the system, the number of its properties and the number of relationships between these elements or properties." Betz (2006, p. 81). This characterization can be valid with regard to the complex structure of a system, but it does not encompass, in my judgment, the whole realm of complexity, since it does not take into account the dynamics of a system.

giving a precise definition of "complex systems."[46] Generally, he considers that a complex system is "one made up of a large number of parts that have many interactions" (1996, pp. 183–184). In these systems, the whole is more than the sum of the parts, such that "given the properties of the parts and the laws of their interaction, it is not a trivial matter to infer the properties of the whole" (Simon 1996, p. 184). Thus, in a broad sense, complexity can be understood as opposite to simplicity (Simon 2001), although complexity might appear in simple systems (Sprott 2006).

In my judgment, such an approach to complexity is too general. On the one hand, structural complexity cannot be only an organized complexity, but disorganized complexity can also be found in complex systems (for example, in chaotic systems). At the same time, the notion of "hierarchy" is not good enough to cover the whole organized complexity, since it is also possible to have a "poli-hierarchy." In this respect, the need for "coordination" has been highlighted (Rescher 1998b, pp. 10–11).[47] On the other hand, Simon's approach to complexity is too focused on the structural perspective.[48] In this regard, in order to grasp the whole complexity, a deeper conception open to dynamic complexity, both of science considered in itself and of the studied reality (above all, when it is a social or an artificial reality) seems necessary.

Rescher's approach to complexity goes further, in my judgment, than Simon's proposal. In effect, Rescher considers that Simon's conception of complexity is too generic, since there are few real things that can escape complexity when it is understood in terms of the relations between the parts and the whole (Rescher 1998b, p. 22, n. 44). Thus, although Rescher is also mainly focused on "structural complexity," he highlights that there are numerous features of complexity. He does this through a breakdown of the modes of complexity according to two basic levels: the epistemic and the ontological.

Epistemic complexity, as Rescher sees it, is related to formulas. Within this first level of complexity, he proposes three epistemic or cognitive modes of complexity: (i) descriptive complexity, which deals with the level of difficulty that involves describing a system in a proper way; (ii) generative complexity, which reflects the number of steps that are required to give rise to the complex system; and (iii) computational complexity, which is associated with the number of resources required to solve problems related to the system (Rescher 1998b, p. 9).

There is another level of complexity in Rescher, which is of an *ontological* kind (the complexity in reality itself). In this regard, he distinguishes three ontological modes of complexity. Each has several possible options: (a) compositional complexity (constitutional and taxonomic); (b) structural complexity (organizational

---

[46] It should be pointed out that there is not a definition of "complexity" or "complex system" that is generally accepted. Cf. Chu et al (2003, p. 19).

[47] A comparison between Rescher and Simon's approaches to complexity is offered in Gonzalez (2012b, pp. 76–79, and 2012c, pp. 151–153).

[48] Certainly, the dynamic trait is also present in Simon's analysis of the sciences of design as sciences of complexity, but he is more interested in the structure of the complex systems than in the dynamic complexity. Cf. Simon (1996).

and hierarchical); and (c) functional complexity (operational and nomic) [Rescher 1998b, p. 9].[49] With this collection of the epistemic and ontological modes of complexity, Rescher offers a framework of complexity that is certainly more systematic and broader than other approaches.

His analysis of ontological complexity can be used to address the problem of scientific prediction. In this regard, when scientific prediction is analyzed from an ontological perspective, Rescher's proposal on the ontological modes of complexity is certainly of philosophical relevance because both the reality predicted and the reality used for prediction could be complex realities. So the study of this ontological complexity (that has repercussions on the epistemological and methodological levels) can have a key role in characterizing the prediction to be made. On the one hand, it allows us to report some obstacles to predictability; and, on the other, it can help to overcome those obstacles.

Thus, Rescher's proposal on complexity is of real interest, but is excessively focused on the structural complexity and with scarce attention to the dynamic trait. A broader approach to complexity should, I believe, pay attention to dynamic complexity too, which is connected with the notion of historicity. This is especially clear when scientific prediction is considered, since it is oriented towards a possible future, so has to deal with the changes that occur in systems over time.[50]

7.4.2   Ontological Modes of Complexity and Their Incidence on Prediction

A system can be complex due to its own ontological characteristics. This issue underlies Rescher's identification of the "modes of complexity." His analysis is focused on the epistemic modes of complexity (those related to knowledge) and the ontological modes of complexity (that are related to reality itself). The *methodological modes* of complexity can be added to his proposal, since the complexity related to reality and to knowledge has repercussions on the scientific processes (for example, in the case of the predictive models).

All these modes of complexity (epistemic, ontological, and methodological) have repercussions on science, in general, and on scientific prediction, in particular. In this regard, within a pragmatic idealism such as that defended by Rescher, ontological complexity is especially relevant, insofar as it is the support for the epistemological level (the achievable knowledge) and the methodological side (the scientific processes). From an ontological perspective—which has to do with the reality itself researched (natural, social or artificial)—a system can be complex (for

---

[49] He points out just two possible options within the three ontological modes, but they can be enlarged if we take into account dynamic aspects.

[50] This is relevant not only for the social sciences and the sciences of the artificial, but also for natural sciences such as biology. See Gonzalez (2015b).

him) in accordance with three modes of complexity: (a) compositional complexity; (b) structural complexity; and c) functional complexity (Rescher 1998b, p. 9).

Compositional complexity is related to the elements that compound the system and the variety of those elements. Thus, Rescher distinguishes two forms of complexity related to system's composition: *constitutional complexity* and *taxonomic complexity* or *heterogeneity*. Constitutional complexity is associated with the number of elements that compound a system, while taxonomic complexity or heterogeneity has to do with the variety of those elements (1998b, p. 11).

Structural complexity has in Rescher a restricted sense, since it is circumscribed to the organization and relation between the elements of the system, instead of encompassing all the structural factors of complexity. Within what he calls "structural complexity," there are also two possibilities: *organizational complexity* and *hierarchical complexity*. The former is related to the forms of organization that can be seen in the system (i.e., it has to do with the possible forms of interrelations between its components); the latter has to do with the possible ways of relating the elements of the system (Rescher 1998b, pp. 11–12).

Meanwhile, functional complexity is related to the behavior of the system that clearly involves a temporal dimension. It is diversified, in turn, in two modes of complexity: *operational complexity* and *nomic complexity*. Operational complexity studies the variety of the modes of operation and the types of functioning that appear in the system, an issue that he sees in terms of processes. Nomic complexity is related to the patterns that regulate the relations between the elements of the system (Rescher 1998b, pp. 12–13).[51]

Through this framework about the ontological complexity (which pays attention to three features: the composition, the structure, and the function of the complex systems) Rescher highlights that, when we try to encompass the complex reality of a complex system (natural, social or artificial) there are a wide variety of factors to take into account. Thus, the ontological complexity of a system means that we have to take into account the components (their number and variety), the structure (regarding the organization and the relations between the components), and the functions (the types of operation and the working patterns).

From this perspective, it can be clearly seen that prediction of a complex system with ontological complexity has to deal with a wide variety of factors. This is because, although a system does not need to be complex in all the aforementioned senses, "the different modes of complexity do tend to run together in practice" (Rescher 1998b, p. 15). Thus, when a system is complex with regard to its composition and structure, it is usually complex regarding its function too (Rescher 1998b, p. 15). Certainly, this makes it difficult to predict, and has repercussions on issues such as the reliability or the degree of accuracy and precision that can be achievable.

---

[51] This functional complexity involves some kind of openness to a teleological component insofar as the operations are oriented to ends. But this aspect, as well as its dynamic consequences, is not the focus of attention of Rescher's approach to complexity.

But it happens that, regarding complex systems, there are still more factors to take into account than just the epistemic and ontological modes of complexity. Thus, although Rescher's approach to complexity is really detailed, his attention is mainly on the structural dimension of complexity (understood in a broader sense, which also encompass the compositional complexity). For that reason, although his approach is open to the dynamic trait—which has to do with the changes that occur in the system over the time—it is a perspective that, in my judgment, is not exhaustively developed in Rescher.

In effect, when dynamic complexity is considered, we must take into account, initially, two dimensions: the "internal" dynamics and the "external" dynamics (Gonzalez 2012a, pp. 13–14). The internal dynamics is related to the activity of the system itself, whereas the external dynamics deals with the relations to the environment. Rescher pays special attention to the "internal" dynamics of the complex system, which he considers in accordance with the notion of "process."[52] But he does not pay attention *de facto* to the "external" dynamics, where there are sources of complexity regarding the relations of the system with the environment.

In this regard, the concept of "historicity" has been proposed to characterize the change in the complex dynamics ("internal" and "external") and its repercussions on scientific prediction (Gonzalez 2012b, pp. 79–88). It is a notion that goes further than the mere temporality, since it is difficult to think of a reality that escapes from the temporal dimension. However, historicity is a characteristic of the human realm: it links with human activity that is *eo ipso* historic, in the sense that it changes over time, and some changes are particularly relevant (Gonzalez 2003, p. 88). In addition, historicity is also in the context that surrounds human activity, which can have relevant changes as well.

Thus, historicity is related to three levels of analysis: science, the agents, and the reality itself that is researched. This was pointed out by Wenceslao J. Gonzalez: "(1) Historicity (*Geschichtlichkeit/ historicidad*) is a trait of science, in general, and each science, in particular. This facet can be found in the whole set of constitutive elements of science, such as language, structure, knowledge, method, activity, ends, and values. (2) Historicity configures the agents themselves involved in the development of scientific research, insofar as they are human beings within a historical context. 3) Historicity is a characteristic of the reality itself that is researched (above all, in the social and artificial realms)" (2011a, p. 43).

When these three levels of analysis are considered, it seems clear that the notion of *historicity* is relevant when the problems of complexity, in general, and its repercussions on scientific prediction, in particular, are considered. Firstly, historicity is a feature that modulates science. This happens at three successive levels: the constitutive elements of science (that lead to the structural dimension of complexity), the configuration of the aims, processes, and results (the internal dynamics of scientific activity), and the relations with the changing environment (the external dynamics).

Secondly, historicity configures the agents who develop the scientific activity, who are in a concrete socio-historical context. This feature can add more complexity

---

[52] On the notion of "process" in Rescher, see Gonzalez (2012b), pp. 79–80.

(for instance, through human creativity, both individual and social). Thirdly, historicity can be a feature of the reality itself studied by science, above all when it is a social or artificial reality. From this perspective, historicity can be the basis to understand why prediction is more difficult in social sciences and the sciences of the artificial, where the complex systems are historical.

7.4.3 Complexity and Historicity in the Social Sciences and the Sciences of the Artificial

A key issue for the reflection on social sciences and the sciences of the artificial is the problem of complexity. Even more, these sciences can be analyzed as sciences of complexity, both with respect to their structure and with regard to their dynamics (internal and external). In effect, they are sciences with structural complexity. This can be seen in relation to their constitutive elements, such as language, internal articulation, knowledge, method, activity, ends, and values (Gonzalez 2012b, p. 74). Additionally, they have a complex dynamics (internal and external), above all when they are configured as applied sciences (Gonzalez 2012a, pp. 8–9). Thus, in their internal dynamics, there is an articulation of aims and processes, which generate a result. At the same time, they involve an external dynamics, since their activity is developed in a changing environment.

Both the structural dimension and the dynamic trait of complexity in social sciences and the sciences of the artificial are modulated by historicity. This is a feature that they have in common with the natural sciences, since every science is the result of the human activity, which is historic in itself. But historicity is also an ontological feature of the reality researched by social sciences and the sciences of the artificial, since they are disciplines that study the social reality and the human-made realm, respectively. The social ontology and the artificial ontology are both under the influence of historicity.

From this perspective, it seems clear that there is also a contextual component in the scientific undertakings. "Because the agents—individual and social—act usually with regard to contexts (historical, cultural, political, …), which are changing" (Gonzalez 2012a, p. 14). So historicity allows us to understand the huge variability of the reality (above all, social and artificial), because it offers the appropriate framework to address factors like human creativity or the choices of the agents that are a source of complexity for predictions in social sciences and sciences of the artificial.

It is usual to consider that prediction in social sciences is more difficult than in natural sciences, both in term of possibility and in terms of results. At the same time, it is also usual to think that this difficulty of social prediction is because there is more complexity in the social systems (or in the artificial systems) than in the natural systems (Gonzalez 2011b, pp. 319–321). Thus, from a structural viewpoint, more factors might intervene in them, with more complex interactions; and, from a dynamic perspective, the variability is usually higher than in natural systems.

However, the studies on complexity, both in social sciences and in the sciences of the artificial, do not take usually into account *historicity*. Commonly, they focus on notions like "evolution" or "process," which do not achieve the depth that historicity achieves (Gonzalez 2012b, pp. 79–84). In my judgment, a more sophisticated analysis of complexity (structural and dynamic) can be developed through historicity, above all in the social and artificial realms. In the case of economics, this has been highlighted by Gonzalez. Economics is a dual science (it is a social science and also a science of the artificial), so it involves features of both groups of disciplines.

Gonzalez considers that, when the problem of complexity in economics and its repercussions on economic prediction is analyzed, there is an ontological duality that must be taken into account in its configuration. This ontological duality is the differentiation between "economic activity" and "economics as activity" (Gonzalez 1994). It is a distinction that involves a relevant source of complexity for economics, as a human endeavor imbued with historicity. This complexity adds difficulty to economic prediction, which has to take into account the duality of components "economic activity" and "economics as activity."

On the one hand, *economic activity* can be understood, in principle, as autonomous regarding other human activities, since it has its own characteristics. Thus, it involves the activities carried out by human beings, which include supply and demand, the transactions of good and services, etc. On the other hand, *economics as activity* is a human endeavor that is related with other human activities (cultural, political, social, ecological, etc.). Commonly, it is interconnected with them in many ways. From this perspective, there are relations with the environment, so economics as human activity—which is interrelated with other activities—is not independent of the context (Gonzalez 2012b, pp. 86–87).

It happens that, through *economic activity*, "prediction has to deal with forms of complexity that are genuinely of an economic character" (Gonzalez 2012c, p. 146). At the same time, *economics as activity* "involves that scientific prediction should consider a series of different types of complexity (social, political, cultural, ecological, etc.), that certainly have repercussions on the economic variables" (Gonzalez 2012c, p. 146). In both cases, historicity is a key feature, since economic reality cannot be understood without taking into account the historicity of human activity, which is developed in a social environment.

Therefore, the duality "economic activity" and "economics as activity" means that "the predictability of economic activity—which is, in principle, autonomous—is possible, and the prediction could be reliable; whereas predictability of economics as a human activity among others appears more unreliable, precisely due to the interdependence with other activities" (Gonzalez 2012b, p. 87). But accepting the historicity of the economic reality does not involve, certainly, neglecting its objectivity. Both "economic activity" and "economics as activity" are objective realities. However, whereas economic activity is, in principle, measurable, economics as activity is more difficult to measure, so it is not usually included in the neoclassical econometrical models (Gonzalez 2012b, p. 92).

With regard to the difficulties to predict in social sciences, in general, and in economics, in particular, Rescher insists on the fact that those difficulties are mainly rooted in the presence of cognitive factors. This is because social prediction, *prima facie*, is about the acts and choices of the agents, and it happens that agents do not make decisions on the basis of the reality of the world, but on the basis of what they *think* of that reality. In this way, the acts and choices of the agents depend on factors such as their beliefs, ideas, etc., which are variable.[53] This means that there is a subjective component, and social prediction has to deal with it.

However, if the distinction "economic activity" and "economics as activity" is taken into account, it seems that there are even more sources of complexity that affect economic prediction, so there are more factors than those contemplated by Rescher. This situation, in my judgment, allows us to understand the higher difficulty of economic prediction in comparison with prediction in some natural sciences such as physics. In addition, this distinction—which has as it starting point the acknowledgement of the historicity that configures the social and artificial reality—can be applied to other dual sciences, such as the communication sciences (see, for example, Arrojo 2012).

All these considerations allow us to see how the analysis of complexity from the historicity of human activity can shed light on the problem of scientific prediction, in general, and prediction in the social sciences and the sciences of artificial, in particular. Certainly, Rescher's approach to the epistemic and ontological modes of complexity allows us to understand some of the methodological difficulties in prediction. But this approach should be incorporated into a broader account of complexity. Thus, we have to consider two aspects of complexity: the structural dimension and the dynamic trait (internal and external). In order to do this, historicity should, in my judgment, be explicitly acknowledged, since it is a key feature of science, the agents, and the reality itself that is researched (above all, social and artificial).

References

Arrojo, M. J. 2012. La sobriedad de factores en el análisis de la complejidad en las Ciencias de la Comunicación. El estudio de la Televisión Digital Terrestre. In *Las Ciencias de la Complejidad: Vertiente dinámica de las Ciencias de Diseño y sobriedad de factores*, ed. W.J. Gonzalez, 337–358. A Coruña: Netbiblo.

Betz, G. 2006. *Prediction or prophecy? The boundaries of economic foreknowledge and their socio-political consequences.* Wiesbaden: Deutscher Universitäts-Verlag.

Boden, M.A. 2004. *The creative mind. Myths and mechanisms*, 2nd ed. London: Routledge (1st ed., 1990).

---

[53] Cf. Rescher, *Personal communication*, 10.6.2014.

Chu, D., R. Strand, and R. Fjelland. 2003. Theories of complexity. Common denominators of complex systems. *Complexity* 8 (3): 19–30.

Cilliers, P. 2007. Why we cannot know complex things completely. In *Reframing complexity. Perspectives from the North and South. Exploring complexity*, ed. F. Capra, A. Juarrero, P. Sotolongo, and J. van Uden, vol. I, 81–89. Mansfield: ISCE.

Csikszentmihalyi, M. 1996. *Creativity. Flow and the psychology of discovery and invention.* New York: Harper/Collins (reprinted in 1997).

Dasgupta, S. 1994. *Creativity in invention and design.* Cambridge: Cambridge University Press.

Dummett, M. 1954. Can an effect precede its cause? *Proceedings of the Aristotelian Society* 28: 27–44.

———. 1964. Bringing about the past. *Philosophical Review* 73: 338–359.

Firth, M. 1977. *Forecasting methods in business and management.* London: E. Arnold.

Gonzalez, W.J. 1986. *La Teoría de la Referencia. Strawson y la Filosofía Analítica.* Salamanca/ Murcia: Ediciones Universidad de Salamanca and Publicaciones de la Universidad de Murcia.

———. 1993. El realismo y sus variedades: El debate actual sobre las bases filosóficas de la Ciencia. In *Conocimiento, Ciencia y Realidad*, ed. A. Carreras, 11–58. Zaragoza: Seminario Interdisciplinar de la Universidad de Zaragoza-Ediciones Mira.

———. 1994. Economic prediction and human activity. An analysis of prediction in economics from action theory. *Epistemologia* 17 (2): 253–294.

———. 2003. Racionalidad y Economía: De la racionalidad de la Economía como Ciencia a la racionalidad de los agentes económicos. In *Racionalidad, historicidad y predicción en Herbert A. Simon*, ed. W.J. Gonzalez, 65–96. A Coruña: Netbiblo.

———. 2005. The philosophical approach to science, technology and society. In *Science, technology and society: A philosophical perspective*, ed. W.J. Gonzalez, 3–49. A Coruña: Netbiblo.

———. 2008a. Rationality and prediction in the sciences of the artificial. In *Reasoning, rationality, and probability*, ed. M.C. Galavotti, R. Scazzieri, and P. Suppes, 165–186. Stanford: CSLI Publications.

———. 2008b. El enfoque cognitivo en la Ciencia y el problema de la historicidad: Caracterización desde los conceptos. *Letras* 114 (79): 51–80.

———. 2010. *La predicción científica. Concepciones filosófico-metodológicas desde H. Reichenbach a N. Rescher.* Montesinos: Barcelona.

———. 2011a. Conceptual changes and scientific diversity: The role of historicity. In *Conceptual revolutions: From cognitive science to medicine*, ed. W.J. Gonzalez, 39–62. A Coruña: Netbiblo.

———. 2011b. Complexity in economics and prediction: The role of parsimonious factors. In *Explanation, prediction, and confirmation*, ed. D. Dieks, W.J. Gonzalez, S. Hartman, Th. Uebel, and M. Weber, 319–330. Dordrecht: Springer.

———. 2012a. Las Ciencias de Diseño en cuanto Ciencias de la Complejidad: Análisis de la Economía, Documentación y Comunicación. In *Las Ciencias de la Complejidad: Vertiente dinámica de las Ciencias de Diseño y sobriedad de factores*, ed. W.J. Gonzalez, 7–30. A Coruña: Netbiblo.

———. 2012b. La vertiente dinámica de las Ciencias de la Complejidad. Repercusión de la historicidad para la predicción científica en las Ciencias Diseño. In *Las Ciencias de la Complejidad: Vertiente dinámica de las Ciencias de Diseño y sobriedad de factores*, ed. W. J. Gonzalez, 73–106. A Coruña: Netbiblo.

———. 2012c. Complejidad estructural en Ciencias de Diseño y su incidencia en la predicción científica: El papel de la sobriedad de factores (*parsimonious factors*). In *Las Ciencias de la Complejidad: Vertiente dinámica de las Ciencias de Diseño y sobriedad de factores*, ed. W. J. Gonzalez, 143–167. A Coruña: Netbiblo.

———. (ed.). 2012d. *Freedom and determinism: Social sciences and natural sciences.* Monographic issue of *Peruvian Journal of Epistemology*, vol. 1. Cambre : Imprenta Mundo.

———. 2013a. The roles of scientific creativity and technological innovation in the context of complexity of science. In *Creativity, innovation, and complexity in science*, ed. W.J. Gonzalez, 11–40. A Coruña: Netbiblo.

————. 2013b. The sciences of design as sciences of complexity: The dynamic trait. In *New Challenges to Philosophy of Science*, ed. H. Andersen, D. Dieks, W.J. Gonzalez, Th. Uebel, and G. Wheeler, 299–311. Dordrecht: Springer.

————. 2014. The evolution of Lakatos's repercussion on the methodology of economics. *HOPOS: The Journal of the International Society for the History of Philosophy of Science* 4 (1): 1–25.

————. 2015a. *Philosophico-methodological analysis of prediction and its role in economics.* Dordrecht: Springer.

————. 2015b. Prediction and prescription in biological systems: The role of technology for measurement and transformation. In *The future of scientific practice: "Bio-techno-logos"*, ed. M. Bertolaso, 133–146 and 209–213. London: Pickering and Chatto.

————. 2016. Rethinking the limits of science: From the difficulties for the frontiers to the concern on the confines. In *The limits of science: An analysis from "barriers" to "confines,"* Poznan studies in the philosophy of the sciences and the humanities, ed. W.J. Gonzalez, 3–30. Leiden: Brill.

Granger, C. W. J. 1989. *Forecasting in business and economics*, 2nd ed.. San Diego: Academic (1st ed., 1980).

Guillán, A. 2013. Analysis of creativity in the sciences of design. In *Creativity, innovation, and complexity in science*, ed. W.J. Gonzalez, 125–139. A Coruña: Netbiblo.

————. 2016a. The limits of future knowledge: An analysis of Nicholas Rescher's epistemological approach. In *The limits of science: An analysis from "barriers" to "confines"*, Poznan studies in the philosophy of the sciences and the humanities, ed. W.J. Gonzalez, 134–149. Leiden: Brill.

————. 2016b. The obstacles to scientific prediction: An analysis of the limits of predictability from the ontology of science. In *The limits of science: An analysis from "barriers" to "confines"*, Poznan Studies in the Philosophy of the Sciences and the Humanities, ed. W.J. Gonzalez. Leiden: Brill.

Hacking, I. 1983. *Representing and intervening.* Cambridge: Cambridge University Press.

Mitchell, M. 2009. *Complexity: A guided tour.* Oxford: Oxford University Press.

Neira, P. 2012. Complejidad en Ciencias de la Comunicación debido a la racionalidad: Papel de la racionalidad limitada ante la creatividad e innovación en Internet. In *Las Ciencias de la Complejidad: Vertiente dinámica de las Ciencias de Diseño y sobriedad de factores*, ed. W.J. Gonzalez, 205–230. A Coruña: Netbiblo.

Niiniluoto, I. 1984. *Is science progressive?* Dordrecht: Reidel.

————. 1993. The aim and structure of applied research. *Erkenntnis* 38 (1): 1–21.

————. 1995. Approximation in applied science. *Poznan Studies in the Philosophy of Sciences and the Humanities* 42: 127–139.

Rescher, N. 1978. Scientific progress. In *A philosophical essay on the economics of the natural science*. Oxford: Blackwell.

————. 1987. *Scientific realism.* Dordrecht: Reidel.

————. 1990. Our science as o-u-r science. In *A useful inheritance. Evolutionary aspects of the theory of knowledge*, N. Rescher, 77–104. Savage: Rowman and Littlefield.

————. 1992a. *A system of pragmatic idealism. Vol. I: Human knowledge in idealistic perspective.* Princeton: Princeton University Press.

————. 1992b. Realism and idealism. In *A system of pragmatic idealism. Vol. I: Human knowledge in idealistic perspective*, N. Rescher, 304–327. Princeton: Princeton University Press.

————. 1995. Predictive incapacity and rational decision. *European Review* 3 (4): 327–332.

————. 1997. *Objectivity. The obligations of impersonal reason.* Notre Dame: University of Notre Dame Press.

————. 1998a. *Predicting the future. An introduction to the theory of forecasting.* New York: State University of New York Press.

————. 1998b. *Complexity: A philosophical overview.* N. Brunswick: Transaction Publishers.

————. 1998c. Objectivity and communication. How ordinary discourse is committed to objectivity. In *Communicative pragmatism and other philosophical essays on language*, N. Rescher, 85–97. Lanham: Rowman and Littlefield.

———. 1999a. *Razón y valores en la Era científico-tecnológica*. Barcelona: Paidós.

———. 1999b. *The limits of science*, revised edition. Pittsburgh: University of Pittsburgh Press.

———. 2005a. *Realism and pragmatic epistemology*. Pittsburgh: University of Pittsburgh Press.

———. 2005b. *Reason and reality. Realism and idealism in pragmatic perspective*. Lanham: Rowman and Littlefield.

———. 2009a. *Free will: A philosophical reappraisal*. New Brunswick: Transaction Publishers.

———. 2009b. *Ignorance. On the wider implications of deficient knowledge*. Pittsburgh: University of Pittsburgh Press.

———. 2009c. *Unknowability. An inquiry into the limits of knowledge*. Lanham: Lexington Books.

———. 2010. *Reality and its appearance*. London: Continuum.

———. 2012a. The problem of future knowledge. *Mind and Society* 11 (2): 149–163.

———. 2012b. Political pragmatism. In *Pragmatism. The restoration of its scientific roots*, N. Rescher, 205–215. New Brunswick: Transaction Publishers.

Sen, A. 1986. Prediction and economic theory. In *Predictability in science and society*, ed. J. Mason, P. Mathias, and J.H. Westcott, 103–125. London: The Royal Society and The British Academy.

Simon, H.A. 1990. Prediction and prescription in systems modeling. *Operations Research* 38: 7–14. Compiled Simon, H.A. 1997. *Models of bounded rationality. Vol. 3: Empirically Grounded Economic Reason*, 115–128. Cambridge, MA: The MIT Press.

———. 1991. *Models of my life*. New York: Basic Books (reprinted in Cambridge, MA: The MIT Press, 1996a).

———. 1995. Problem forming, problem finding, and problem solving in design. In *Design and systems: General applications of methodology*, ed. A. Collen and W.W. Gasparski, vol. 3, 245–257. New Brunswick: Transaction Publishers.

———. 1996. *The sciences of the artificial*, 3rd ed. Cambridge, MA: The MIT Press (1st ed., 1969; 2nd ed., 1981).

———. 2001. Science seeks parsimony, not simplicity: Searching for pattern in phenomena. In *Simplicity, inference and modeling. Keeping it sophisticatedly simple*, ed. A. Zellner, H.A. Keuzenkamp, and M. McAleer, 32–72. Cambridge: Cambridge University Press.

———. 2002. Forecasting the future or shaping it? *Industrial and Corporate Change* 11 (3): 601–605.

Sprott, J. C. 2006. Complex behavior of simple systems. In *Unifying themes in complex systems, Vol. IIIB, New research*, ed. A.A. Minai and Y. Bar-Yam, 2–11. Proceedings of the Third International Conference on Complex Systems. Berlin/Heidelberg/New York: Springer.

Chapter 8
Axiological Elements of Scientific Prediction

Abstract The axiological elements of scientific prediction are relevant. In this regard, this chapter follows several steps. (1) Nicholas Rescher's proposal regarding the axiology of research is considered in order to see how he modulates the axiological features of prediction. Thus, values are seen as a system, within a double perspective of analysis: internal and external. The internal perspective sees science as activity in itself; while the external viewpoint deals with the relations of science with the context. (2) The axiological characters of prediction are investigated. This is twofold: on the one hand, the research in prediction as a value of science is addressed; and, on the other, the values which accompany prediction are analyzed.

Regarding these problems, there are two dimensions of analysis: the structural perspective and the dynamic component. Rescher's proposal is preferentially structural. In this way, it is possible to broaden his proposal through attention to the dynamic component. Consequently, the study of how prediction and the connected values modulate the aims, processes, and results of the scientific research (basic, applied, or of application) is required, from both an internal and external perspective.

Keywords Axiology of science • Scientific prediction • Prescription • Basic science • Applied science • Application of science • Historicity

Axiology of scientific research is one of Rescher's main interests in the realm of the philosophy of science. This can be seen in the number and quality of his work in this thematic field. Thus, the second of his three volumes devoted to *A System of Pragmatic Idealism* (1993a), where he seeks to make the coordinates of his system of thought explicit, has as its subtitle *The Validity of Values: Human Values in Pragmatic Perspective*. This is relevant, since the book is entirely focused on the reflection on human values, in general, and scientific values, in particular. Furthermore, the importance of the axiology of research can be seen when Rescher addresses the problem of scientific prediction, since axiological elements are especially relevant in this topic (1998a, ch. 7, pp. 113–131).

A. Guillán, *Pragmatic Idealism and Scientific Prediction*, European Studies in Philosophy of Science 8, DOI 10.1007/978-3-319-63043-4_8

To carry through the analysis of the axiological elements of scientific prediction, it should be noted that axiology of research is twofold: it has a structural and a dynamic trait. When the structural trait is addressed, the attention is focused on the constitutive elements of science (language, structure, knowledge, method, etc.), so above all the internal side of analysis is stressed (Gonzalez 2005). Thus, when the study of the axiological elements of scientific activity is made from a structural dimension, the focus is usually on the values (semantic, logical, epistemological, methodological, etc.) that modulate scientific activity *as such*.

Meanwhile, the perspective changes when the dynamic trait is considered: the *historical character* of scientific activity—which is modulated by several factors— is then emphasized. This activity involves, on the one hand, paying attention to the aims, processes, and results ("internal" dynamics) [Gonzalez 2013a] and, on the other, to the relations with the context—social, cultural, politic, economic, etc.—, which is changeable ("external dynamics") [Gonzalez 2012].

In turn, aims, processes, and results—and the connected values—change in accordance with the type of activity: basic science, applied science, or application of science.[1] In this way, when values are contemplated from the dynamic trait, external values of science—i.e., the values that modulate science as an activity related to other human activities (social, cultural, politic, economic, etc.)—can be seen more clearly.[2]

Within this thematic framework about values, which emphasizes the twofold character of the axiology of science—the structural dimension and the dynamic trait—and the double perspective of analysis—the internal aspect and the external viewpoint—Rescher's axiology of science defends a *holism* of values. He suggests this holism from the idea of a system, so the different values (internal and external) form a network of interdependences. Due to the attention to values as a system, his axiological approach stresses the structural dimension, where the internal values (especially, epistemological and methodological) have primacy.

But, to the extent that his axiology is holistic, the dynamic trait should be also considered, which deals with the aims, processes, and results ("internal" dynamics), as well as to the relations to the changing context ("external" dynamics). However, because Rescher has little interest in the *historicity* of human activity, he does not develop the dynamic trait of the relation between science and values. This is clear when his attention goes to the axiology of scientific prediction; because, for him, prediction is mainly a cognitive content with methodological import, so the structural factors have primacy (especially, the epistemological and methodological features).

Within these coordinates, this chapter has two main aims: (1) to clarify the framework of Rescher's axiology of science; and (2) to offer a critical analysis of his axiology of scientific prediction. To do this, several steps are followed. First, the research

---

[1] On the differences between basic science and applied science with regard to the aims, processes, and results, see Gonzalez (1999b, pp. 139–171; especially, pp. 158–159). On the distinction between applied science and the application of science, see Niiniluoto (1993) and Gonzalez (2013b, pp. 17–18). See also Chap. 5, Sect. 5.2 herein.

[2] This happens, especially, when the research is oriented towards applied science or the application of science, because the nexus with the context (social, cultural, economic, etc.) is clearer than in the case of basic science.

is oriented towards his holistic conception of values, which sees science as a human activity. This is related to his approach to rationality, which includes an evaluative rationality or rationality of ends, so scientific rationality is not merely an instrumental rationality (Rescher 1988). Secondly, his approach to the *system* of values of science is stressed, where the internal component has primacy. In this way, it can be seen how the main values are, for Rescher, epistemological and methodological values.

Thirdly, the attention is focused on the axiological elements of scientific prediction. In this regard, it can be pointed out that axiological perspective of scientific prediction is initially twofold: (a) the value of prediction as such; and (b) the values that are or might be characteristic of scientific prediction. In this case, there is a duality that should be considered—structural and dynamic—that has to do with the values of prediction.

On this basis, the analysis offered about the axiological elements of scientific prediction takes into account three aspects: (I) prediction as a *value* of science, that is related to the problem of predictivism from an axiological viewpoint; (II) the values of scientific prediction from the *structural* perspective, which is the dominant viewpoint in Rescher's approach; and (III) the values of prediction from the *dynamic* point of view, which allows us to broaden his axiological proposal through attention to historicity.

8.1   Rescher's Axiology as a System of Science and Values

As the background of his "pragmatic Kantism," Rescher defends a holism of values, which is related to the idea of a *system* (1993b, c). On the one hand, there is in his approach a clear predilection for the internal values of science (which he sees above all from the structural perspective). But, on the other hand, he develops a *broad* axiology of research (Gonzalez 2008b). Thus, even when it is certainly possible to establish distinctions regarding values (for example, internal values in contrast to external values), strictly speaking, a complete separation between them is not possible (Gonzalez 1999a, p. 22). This is the case because, *de facto*, values of science (internal and external) are related to the full set of human values and, for Rescher, they are rooted in human needs, which are of universal character. Hence he considers that human values, in general, and scientific values, in particular, are objective (Rescher 1999a, pp. 74–76).

Together with the objectivity of values, as based in human needs, Rescher develops an approach that implies that scientific rationality is not merely an instrumental rationality or rationality of means, but rather it also includes an evaluative rationality or rationality of ends (Rescher 1988). Such an approach to the axiology of research allows us to overcome proposals of instrumental reductionism, where values just have a role for the choice and evaluation of the means with regard to given ends. Moreover, Rescher's proposal contributes to clarifying the relations of scientific rationality with technological rationality, where the instrumental component usually has a more prominent role.

8.1.1 A Holistic View of Values as a Human Activity

Rescher's axiology of science can be characterized as a combination of pragmatic elements (the primacy of practice), Kantian features (the coordination of a system), and realist aspects (the emphasis on objectivity). In this way, his axiology of scientific research is in the same direction as his philosophy and methodology of science, which configures a system of pragmatic idealism open to elements of realism. Thus, his axiology is pragmatic insofar as he sees scientific activity as a practice connected to the entire human experience. This leads him to a holism of values, which is linked to his Kantism, where values configure a system. Moreover, the most important values are, in his judgment, the internal values to science (above all, epistemological and methodological values).

This "pragmatic Kantism" that modulates his axiology of scientific research is rooted in a realist basis which appeals to the objectivity of values, so Rescher considers that the rational deliberation about evaluative issues is possible.[3] This objectivity leads him to distinguish between values rooted in human needs and values of optional character. Furthermore, objectivity leads to the idea of "impartiality;"[4] that is, an unbiased or not particular character, open to a universal feature (i.e., present in the diverse human agents).

The idea that human values, in general, and the values of science, in particular, are related to *human activity* is Rescher's starting point. In this way, values "do not need an 'isolated world'—in the Platonic way—in order to base them, neither are they about a transcendental subject or a pure consciousness, but [they are] about people" (Gonzalez 1999a, p. 11). This attention to human activity as the axis of his axiology of science leads him to a holistic view of values. Because scientific activity can be thought as a human activity *among others*, so it is not isolated from other realms of human experience (social, cultural, economic, ecological, etc.).

In effect, "Rescher conceives scientific goals (mainly, cognitive ones) as related to the rest of our goals (social, cultural, economic, etc.). Thus, besides the teleological character of science, this viewpoint insists on science as a human undertaking in a contextual setting rather than in a purely cognitive project or an isolated doing. In other words, science belongs to a human network" (Gonzalez 2008b, p. 90). From this perspective, science is modulated by a plurality of values (internal and external) that configure a *system*, so there is an interdependence network among them. But the idea of a system is compatible with a hierarchy of values. Once again the Kantian influence on Rescher's thought is considerable, since, among the values of the system, the internal values (mainly, epistemological and methodological) are the most important ones.

---

[3] "Value issues should also be seen in a 'realistic' light. Matters of value too can and should be regarded as objectively factual—the difference is just that we are dealing with *evaluative* rather than simply *informative* facts," Rescher (2014b, p. 77).

[4] On his approach to objectivity, see Rescher (1997).

Besides the holism of values—which is compatible with an articulation that gives priority to internal values—the objective character of Rescher's axiology should be highlighted. This is another main aspect in his characterization of values, in general, and the values that modulate scientific activity, in particular. For him, values are not a merely subjective or intersubjective matter, but the rational deliberation about the *validity* of values is possible. This means that, in contrast to the mere "tastes" or "preferences," the notion of "value" is linked to what is *preferable*—not simply preferred—in an objective way (Rescher 1999a, pp. 74–76 and 90–93). In his judgment, the objectivity of values (scientific and non-scientific) has an ontological basis that is rooted in human needs.[5]

These coordinates that articulate Rescher's proposal about the relation between science and values allow him to configure a *broad* axiology of research, where the internal dimension and the external component are interrelated. Thus, even when he shares with authors such as Larry Laudan a pragmatic view of knowledge and certain predilection for the internal values (above all, the cognitive ones),[6] Rescher's proposal certainly goes further than the simple attention to the role of the cognitive values in the configuration of the aims of research.[7] This greater richness can be seen in both the internal and external aspects.

On the one hand, the holism of values maintained by Rescher involves assuming that axiology of research can follow, in principle, two different (although connected) thematic orientations: the internal and the external. From the internal perspective, the attention goes to the values of science by itself (cognitive values, methodological values, …), while, from the external perspective the values studied are those that go with science insofar as it is a human activity connected with other human activities (social, cultural, economic, ecological, etc.) [Gonzalez 1999b, pp. 148–149].

Moreover, Rescher's proposal assumes that values (internal and external) have a role in science: firstly, they have to do with the selection of the *aims* of the scientific activity; and, secondly, they have incidence on the selection of the most adequate *means* to achieve those aims. Thirdly, values intervene in the evaluation of the *results* of the research (and their possible consequences) [Gonzalez 2008b, pp. 93–96]. All

---

[5] "Our values themselves are not—and should not be—arbitrary and haphazard. For in the final analysis, they pivot not on mere wants and the vagaries of arbitrary choice in fortuitous preference, but on our best interests and real needs—on what is necessary to or advantageous for a person's well-being. (…) For the fact is that values are valid just exactly to the extent they serve to implement and satisfy our needs and our correlatively appropriate interests," Rescher (2012b, p. 38). On human needs as the support of objectivity of values, see Rescher (1999a, pp. 85–90).

[6] Both Laudan and Rescher were students of Hempel. Furthermore, Rescher claims that both he and Laudan are part of the "younger generation" of the legacy of the Berlin School. Cf. Rescher (2006). The influence of the received view in both authors can be seen in the fact that, when they think of the relation between science and values, internal criteria prevail, above all, cognitive criteria. However, Laudan was also influenced by Thomas Kuhn and especially by Imre Lakatos, so his concern for the historicity of science is much higher than Rescher's.

[7] In effect, Laudan's proposal about the relation between science and values is focused on cognitive values and their role with regard to the aims of research. Cf. Laudan (1984). In this regard, see Wenceslao J. Gonzalez's comparison between the axiology of scientific research of Laudan and Rescher's approach: Gonzalez (2008b).

these considerations have repercussions for the characterization of scientific rationality. Thus, Rescher considers that the deliberation on *value* matters is a rational deliberation that has objective bases. He maintains that human rationality, in general, and scientific rationality, in particular, is not only an instrumental rationality or rationality of means, but that there is also an evaluative rationality or rationality of ends, which deals with what we should estimate or value.[8]

8.1.2   From the Internal Dimension to the External Aspect: Importance of Evaluative Rationality and Relation to Technology

Rescher thinks that values are mainly related to the aims of scientific activity. In his axiology of research, "values have a central role in science and (…) this role is not something arbitrary or added, but it is inherent to the goal structure that is definitive of science as rational search" (1999a, p. 95). The means are not all that matter, because—to the extent that science is a rational activity—there should be a rational deliberation about the validity of the aims of the scientific activity. This is because, in his judgment, "true rationality calls for the pursuit of *appropriate* ends based on valid human interests, rather than following the siren call of unexamined wants or preferences" (Rescher 1988, p. 107).

In effect, he admits three types of rationality in accordance with their object of deliberation: (i) cognitive rationality, which deals with what can be accepted in the realm of knowledge; (ii) practical rationality, which is about what actions can be taken; and (iii) evaluative rationality, which has to do with what should be considered as valuable or preferable (Rescher 1988, p. 3). In the case of scientific rationality, the repercussion for the axiology of science is clear: scientific rationality is not just an instrumental rationality, because it involves a rationality of ends. This selection of the ends of scientific activity must be made in accordance with values (Rescher 1999a, pp. 93–96).

Thus, first, values have a role for science insofar as they modulate the *aims* of the research. Secondly, values intervene in the selection of the best *means* to achieve those previously selected ends. Thirdly, the achieved *results* can be valued by taking into account whether or not they meet the ends sought. In this way, the framework proposed by Rescher does not just remain at the instrumental level, because there is a selection of aims, which is prior to the rational deliberation about the means and is also before the potential results (Gonzalez 1999b, pp. 147–148).

This involves an improvement on Simon's approach to rationality as a matter that only has to do with the means (see Sect. 4.1). It is also an advancement with respect to any other account of instrumental reductionism in social sciences and in the

---

[8] These ideas are developed by Rescher in several works. Among them, they can be seen Rescher (1988, 1995, 2004, 2012c).

sciences of the artificial (such as, for example, the approach that has been dominant in neoclassical economics) [Gonzalez 2003, pp. 70–72]. Obviously, the acknowledgement of the role of values is higher in Rescher than what was accepted by instrumentalist authors in the line of Friedman (1953).

Concurrently, to the extent that scientific rationality is, for Rescher, mainly a rationality of ends, there are clear differences with regard to technological rationality. In effect, science and technology are different activities, at least when they are analyzed from a conceptual perspective. This is because science seeks a variety of aims through cognitive means, either in order to increase our knowledge (basic science) or to solve concrete problems (applied science) [Niiniluoto, 1993, 1995]. Meanwhile, technological rationality is oriented towards a creative transformation of reality in order to generate new results (usually, an artifact).[9] Consequently, "science" and "technology" are different human activities with different types of rationality, although there are frequent and intense relations between them,[10] which are especially important in the case of the sciences of design, where the level of interaction is very high.

Certainly, the differences between scientific activity and technological doing involve differences in the type of rationality. Thus, the instrumental component—the rationality of means—prevails in technological rationality, insofar as technology is a *doing that transforms reality*. From this point of view, the most important values in technology are those related to processes (efficacy, efficiency, etc.). This feature is even clearer in engineering, which is basically praxeological and, therefore, the epistemological aspect is secondary (Neira 2003, p. 151). Concurrently, the importance of the external values (social, political, economic, ecological, etc.) is clearer in the case of technology than in science, because the effect of the technological artifacts on the society and the environment is usually more direct than in the case of science (above all, in basic science).[11]

Among the features of science, it is oriented towards the search of *aims*, which are generally cognitive. These aims of knowledge usually guide the research both in formal sciences and in empirical sciences (Gonzalez 2005, pp. 10–11). In addition, these aims are modulated by values, which can be either internal (for example,

---

[9]On the differences between science and technology, see Gonzalez (2005, pp. 3–49; especially, pp. 11–12). Obviously, this conceptual difference, which has to do with the aims, processes, and results of science and technology, is compatible with the acknowledgement of a clear practical interaction, above all in certain companies, between science and technology. This intense interaction is one of the reasons that have led to think of "technoscience" for several years. There is a line of thought about this topic, where diverse senses of "technoscience" can be considered. Javier Echeverría has contributed to this realm (2001, 2003).

[10]Since Rescher is mainly focused on the natural sciences, especially when they are oriented towards basic research, the relation with technology is centered, above all, in the need for technological artifacts in order to develop the scientific activity. See, for instance, Rescher (1999a, p. 100).

[11]On values in technology, see Gonzalez (1999c). The relation between technology and values (with especial attention to the ethical values) is analyzed from diverse perspectives in the papers of the book Gonzalez (2015b).

cognitive values such as truth) or external (social, cultural, political, etc.). Once the rational analysis of those aims—they are selected according to values—is made, there is also a rational deliberation about the most suitable *processes*. The use of these processes will eventually produce a result. This attention to the aims, processes, and results involves a dynamic perspective in order to address the axiology of research. It is an approach that is contained in the axiological framework proposed by Rescher, but he does not develop it, because he gives primacy to the structural perspective.

8.2 Axiology of Scientific Research in the Pragmatic Idealism

Rescher defends a holism of values, so there is an explicit acknowledgement of the presence of values—internal and external—in scientific activity. He addresses this holism of values from the idea of a *system*, which is mainly articulated as a structural approach to the axiology of scientific research, so the dynamic trait is habitually avoided.

According to the structural approach, values are above all related to the constitutive elements of science (language, structure, knowledge, method, etc.), so the internal dimension of analysis has primacy. In this regard, the Kantism that characterizes Rescher's thought can be clearly seen, because he gives more importance to the *content* of science than to the relations with the social milieu. From this point of view, the values that prevail are, in his judgment, internal values to science; especially, the epistemological and methodological values.

8.2.1 Values as a System in the Case of Science: The Primacy of the Internal Dimension

Like many idealist authors, Rescher develops a conception of a *system*.[12] In his approach to the axiology of research, the relation between science and values can be seen from several viewpoints: (i) insofar as science is oriented towards aims such as information and truth; (ii) since there are economic values that modulate the search of those aims and that are related to the pattern of cost-benefit, so there is an "economics of research" that modulates scientific progress; (iii) from the feature of science as a social activity that is based on a process of human cooperation; (iv) in accordance with the consequences of the scientific activity, which involve the evaluation of the applications and uses of science (Gonzalez 1999a, p. 16). Each of these steps requires more attention.

---

[12] The idea of a "system," which serves to characterize the three volumes where Rescher synthesizes his contributions (1992a, 1993a, 1994), appears also in many other publications. See, for example, Rescher (1973, 1977, 1978, 1988, 1999b).

1. There is a teleological starting point in Rescher's axiology of research which consists in the characterization of the scientific activity as a *goal-oriented* activity. For him, "values have a central role in science and (…) this role is not something arbitrary or added, but it is inherent to the goal structure that is definitive of science as rational search" (Rescher 1999a, p. 95). When those goals or aims are considered, we have that "certain factors of values represent the *roles of scientific research* itself considered: concretely, effective description, explanation, prediction, and control over nature" (1999a, p. 94).[13] Therefore, these goals are valuable in themselves.

   In addition to these values of the investigation itself, there are values that modulate scientific theories (Rescher 1999a, p. 94). He gives them an *instrumental* character, in the sense of specification: their search contributes to the achievement of the aims of scientific research (description, explanation, prediction, and control). In this regard, the values he points out are cognitive ones: coherence, consistency, generality, simplicity, accuracy, precision, etc. Besides cognitive values there are also methodological values, since the values related to the standards of proof and rigor also intervene here. They specify what is worthy in the processes. Thus, they are related to the quantity and the type of proofs that are needed in order to establish a statement as a scientific claim.[14]

   Those values that go with scientific theories can be different in formal sciences and empirical sciences. In turn, within empirical sciences, values can vary in natural sciences, social sciences, or the sciences of the artificial.[15] From this perspective, although Rescher usually focuses on the natural sciences, he admits that there are differences regarding the values among the different groups of sciences. Therefore, his axiological proposal is not naturalistic. However, he does not address how those scientific values can vary over time. Thus, even when he pays attention to the context, he does not take into account historicity.

2. Rescher insists on the "economics of research." Thus, he considers that, from a methodological viewpoint, science requires an economy of means. In this way, he highlights the importance of the *economic values* in "our science." He emphasizes then values such as profitability, efficacy, efficiency, etc. "He places those values in several successive levels (cognitive, methodological, social and related to policy), that correspond with elements related to science from an internal perspective (the first two) or external (the second ones)" (Gonzalez 2010, p. 270). Therefore, these values modulate scientific research and they have a key role for scientific progress. This is because, in Rescher's conception, the cost-benefit relation is especially relevant for the analysis of scientific progress (see Sect. 1.3.2 of this book).

---

[13] In this regard, his higher interest in basic science than in applied science can be seen in the fact that he does not mention expressly *prescription* as one of the aims of scientific research.

[14] This directly affects hypotheses and, therefore, scientific theories, especially if they are contemplated from statements that convey contents regarding real things.

[15] Expressly, he mentions the case of formal sciences: "We do not ask social scientists for the same criterion of rigor that a mathematician imposes himself," Rescher (1999a, p. 94).

On the one hand, the achievement of knowledge involves a benefit that can be theoretical or practical. Thus, there is scientific progress to the extent that science increases its capability of explaining and predicting phenomena (basic science) and when it solves more concrete problems or solves them in a more efficient way (applied science). But on the other hand, there are costs that make the advancement of science difficult (Gonzalez 2008b, p. 93). Those can be due either to the increasing complexity of the studied phenomena or to an increase in the resources needed (Rescher 1998b, p. 85). Regarding this issue, Rescher highlights the relation between science (above all, natural sciences and sciences of the artificial) and technology, because the advancement of science requires more and more costly technology (Rescher 1999a, pp. 100–103).

3. There are also values in science insofar as it is a *social activity*. This is reasonable because science is social in its goals or aims, to the extent that it does not start from scratch and the new goals require the cooperation of the researchers. It is also social in its means, because scientific progress needs social support in order to have adequate research tools. Furthermore, science is social in its results, because scientific evaluation is a task made by persons and organizations devoted to these ends.

As social values of science, Rescher points to values such as truthfulness, honesty, or perseverance. They are values that go with the aims of research and the processes that contribute to achieving those aims. Besides these values, he admits the presence of desires that have to do with the self-promotion of the individuals or groups: influence, power, prestige, etc. (Rescher 1999a, p. 76). These desires do not have the character that he attributes to "values," since they lack an objective basis. In effect, they are due to the preferences of the concrete individuals or groups, instead of being something objectively desirable for science.

4. This framework is completed by the set of values of science *as application*. In Rescher's judgment, these values represent the benefit of the products of science, and they are "related mainly with the application of science to human *desiderata*, such as wellbeing, health, longevity, comfort, etc., in medicine, engineering, agriculture…" (Rescher 1999a, p. 95). From this perspective, the results of the research and their possible uses and consequences are explicitly considered. Furthermore, this leads to a relation between the values of science and the values of technology, where ethical values have an important role (for example, in biomedicine).

Once again, these values of the applications of science arise from knowledge (it is about their "consequences"), where practical knowledge gains special prominence, because it adapts to the context where it is used. These values are related to the use of the cognitive content and the processes of application, so what is worthy is not something momentary or transitory. To sum up, it is an "intellectual" approach to values, so it is not a "volitive" account of values: *content* prevails over the decision regarding what is "worthy."

Therefore, when Rescher's axiology of research addresses the four realms mentioned—the aims, the economic values, the social component, and the consequences of scientific activity—it gives priority to the internal dimension. The values that characterize the human activity of making science are values "internal" to scientific activity by itself (semantic, logical, epistemological, methodological, ontological values, etc.). Thus, they modulate the aims of the research—description, explanation, prediction, and control (to which prescription should be added)[16]—that seek to increase the available knowledge (basic science) or to solve concrete problems (applied science).

Although he insists on the "internal" dimension, "external" values also have an important role. Rescher admits this when he notes that scientific knowledge is a good *among others*: "It is only one element of the constellation of human desiderata —one valuable project among others, whose cultivation is only one component of the wider framework of human purposes and interest" (Rescher 1999b, p. 243). In this way, the role of the external values (cultural, political, economic, ecological, etc.) is not merely secondary, but can also be relevant, due to a holistic view of values.

Now, this interrelation between the internal and the external component highlights that scientific activity can be subject to external limitations. This is because the whole set of human values serves human needs, so the search for knowledge should not interfere with other legitimate goods (health, wellbeing, etc.). In that case, the values "internal" to scientific activity (semantic, logical, epistemological, methodological, etc.) are interrelated with the "external" values (social, cultural, ecological, etc.).

Rescher avoids an "atomism" of values: they are not "isolated" values, because they are interrelated. Consequently, the search for knowledge as a good "in no way hinders the cultivation of other legitimate goods; on the contrary, it aids and facilitates their pursuit, thereby acquiring an *instrumental* value in addition to its value as an absolute good in its own right. Whatever other projects we may have in view—justice, health, environmental attractiveness, the cultivation of human relations, and so on—it is pretty much inevitable that their realization will be facilitated by the knowledge of relevant facts" (Rescher 1999b, p. 243).

Certainly, he addresses this double perspective of analysis—internal and external—from a structural viewpoint that appeals to values as a *system*. For this reason, although his approach takes into account the aims, processes, and results—which are the three moments that configure the internal dynamics—and the relations with the changing context (the external dynamics), Rescher does not develop a satisfactory dynamic approach. In my judgment, the dynamic approach must assume the *historicity* (of science, agents, and the reality itself that is researched) [Gonzalez 2008a].

In contrast to a conception of this type (open to historicity), Rescher's proposal is more static. In effect, the dynamic approach, whose starting point is the acknowledgement of the historical character of scientific research, involves several aspects. First, it requires paying attention to the changes that occur in the aims, processes and results of the research (and the connected values); and, secondly, it involves the

---

[16] Prescription is required in order to encompass applied science.

study of the relations with the context, which is also changeable. Meanwhile, values in Rescher are always structural, so he does not think of values that can be modulated by historicity.

8.2.2   Main Values: Epistemological and Methodological

Within this framework of a holism of values, Rescher accepts the presence of a plurality of values in science. However, he considers that values internal to scientific activity (above all, epistemological and methodological) are more important than the external values (social, cultural, economic, ecological, etc.). In this regard, Wenceslao J. Gonzalez notes that "his Kantism is obvious, to the extent that the content of science is more relevant than the socio-historical context" (2010, p. 269).

His holism of values admits a hierarchy or scale of preferences among them. Thus, in Rescher's judgment, all the elements of value that are present in science have not the same status. In this holistic axiology of research, the internal values (epistemological, methodological, etc.) are not the only values that modulate research, but they are, in principle, more important than the external values (social, cultural, political, …). So, in the first place, his conception that broadens the realm of values in contrast to other proposals that are *exclusively* focused on the internal dimension;[17] and, in the second place, it is clearly different from the studies of science, technology, and society, which usually highlight the external component.[18]

On the one hand, internal values have more weight in Rescher than external values. He considers that they characterize scientific research in a better way. Above all, he focuses on epistemological values such as truth, accuracy, or coherence; and methodological values like explanatory and predictive power of theories. In turn, the results of science (mainly the development of scientific theories) that have those values can contribute decisively to the achievement of other human values, such as personal dignity, health, or sustainability. On the other hand, the primacy of the internal dimension has to do with a conception that emphasizes the structural aspects with regard to science. This is because he focuses on what is worthy regarding science as *content*, so he gives priority to the epistemological and methodological realms.

By accepting that science is *our* science, Rescher assumes that values are in science from the beginning of the activity. Thus, they are in the realm of the aims or goals of scientific research. According to the teleological character of science, two kinds of values have priority in his judgment: (a) values that characterize science, as

---

[17] This can be seen in the comparison made by Wenceslao J. Gonzalez between Rescher's and Laudan's proposals. Cf. Gonzalez (2008b).

[18] Regarding the greater importance of the internal dimension in Rescher's axiology of scientific research, there are convergence points between his proposal and the logical empiricism and naturalism, which are two of the most influent philosophical traditions in the North American philosophy of science in the 20th century. Rescher himself notes his relation with the legacy of the logical empiricism and considers that he is a member of the "young generation" of the Berlin School, because he was a student of Hempel. Cf. Rescher (2006, p. 282).

they go with *the type of knowledge* that science gives (epistemological values); and (b) values that modulate *the processes* that lead to the achievement of that knowledge (methodological values). Within methodological values, economic values are highlighted, since science requires, for him, an "economy of means" (Gonzalez 1999a, p. 16).

Within epistemological values, Rescher highlights values such as coherence, consistency, generality, understandability, simplicity, accuracy, and precision (Rescher 1999a, p. 94). These values modulate the aims of scientific research, insofar as by searching for them we can effectively achieve the aims of science (among them, prediction). Moreover, they have a role in evaluating the results of the research (for example, accuracy and precision might be key factors in evaluating a scientific prediction) [Gonzalez 2015a, pp. 243–245]. Rescher considers that "in this realm, we also find the values included in the management of the cognitive risk; especially, in the standards of proof and rigor" (Rescher 1999a, p. 94).

Rescher acknowledges the diversity of values in accordance with the thematic realms. Thus, he admits that there are values that can vary in the different groups of sciences. Furthermore, the values that have priority in social sciences and in natural sciences are not necessarily the same. In this way, he assumes that there can be different scales of preferences in sciences with regard to the epistemological values. In effect, it seems that a value such as precision can be more relevant in astronomy than in linguistics or social anthropology, where other values such as coherence, might be more important.

But, in my judgment, Rescher's proposal should be broadened because there can be variations within each group of sciences (natural sciences, social sciences, or the sciences of the artificial) and within each science (astronomy, economics, physics, sociology, etc.) depending on socio-historical factors. These variations, which highlight the *historicity* of scientific activity, are compatible with the *objectivity* of values. In this regard, two factors should be distinguished in order to join the objectivity of values and their historical character: (a) values as such; and (b) the appraisal of values.

When values are considered as such, it can be seen that they have an objective basis, so it is possible to claim that coherence, accuracy, precision, simplicity, etc. are worthy factors that should accompany scientific theories. Therefore, the deliberation about what the elements of value that should guide the aims of the research are is not a merely subjective or intersubjective matter; rather there are rational bases that support the evaluation. Meanwhile, the appraisal of values is something that can vary in accordance with different socio-cultural contexts and different historical moments.

As Javier Echeverría notices, "the criteria of evaluation of scientists themselves have varied according to the contexts, disciplines, periods, and concrete situations in which they act" (2002, p. 185).[19] Even more, the *historicity* of values cannot be just a secondary matter to the axiology of scientific research, since "the most pressing issue consists in analyzing the changes in values over the history of science and within a discipline, or even during one evaluative process" (Echeverría 2002,

---

[19] This feature of values can be clearly seen in the case of brain studies. Cf. Gómez (2005).

p. 181). In effect, it is a crucial issue for the history of science, because "fundamental scientific controversies and large changes in the way science is practiced typically involve conflicts over, and transformations of, the value commitments that dominate genuine knowledge production" (Doppelt 2007, p. 198).

This component of change and variation that modulates the criteria of evaluation is difficult to adapt to Rescher's structural approach. Certainly, he emphasizes the issue of contextuality, so he is aware that changes occur in science, and also in the realm of values.[20] But it seems to me that his approach does not develop the three main realms where historicity can intervene: science, the agents, and the reality researched (above all, social and artificial). In my judgment, the component of change can be seen more clearly within the framework of the approaches that highlight the historicity of scientific research, which is a wider notion than contextuality.[21]

In effect, establishing the criteria to assess the acceptability of theories depends on the validities accepted by scientific community. These criteria might vary from one socio-cultural context to other (here the level of acceptance of the autonomy of science has influence) and they might depend on concrete historical moments. Certainly, the acceptance of the variability of values according to socio-historical factors does not lead us to assume an axiological relativism, because historicity is compatible with the objective character of evaluative issues.

Together with the epistemological values, Rescher emphasizes methodological values, which are values that have to do with scientific processes. Within these values, his attention is focused on economic values, as they are an expression of the "economy of means," because, in his judgment, "rationality and economy are inextricably interconnected. Rational inquiry is a matter of epistemic optimization, of achieving the best overall balance of cognitive benefits relative to cognitive costs" (Rescher 1989, p. 13). From this perspective, methodological values such as efficacy or efficiency have priority in his conception, since research requires "profitability" in terms of time and effort.

Regarding methodological values, Rescher's approach is also focused on the structural dimension, where the internal perspective has priority over the external aspect. Thus, even when he is interested in the nexus of science with the context (so there are external values that modulate science with regard to the aims, processes, and results), he does not develop this nexus within a framework of *historicity*. For this reason, he insists on methodological values of economic character such as efficacy and efficiency, which are related to the fulfillment of other criteria, like sim-

---

[20] "One has to accommodate general considerations with particular contexts and with variability of contexts. This has impact on normative conceptions (values, duties, etc.)," Rescher, *Personal communication*, 26.5.2015.

[21] Since the 1960s, there has been a series of proposals that have acknowledged the importance of the historicity of science. Thus, the Lakatosian "research programs," Laudan's "research traditions," and other conceptions of the advancement of science have insisted on scientific activity related to historicity (for example, Paul Thagard). On this issue and the debate about "scientific revolutions," see Gonzalez (2011).

plicity, uniformity, regularity, etc. (Rescher 1989, p. 96), instead of paying more attention to the economic values that modulate science as social activity.[22]

There are two coordinates that modulate his axiology of scientific research and that are important for scientific prediction: (i) the primacy of the internal dimension, within a structural configuration when he address the *system* of values; and (ii) the priority of the epistemological and methodological aims over the social and operative ones. Within this framework, Rescher addresses the study of the axiological elements of scientific prediction. In this regard, his approach goes in two different (although connected) directions. On the one hand, he addresses the *value of prediction* as such; that is, its importance within scientific activity (basic, applied, or of application); and, on the other, he analyzes what *values* are or might be *characteristic of prediction* (accuracy, precision, credibility, etc.).

8.3 Prediction as Value of Science: The Problem of Predictivism

Axiologically, the problem of scientific prediction has two different dimensions. First, scientific prediction can be analyzed *as value of science*; that is, the importance it has within the scientific context. Secondly, there is the problem of the *values* that should go with *scientific prediction* (accuracy, precision, etc.), which are the features that modulate its scientific appraisal. The first problem has to do, above all, with the role of prediction as one of the main goals of scientific activity. Meanwhile, the second issue has direct repercussions on the values that modulate this goal, which makes prediction worthy.

This double dimension that affects the axiological perspective of scientific prediction—*prediction as value* of science and the *values of prediction*—can be seen in Rescher's approach. On the one hand, he defends a conception of (moderate) predictivist character, since he considers that prediction has a high scientific value (Rescher 1998a).[23] On the other, he explicitly addresses the values that go with prediction (accuracy, precision, correctness, robustness, etc.), which he preferably analyzes from a structural perspective, where the internal values have primacy (Gonzalez 2010, pp. 269–270).

The first dimension—prediction as value of science—connects with the problem of predictivism. In this regard, two main issues can be contemplated. First, there is the topic of the scientific value of prediction regarding the set of the goals of science. This involves considering explanation and prediction in basic science and prediction and prescription in applied science. Secondly, the value of scientific prediction can be considered with regard to the processes and results. This connects with the predic-

---

[22] The internal dimension and the external perspective with regard to economic values are developed in Gonzalez (2008b).

[23] On the different versions of predictivism, see Harker (2008) and Barnes (2008).

tion-accommodation controversy, which considers which has more methodological weight: the prediction of novel facts or the accommodation to already known facts.

8.3.1 Prediction as Value with Regard to the Aims of Science

When the value of prediction is considered with regard to the aims of science, Rescher opts for adopting a moderate position because he notes that "successful prediction is one of the cardinal aims of science, though not its sole objective" (Rescher 1998a, p. 159). Thus, he considers that prediction is an important aim of scientific activity *among others*. But, at the same time, he thinks that it is an especially important aim, since "prediction is the very touchstone of science in that affords our best and most effective test for the adequacy of our scientific endeavors" (1998a, p. 161).

Due to the special importance that Rescher attaches to prediction, his proposal can be classed, in my judgment, among the predictivist tradition. It is the orientation that, initially, philosophers such as F. Bacon, G. W. Leibniz and W. Whewell defended. They emphasized the value of *prediction* as a prominent aim of science. Afterwards, in the 20th century, philosophers such as H. Reichenbach or economists like M. Friedman continued this tradition. Reichenbach and Friedman had in common that they considered prediction as the highest scientific value, so other aims can be subordinated to the aim of predicting.

Karl Popper, who has been especially influential in conceptions of general philosophy and methodology of science and also in methodological issues regarding economics, developed a dual approach to prediction. With regard to the general realm—with predilection for physics as the most valued science—his falsifiability emphasized the role of prediction and its role for refutation. Thus, he put the example of Einstein and his critical attitude: the failure of prediction can lead to neglecting a scientific theory. Meanwhile, regarding socio-historical prediction, Popper addressed the problem of historicism. Thus, he rejected the scientific character of socio-historical predictions in the long run and with a broad scope, while he finally admitted economic predictions that are about certain kind of phenomena.[24] To a large extent, Popper showed a predictivist approach at the general methodological level and criticized predictivism at the special methodological level (Popper 1957).

Afterwards, within the "historical turn" in philosophy of science, there were also philosophers who highlighted the value of prediction. Thus, according to Thomas Kuhn, "probably the most deeply held values concern predictions: they should be accurate; quantitative predictions are preferable to qualitative ones; whatever the margin of permissible error, it should be consistently satisfied in a given field; and so on" (Kuhn 1970, p. 185).[25] Predictivism is even clearer in Imre Lakatos, who

---

[24] A detailed analysis of his approach to prediction is in Gonzalez (2004).

[25] A study of Kuhn's proposal on scientific prediction is in Gonzalez (2010, pp. 127–159).

considered that prediction of novel facts is the key factor in seeing a research program as progressive.[26]

On this problem, Rescher's proposal is more in tune with Leibniz's than with the approaches of contemporary authors. In effect, Rescher quotes Leibniz as support[27] to maintain that "prediction is the very touchstone of science in that it affords our best and most effective test for the adequacy of our scientific endeavor" (Rescher 1998a, p. 161). Thus, he mainly relates the value of prediction to its role as test of theories. But prediction has other important roles in science. For this reason, besides basic science, the role of prediction in applied science and the application of science should be considered.

In effect, the value of prediction can be related to the different roles it has: (1) In basic science, prediction can be used as a test for theories, in general, and hypotheses, in particular. (2) In the case of applied science (pharmacology, medicine, applied economics, etc.), prediction habitually is the previous step to prescription.[28] 3) When the problem of the application of science is considered, prediction has also a role, because it can be the support for decision-making procedures (Gonzalez 2010, p. 11).

As is usual in his philosophy of science, when Rescher thinks of *prediction as value* with regard to the aims of science, he considers its role in basic science (above all, in natural sciences). He rarely contemplates the use of prediction in applied sciences as a guide for prescription. Furthermore, prescription does not appear in an explicit way when he lists the goals of science. Moreover, he barely takes into account, strictly speaking, its role for the application of science, although he is interested in the use of scientific prediction by agents in matters related to everyday life (see, for example, Rescher 2003). In this way, the value of prediction appears related to its use as an indicator of scientific progress in basic science, insofar as it is a test to determine the scientific character of theories.

Regarding the value of prediction, Rescher distances himself from the type of predictivism defended by authors such as Reichenbach or Friedman, who think of prediction as the most important aim of science. For them, predictivism is related to methodological instrumentalism with regard to prediction. Thus, they think that predictive success is the most important thing for science, so scientific processes should be oriented towards the aim of successfully predicting the possible future (Friedman 1953 and Reichenbach 1949). Meanwhile, the kind of predictivism defended by Rescher can be characterized as a *moderate predictivism*, which highlights the importance of prediction but does not subordinate other goals of science to the aim of predicting.

---

[26] Cf. Lakatos (1970, pp. 91–196). For an analysis of the role of scientific prediction in Lakatos' methodology of scientific research programs, see Gonzalez (2010, pp. 161–192). On his influence on the methodology of economics, see Gonzalez (2014, 2015a, pp. 103–124).

[27] "It may even turn out that a certain hypothesis can be accepted as physically certain [*pro physice certa*] if, namely, it completely satisfies all the phenomena which occur, as does the key to a cryptograph. Those hypotheses deserve the highest praise (after the truth), however, by whose aid predictions can be made, even about phenomena or observations which have not been tested before; for a hypothesis of this kind can in practice be accepted as truth," Leibniz (1969, p. 188). This text is quoted in Rescher (1998a, pp. 160–161).

[28] This is especially clear in the case of economics. See Simon (1990), and Gonzalez (2015a, pp. 321–345).

Furthermore, Rescher expressly criticizes expressly methodological instrumentalism: "Some philosophers take this matter of the predictive utility of good theories too far by adopting a wholly 'instrumentalistic' view of the theories of natural science as mere predictive instruments, altogether dismissing the issue of describing and explaining the world's occurrences" (1998a, p. 164). In this regard, it seems to me that Rescher's predictivism offers a more suitable alternative than methodological instrumentalism. On the one hand, he sees prediction as a value of science, insofar as it is one important aim *among others*; and, on the other, he emphasizes its value with regard to the processes (as an indicator of scientific progress) and regarding the results (due to the knowledge it provides about the possible future).

8.3.2　The Value of Prediction in Processes and Results. The Discussion Between Accommodation and Prediction

The value of prediction can be considered with regard to the processes and results of science. It is an issue that connects with the problem of scientific progress. This is because scientific prediction might have a role, when we consider what the measure of the advancement of science is. In this regard, the discussion between prediction and accommodation should be emphasized. It is a controversy about the methodological weight of each one of them; that is, what is more important for the evaluation of scientific contents; the prediction of novel facts or the accommodation to already known facts.[29]

This discussion has had special repercussions in some sciences, among them economics, which has an important predictivist tradition. In this regard, Friedman's proposal—which has continuity among the authors of the Chicago School—has been especially influential. In his judgment, the results of predictions are the most important feature in assessing scientific progress (Gonzalez 2015a, p. 152). In this case, results have primacy (predictive success) over the activity (the processes), so economic models should be oriented towards achieving predictive success.

Primacy of results leads Friedman to reject the realism of the assumptions in predictive models. Thus, he maintains that "the only relevant test of the validity of a hypothesis is comparison of its predictions with experience. The hypothesis is rejected if its predictions are contradicted (…); it is accepted if its predictions are not contradicted; great confidence is attached to it if it has survived many opportunities for contradiction" (Friedman 1953, pp. 8–9). It is an approach that gives prediction an excessive value, since "neither scientific theories, in general, nor economic theories, in particular, can be reduced to the single goal of making predictions" (Gonzalez 2015a, p. 152).

---

[29] On this methodological controversy, see Gonzalez (2010, pp. 283–288). Periodically, the prediction-accommodation controversy raises publications. In recent years, this topic has been an object of attention in Alai (2014), Carrier (2014), Harker (2008), Hudson (2007) and Shmueli (2010).

An especially critical proposal of Friedman's predictivist thesis is that of Herbert A. Simon. Instead of results, Simon's interest is in the *processes* of decision-making, so rationality—which he addresses in terms of bounded rationality—also includes a *procedural* component (it is not merely a substantive rationality). This leads him to defend the primacy of *understanding* rather than prediction: "his interest lies in *understanding* the mechanisms that explain past and present economic phenomena rather than in the predictability of economic behavior" (Gonzalez 2015a, p. 205). He rejects that prediction is the main aim of science; and, consequently, it cannot be the only measure of scientific progress (Simon 1989).

Currently, in philosophy and methodology of economics, the prevailing tendency is to consider prediction as a relevant test for economics as a science, rather than being the preferential or prevailing test for economics. This tendency sees prediction as an important aim, but one *among others*. Thus, it is more in tune with Simons' approach—not focusing the attention *only* on predictive success—than with Friedman's radical predictivism.

With regard to the role of prediction in scientific progress—its methodological value for the advancement of science—this means that "prediction of novel facts appears now more as a sufficient condition, rather than a necessary condition, for scientific progress. Hence, it is still a relevant criterion for science, although the emphasis now is on considering prediction of novel facts as a sufficient condition of scientific evaluation of scientific research programs" (Gonzalez 2014, p. 16).

On this issue, Rescher's main interest is not in economics but in natural sciences. However, there is a criticism of Friedman's methodological instrumentalism, which he addresses in two different directions: first, with regard to the necessity of realism of the assumptions in scientific models, in general, and economic models, in particular (Rescher 1998a, p. 109); and, second, with regard to the primacy of the predictive success as a criterion to evaluate scientific progress (1998a, pp. 194–196). Rescher is against Friedman's radical predictivism, since he defends, in my judgment, a moderate version of predictivism, which is within the framework of his methodological pragmatism.

It seems clear that Rescher highlights the methodological role of prediction as an indicator to assess scientific progress. He considers this problem within a methodological conception that is pragmatic. Within this framework, he considers that the best criteria we have to establish the adequacy of scientific theories are the capability of the methods to obtain successful predictions and to achieve an effective control over phenomena. Thus, he does not think that successful prediction is the *only indicator* to assess the comparative theoretical adequacy of theories. Instead, he thinks that prediction is *the best criterion* we have (Rescher 1998a, p. 164).

In his judgment, the especial importance of prediction is rooted in its objectivity: prediction provides an objective criterion to confirm or disconfirm a theory. Moreover, predictive success allows us to compare the adequacy of a concrete theory with regard to contrary or alternative theories. In that case, prediction appears as a *result*: it is a scientific statement that can be supported by experience. For this reason, besides the attention to the processes, Rescher considers that also results—

above all, predictions as statements about the future—have value for the advancement of knowledge.

Consequently, Rescher takes into account the methodological value of prediction with regard to aims, processes, and results. In this way, his approach to this problem is broader than the proposal of other authors, insofar as he considers the diverse elements at stake. Moreover, even when his approach is certainly *predictivist*, he clearly rejects a methodological instrumentalism. Thus, he defends a *moderate* conception of predictivism, where prediction appears as sufficient condition for scientific progress.

Concurrently, due to the fact that his approach to science is a *system*, this problem of *prediction as a value* with regard to the aims, processes, and results is closely related to the issue that has to do with the *values of prediction*. Thus, he does not consider that predictive success as such is what matters, but that prediction has a prominent place within the values of science insofar as prediction itself has a series of values, that he wants to make explicit (1998a, pp. 113–131). He does this within an axiological approach to prediction that is coherent with his axiology of scientific research, so he considers, above all, the structural dimension, where internal values (above all, epistemological and methodological ones) have priority.

8.4 From the Structural Perspective: Values of Scientific Prediction in Rescher's Proposal

In Rescher's philosophy a view of values of science from the structural dimension prevails. In his judgment, "the cardinal principle of the theory of prediction is that it is not just predictions we want, and not even just correct predictions, but predictions of high quality in relation to the entire spectrum of relevant criteria" (Rescher 1998a, p. 125). For this reason, he considers that it is an especially important issue to clarify what those criteria are or should be. To do this, he develops a structural approach to the values of scientific prediction, where values that have to do with prediction as *content* are highlighted. Because, in his approach, prediction is, above all, a statement about the future that is the result of a rational process. So when he addresses the problem of prediction and its relation with values, he considers prediction as an answer to a question about a future event or phenomenon.[30]

Rescher presents a very rich framework for the structural dimension, where the relation between scientific prediction and values is addressed at several successive levels (Guillán, forthcoming). First, predictive *questions* lead to considering what elements of value the questions we pose about future phenomena or events have. Secondly, future *statements* consider what the values of prediction are, as a statement with cognitive content. Thirdly, the *task* made by the predictors or predictive

---

[30] "To predict is, more or less by definition, to endeavor to provide warranted answers to detailed substantive questions about the world's future developments," Rescher (1998a, pp. 37–38).

methods poses the question relative to the elements that should be taken into account in the evaluation of predictive procedures or methods.[31]

With regard to the merit of prediction, Rescher suggests that it should be evaluated according to two realms of content: (i) by the evaluation of predictive questions; and (ii) through the evaluation of the answers (that is, the statements about the future or predictions) [Rescher, pp. 113–115]. In turn, when the relation between prediction and values is analyzed, he gives primacy to the *internal dimension* over the external one. Certainly, he accepts an external aspect, insofar as he seeks to offer elements for the evaluation of the task made by the predictors.[32] But, due to the primacy of the structural approach, he does not develop exhaustively this external dimension, which can be clearly seen when the axiology of research is addressed from the dynamic trait.

8.4.1   Criteria for the Evaluation of the Predictive Questions and Values of Prediction as an Answer

Within a pragmatic approach, scientific prediction can be seen as the result of an activity that seeks to achieve justified answers to meaningful questions regarding future events (Rescher 1998a, pp. 37–39). From this perspective, the type of predictive question posed has a direct repercussion on the statement of future that can be achieved (generic or specific, conditional or categorical, etc.). Therefore, *the values of predictive questions* should be considered first; and, then, *the values of prediction* should be addressed (that is, the values that the predictive statement has or should have).

Regarding the *predictive questions*—the first realm of content mentioned before—Rescher suggests that they should be evaluated according to several requirements (1998a, pp. 113–118). He emphasizes then four features: (a) *importance*, which is assessed in terms of how much is lost—either in practical or theoretical terms—if we do not have a correct answer to the predictive question; (b) *interest*, which depends on personal or collective inclinations regarding the predictive problem; (c) *difficulty*, which is linked with the importance of the prediction,

---

[31] Rescher thinks that a prediction's *quality*—if it is scientific or not—depends on the predictive questions and on the kind of processes used in order to predict. This implies that a non-scientific prediction (for example, anticipating the choice of an individual on the basis of his tastes and inclinations) can also have values such as reliability, accuracy, precision, etc. Hence, the main difference between scientific and non-scientific prediction is rooted in the *type of processes*: it consists in the use of non-scientific procedures or properly scientific methods (such as predictive models) in order to achieve predictions. Cf. Rescher, *Personal communication*, 15.7.2014.

[32] Due to Rescher's insistence in a view of science as *our* science—that is, the result of the interaction between the researcher and the researched object—the analysis of predictors can be seen in his approach as an internal element, instead of being an external factor of prediction. In fact, he thinks that the elements for the evaluation of predictors are the same as the features that we should take into account in order to evaluate the methods of prediction. Cf. Rescher (1998a, pp. 113–118). On his conception of science as our science, see especially Rescher (1992b).

because the most important predictions are, generally, the most complex ones (Rescher 2012a, p. 152); and (d) *resolvability*, which is the possibility of verifying or falsifying the answer to a predictive question.

To assess the *importance* of the predictive question—the first criterion that Rescher points out—the context in which the question oriented towards the future is posed must be considered. Thus, importance is not an absolute value, but it has a contextual or relational value. Consequently, the criteria important for its evaluation must be established. These criteria might be different in the realms of basic science, applied science, and the application of science. This is because basic science, applied science, and the application of science are different activities, so prediction can have different roles within each of them.

When the importance of a predictive question is assessed in the realm of basic science, epistemological values of the possible answer—with regard to the cognitive content it can provide—might be emphasized, as well might its methodological value (to what extent the answer to this question can be a test for the scientific character of a theory). Meanwhile, in applied science the main values to assess the importance of a predictive question are practical criteria, which have to do with its importance for making the subsequent prescription. Moreover, in the case of the application of science, practical values have, in principle, more weight than the cognitive values, insofar as the answer to a predictive question can be the basis for decision-making.

In contrast to importance (which can be assessed on the basis of objective criteria in different contexts), *interest* (the second criterion) provides a weaker evaluation, because it depends on the inclinations (either personal or collective) towards the problems posed. Furthermore, to the extent that Rescher defends the objectivity of values, it can be considered that interest is not, strictly speaking, a value of predictive questions, since it depends on subjective or intersubjective factors. Thus, even when he expressly mentions interest as one of the criteria to evaluate predictive questions, it does not seem that it can have the same status as other criteria that do have an objective basis.

Regarding *difficulty* as a factor for evaluating predictive questions—the third criterion in his list—there are several issues at stake. (a) Within the framework of values that Rescher's proposes, difficulty is related to importance. This is because he usually relates the importance of a prediction to its informativeness (accuracy, precision, etc.) and maintains that the most informative predictions are generally the most difficult to achieve (Rescher 2012a, p. 152). (b) Evaluating the difficulty of a predictive question involves, in my judgment, taking into account the economic values, since the more difficult a predictive question is, the more resources we need, in principle, in order to solve it.

With regard to the fourth criterion to value a predictive question—*resolvability*—it is a criterion Rescher associates with "its definitive verifiability or falsifiability in due course" (1998a, p. 115). In principle, it seems that it is, in effect, a relevant criterion when a predictive question is posed in the realm of basic science, because it is required for the role of prediction as a test of theories. However, it has lesser relevance, in my judgment, in applied science and in the application of science.

Both in applied science and in the application of science prediction serves as a guide for a subsequent action. Thus, on the one hand, prediction is the previous step to prescription in applied science; and, on the other, it is the starting point of practical decision-making in the application of science. In this case, the possibility of confirming or falsifying the prediction is not pressing. Moreover, it could happen that, on the basis of the obtained prediction, an action is designed in to avoid the predicted thing happening. Thus, for example, in the case of design sciences, it is usual to use prediction in order to anticipate the possible problems of design, so prescription is oriented toward avoiding that the predicted problem occurs in the future.

Rescher also notices values with regard to the relevance of the *answers* or *statements about the future* (the second level of content that he contemplates), to the extent that they have or do not have the different worthy factors (1998a, pp. 119–125). He points out six criteria: (1) relevance with regard to the predictive answer; (2) correctness, which cannot be guaranteed at the moment the prediction is made,[33] but that can be estimated according to its credibility; (3) accuracy, which consists in the level the content of prediction reflects specific facts in the future; (4) precision, which has to do with the level of detail the prediction reaches;[34] (5) credibility, which is based on the evidence and the probability that supports the predictive statement; and (6) robustness, which is the cohesion between the prediction and other predictions achieved or methods used within a concrete field.

According to these values suggested by Rescher, his main interest is in the *content* of prediction. This has direct consequences for his proposal about the values of scientific prediction. First, the internal dimension clearly has primacy over the external axiological perspective, to the point that external values (social, political, economic, ecological, etc.) are not explicit in the framework he proposes about the values of scientific prediction. Secondly, among the internal values, the main values are the cognitive ones (accuracy, precision, correctness, etc.), which are related to scientific prediction as cognitive content (and have incidence on the methodological processes).

Thirdly, on the basis of the primacy of internal values and the priority of cognitive values, it seems to me that Rescher is generally thinking of the evaluation of the predictive statements in a context of basic science. Consequently, there can be problems if we try to transfer his proposal to the realm of applied science or to that of application of science. This is because, in basic science, prediction is evaluated on the basis of the increase of the available knowledge, so it is important that the cognitive content of the prediction can be tested. Thus, prediction can serve as a test for theories. To do this, values such as accuracy, precision, correctness, or robustness—which are the values that Rescher habitually notes—are especially important.

But, in applied sciences, prediction is usually the previous step to prescription. Its main role is to anticipate problems and to contribute to designing possible solu-

---

[33] "In the end, only the course of experience can inform us about the extent to which the phenomena of a particular domain are predictable," Rescher (2014a, p. 44).

[34] Regarding the level of precision of predictions, there is an important problem: the more informative a predictive statement is, the less secure it is. Thus, generally, the more generic predictions are also the more secure ones. Cf. Rescher (1998c, pp. 1–48; especially, pp. 19–24).

tions. For this reason, the evaluation of prediction in this realm has a more pragmatic dimension: it is made on the basis of its contribution to the solution of the
concrete problem posed. Thus, on the one hand, it is usual to consider cognitive
values in applied scientific activity as less important than in basic science; and, on
the other, the presence of external values (social, political, economic, ecological,
etc.) is clearer than in basic science.

From this perspective, criteria suggested by Rescher (correctness, accuracy, precision, etc.) are only some values of prediction *among others*. Thus, it seems clear
that cognitive values are not the only values in scientific activity, and they are not
necessarily the most important ones. Furthermore, the problems that applied science
seeks to solve (for example, in applied economics, pharmacology, medicine, etc.)
are usually related to the wider human experience, so the external axiological
dimension should be taken into account. This is even clearer in the application of
science, where prediction serves aims that, in turn, are modulated by the context,
which is changeable.

8.4.2 Elements for the Evaluation of Predictors and Predictive Methods

Besides the values that have to do with the predictive questions and predictive
answers (that is, predictions as statements of future), there is another level that
Rescher considers: the task made by predictors and the methods of prediction. He
proposes a variety of criteria for the *evaluation of predictors*. In this regard, he notes
that "predictive *methods* can themselves also be looked upon as being predictors
and can be evaluated as such by exactly the same standards" (1998a, p. 130). This
issue is related to his approach to predictive processes, which he classifies in two
large types (1998a, ch. 6, pp. 85–112; especially, pp. 86–88): (i) *unformalized or
judgmental procedures* and (ii) *formalized or discursive methods* (see, in this regard,
ch. 6 herein).

Unformalized or judgmental processes are based on the personal estimations of
experts (for example, the Delphi procedure), while formalized methods follow a
series of rules or inferential principles (as in the case of the current predictive models). Hence, when Rescher focuses on the axiology of prediction, his proposal is
valid both for the estimative procedures—where the evaluation is focused on the
agents—and for the formalized methods, where the evaluation is made with regard
to scientific procedures, without taking into account the subjects.[35]

---

[35] According to Rescher, the quality of predictions does not depend on their scientific character,
because there might be non-scientific predictions that are more reliable than some scientific predictions. Thus, he considers that the difference between scientific and non-scientific predictions is
mainly methodological; that is, it is rooted in the type of processes that are used in order to predict
the future. For this reason, in order to evaluate the quality of predictions, the values of predictive
questions (importance, interest, …) and the values of the answers or statements about the future

Rescher considers that there are nine worthy factors for the evaluation of predictors and predictive methods. (1) Reliability, which is established with regard to the record of predictive results of the past and can be estimated either in an absolute way or in a comparative way. (2) Range or versatility, which has to do with the extent of the thematic range. (3) Daring, i.e., the ability to succeed with difficult or complex issues. (4) Perceptiveness, which deals with the detail and definiteness of the achieved predictions. (5) Foresight, which is characterized by Rescher as the timespan over which a predictor or predictive method is able to function.[36] (6) Consistency, which is considered from the perspective of regularity; that is, the capacity to perform at a uniform level of competence over time. 7) Self-criticism, which has to do with the accuracy of the evaluation that the predictor makes about his or her own task. 8) Knowledgeability, which is the ability to provide non predictive information within the thematic realm of the prediction. 9) Coherence, which requires that the predictions achieved by the predictor or the predictive method are compatible amongst themselves (Rescher 1998a, pp. 125–130).

This proposal by Rescher about the values of prediction, which is focused on the structural dimension—where the epistemological and methodological realms have primacy—has repercussions on his preference for the *internal* values of prediction (mainly, cognitive ones) over the *external* values. However, he defends a position of *holism of values*, which is related to his view of science as a *system*. In this way, the distinction between internal values and external values does not allow us to separate them. He considers that values form a set of interdependent elements within that system (Gonzalez 1999a, p. 22).

Now, the primacy of the structural dimension, which in the case of Rescher is linked to the priority of the internal values (where above all the cognitive values are emphasized) poses problems when this approach is transferred to the realms of applied science and application of science, which are different from basic science. In this regard, it seems to me that we should address the dynamic perspective of the axiology of prediction, which takes into account the aims, processes, and results of the three kinds of activities (basic, applied, and of application), so highlighting the historical dimension of scientific activity.

(accuracy, precision, …) are especially important. Moreover, the quality of the predictions achieved will depend on the type of reality prediction is about (for example, predictions about natural phenomena are usually more reliable than predictions about social reality). Rescher, *Personal Communication*, 15.7.2014.

[36] *Sensu stricto*, the notion of "foresight" refers to a scientific prediction which achieves an effective control of the relevant variables. Therefore, it is the most reliable type of prediction, irrespective of its temporal projection (that can be in the short, medium, or long run). Cf. Fernández Valbuena (1990, p. 388). See also Sect. 2.4.2 herein.

8.5 Values from the Dynamic Perspective: Prediction in Basic Science, Applied Science, and the Application of Science

As science is a human activity of a teleological character, it has aims, processes, and results. Concerning scientific knowledge about the future, the evaluation of these aims, processes, and results is the main objective of the axiology of prediction from the dynamic perspective [37]. This perspective involves taking into account the differences between basic science and applied science, because they have to do with the aims, processes, and results of both kinds of activities (Gonzalez 1999b, p. 158; and 2013b, pp. 17–18). In addition, the realm of the application of science can be considered. In this field, prediction can be the basis of decision-making (for example, in the context of policy).

Concurrently, it happens that scientific activity—which articulates aims and processes that can produce some results—is carried out within a context (historical, social, cultural, political, economic, ecological, …), which is changeable. This external dynamics of scientific activity means that, together with the internal values of scientific activity itself (for example, cognitive values), there are also external values. These values affect science according to its nexus with other human activities (social, political, economic, ecological, etc.). This external component especially affects applied sciences, where prediction is usually the previous step to prescription, and also the application of science, where scientific prediction can be the basis of decision-making.

8.5.1 Evaluation of Aims and Processes in the Three Cases

Certainly, the presence of values in science is clear insofar as they modulate the aims that orientate scientific research and the processes that seek to achieve these aims. But the aims and processes of scientific research may be different in basic science and applied science. The third step—the application of science—has, in turn, its own aims and processes, which are modulated by context (Gonzalez 2013a, pp. 1511–1513; and 2013b, pp. 17–18). Thus, the differences between basic research and applied research have repercussions on the configuration of the values in both kinds of sciences, both from an internal and external perspective.[38] This can be seen regarding the values of science, in general, and the values of scientific prediction, in particular.

According to the differences regarding the aims and processes, it seems clear that cognitive values have primacy in basic science; while in applied science there is a pragmatic or instrumental feature, which is more emphasized (Gonzalez 2013a,

---

[37] On this analysis of the axiology of prediction from the dynamic perspective see also Guillán (forthcoming).

[38] The differences between basic science, applied science, and the application of science are dealt with in Chap. 5, Sect. 5.2.

pp. 1511–1513, and Niiniluoto 2014, pp. 11–15). This feature affects the axiology of prediction in two different senses: (i) *prediction as a value* in basic science and applied science; and (ii) *values of prediction* in both types of sciences. With regard to the first sense (prediction as a value of science), prediction can be considered as a valuable aim in itself.[39] This can be seen both in basic science and applied sciences.

Undoubtedly, in Rescher's philosophical account, prediction is a *value* of science according to the teleological character of scientific research (1999a, p. 106). However, the roles of prediction are different in basic science and in applied science. In basic science, aims, processes, and results of the research are oriented towards extending the available knowledge, so cognitive values related to explanation and prediction take priority (Gonzalez 2013a, p. 1513). In this regard, prediction provides knowledge about the possible future that can be used as a test for scientific theories.

Meanwhile, in applied sciences (pharmacology, medicine, applied economics, information sciences, etc.) the aims, processes, and results of scientific investigation are oriented towards solving specific problems within a concrete domain. Thus, "the natural sciences, the social sciences and the sciences of the artificial require prediction of the possible events and prescription on how to perform" (Gonzalez 2013b, p. 17). Consequently, the main values in applied sciences are those related to prediction and prescription, where the practical dimension is important.

Within this framework, in basic science the value of prediction has to do, on the one hand, with the type of knowledge it offers about the possible future; and, on the other, it is related to its use as a test for the scientific character of theories. Meanwhile, prediction is linked to prescription in applied sciences. This is because prediction is required in order to prescribe (that is, in order to suggest how to perform when the aim is the resolution of a specific problem). Thus, in the realm of applied science, prediction is the previous step to prescription. For this reason, prediction can be considered as a methodological tool in applied sciences: we need to predict the possible future in order to establish how to perform next (Gonzalez 2008c, p. 181).

These differences, that have to do with prediction as a value in the context of basic science and in the realm of applied science, have repercussions on the second sense mentioned regarding the axiology of scientific prediction: the values that predictions have or should have. In basic science, the values of prediction with regard to the aims, processes, and results are mainly of an internal character. Since prediction in basic science can be a test for the scientific character of theories, the *values* of prediction have to do above all with the quality of the knowledge about the future (accuracy, precision, etc.). In this way, if the predictive content is confirmed, prediction can provide empirical support for theories.

However, in applied science prediction has also an instrumental value, insofar as it can be the basis for problem-solving. Due to this instrumental component, it is usual to consider the cognitive values of prediction (accuracy, precision, etc.) as less important than in basic science. This can be seen in Simon's approach to the sciences

---

[39] This claim does not have to do directly with the controversy about the methodological weight of prediction versus accommodation. But it should be acknowledged that Rescher manifests a preference for prediction.

of the artificial (1996). For this Nobel Prize winner in economics, prescription does not require predicting the future with accuracy and precision, but the main concern is to shape the future through designs (Martínez Solano 2012, p. 248). In this way, it can be good enough to have a forecast—the least reliable kind of scientific prediction—instead of having a foresight in the strict sense.[40]

From this perspective, the evaluation of prediction in applied science is carried out on pragmatic basis, which takes into account to what extent prediction allows us to anticipate problems and to design possible solutions. However, the more accurate and precise the prediction is, the higher the prescription's probability of success will be (at least in principle). For this reason, it is desirable, I believe, that applied sciences are oriented towards the search for increasingly reliable and informative predictions (that is, detailed, precise, etc.), so it facilitates the task of selecting among alternative courses of action (Guillán 2013, pp. 128–129).

But, besides the internal values of science (accuracy, precision, coherence, efficacy, efficiency, …), the dynamic trait of the axiology of research implies attention to external values (social, political, economic, ecological, esthetic, …). Thus, in addition to the "internal" dynamics of scientific research (aims, processes, and results), there is also an "external" dynamics, which has to do with the relations between science and the socio-historical context (Gonzalez 2012).

Commonly, the presence of values external to science can be seen more clearly than in basic science. The reason for this higher visibility is clear: the results of applied science and their possible consequences have frequently more incidence in a society's life. These external values can modulate the aims of applied research and can even have a role regarding the processes, which eventually lead to some results.

Among the external values there are the exogenous ethical values (that complement the endogenous ethical values). They are values that affect science as a human activity linked with other human activities (for instance, solidarity, social responsibility, etc.). These exogenous ethical values might have a role in the selection of the aims and processes of the research, above all when it is applied research (for example, in medicine, demography, etc.). However, they are usually more relevant for the evaluation of the results and their possible consequences, to the extent that they can have incidence for the individuals, the society, or the environment.[41]

The relation between scientific prediction and the external values (among them, the exogenous ethical values) can also be seen, in my judgment, in a context of application of science; because prediction can be the basis for agents' decision-making (for example, in the realm of policy, regarding the decisions about climate change). This implies that the application of science is always contextual, that is, it depends on a socio-historical context. Furthermore, it is more conditioned than the cases of basic science and applied science, due to its practical repercussions, that are more direct than in the realms of the basic or applied knowledge (this feature also involves relevant connections with national or international legislations).

---

[40] On the different types of predictions (forecast, prediction, foresight, …) see Sect. 2.4.2 of this monograph.

[41] On Rescher's approach to the ethics of science, see Chap. 9. On the relation between science and ethical values (endogenous and exogenous), see Gonzalez (1999b).

8.5.2 Criteria for the Evaluation of the Results in the Three Realms

The orientation of scientific research towards aims leads to select some processes that, in turn, produce some results. The task of evaluating the results of the research, which is carried out according to values, can follow two different orientations: (a) the internal perspective and b) the external viewpoint. The internal perspective addresses to what extent the achieved results correspond to the previously established aims. Meanwhile, from the external viewpoint, the focus is on the relation with the context (social, cultural, political, economic, ecological, etc.), where it is usual that changes occur (Neira 2015, p. 47).

When the internal orientation is followed in the analysis of the results, criteria for the evaluation of these are different in basic science and in applied science. Thus, in basic science, results are evaluated on the basis of the enlargement of the available knowledge, without taking into account—in principle—the possible concrete uses for these results (Gonzalez 1999b, pp. 158–159). For this reason, the values that have primacy in basic science are commonly values related to explanation and prediction. Among them, cognitive values such as accuracy, precision, novelty, simplicity, coherence, etc., are highlighted, and values related to the problem of truth understood as agreement with reality.

In the case of scientific prediction, its use as test for theories is also relevant. In this regard, there might be a hierarchy of values on the basis of the empirical support that prediction provides for theories in basic research (for example, detail is especially important for prediction, since specific predictions are preferable to generic predictions as test for theories). This case of prediction as test has been of great importance in the case of economics.

Meanwhile, in applied sciences there is a more active component. For this reason, the evaluation of results can be made according to cognitive criteria (the adequacy of the obtained knowledge for solving the concrete problem raised) or it can be made following practical criteria (efficacy and efficiency in the resolution of that problem) [Gonzalez 1999b, pp. 158–159]. Thus, besides the cognitive values, there are values such as the economic ones, which also have an important role for the evaluation of the results of applied research. So cognitive values (accuracy, precision, etc.) are usually considered as less important than practical values when prediction in applied sciences is analyzed. From this perspective, the practical value of prediction as a guide for the resolution of concrete problems is emphasized.

In addition, the prescriptive task of applied science—which is usually oriented toward suggesting how to perform in order to deal with issues that we cannot effectively control—should be considered. Here, it is usual to search for the best possible adaptation to the predicted problems (Simon 1990). Therefore, the evaluation of the results will be positive if they facilitate the adaptation to the context. In this regard, the values of prediction are also relevant, because achieving satisficing results in applied science depends, to a great extent, on the achievement of reliable predictions in order to anticipate the problems and consider the feasibility of the possible solutions.

Besides the internal orientation—which evaluates the achieved results, considering if they adjust or not to the previously established aims—results can also be evaluated following an external viewpoint. From the external viewpoint, the focus is on the relation with the context (social, cultural, political, economic, ecological, etc.). These relations with the context are clearer in applied science, which has more incidence on the whole human experience. Furthermore, applied science serves as a guide for social action (for example, in economics), so external or contextual values commonly have more weight than in basic science (Gonzalez 2013a, p. 1513).

The relevance of external values in applied science is especially clear in the case of ethical values, above all when the attention goes to the practical consequences of the research and their possible repercussions in society, instead of highlighting the epistemic content (Gonzalez 2013a, p. 1513). There is a relation between exogenous ethical values and scientific prediction, which appears insofar as the predictions achieved in applied science can be the basis for problem-solving. Those problems might have incidence on the individuals, society or the context (for example, in the case of pharmacology or ecology).

Concurrently, the dimension of analysis that deals with the results opens up another perspective which has to do with the use that agents make of those results in practical contexts of decision-making (political, economic, ecologic, etc.); that is, the application of science (Niiniluoto 1993). Frequently, the applications of science are made by agents in institutions (for example, in hospitals, economic organizations, etc.). Due to this application in variable contexts, the results of scientific research can be used in the wider socio-historical context. In this regard, values are especially important when matters such as risk management appear (Gonzalez 2013a, p. 1516).

From this perspective of the application of science, prediction has to do with the decision-making in professional contexts (as commonly happens, for example, in medicine or in policy issues in cases such as a hurricane, an earthquake, or a tsunami).[42] Due to the importance of the context for the applications of science, the external dimension of the axiology is crucial. Thus, the applications of science in realms such as economics, medicine, pharmacology, etc., have to do with matters of public interest, insofar as they can affect the individuals, the society, or the environment (in either a positive or negative way). Ethical values have here a highlighted role, above all, those values related to social responsibility. Moreover, "insofar as the consequences are more manifest for citizens and society as a whole, the legislation (regional, national or international) can show up" (Gonzalez 2013a, p. 1508).

To sum up, the elements offered by Rescher for the axiology of prediction are coherent with his axiology of scientific research, which is within the coordinates of a pragmatic Kantism. Thus, the structural dimension has primacy, due to his view of values as a system (Rescher 1993b, c, and 1999a). From this perspective, the internal values to scientific activity are emphasized. If this approach is transferred to the axiological study of scientific prediction, epistemological and methodological values of prediction have a manifest primacy over other kind of values.

---

[42] For Rescher, a main issue regarding prediction is the fact that our ideas about what will happen in the future have a clear influence on our present decisions. Cf. Rescher (1998a, p. 64).

In my judgment, Rescher's proposal can be complemented with the attention to the *dynamic trait* in the relation between science and values. Through the dynamic component—which addresses the aims, processes, and results of the research ("internal" dynamics) and the relations with the context ("external dynamics)—it seems clear that, besides the internal values, external values are also relevant. In effect, science is a human activity linked with other human activities, so there are values that have to do with the relations between scientific activity and the context (social, political, economic, ecological, etc.) [Gonzalez 1999b, 2013a].

Although Rescher gives primacy to the structural dimension of science, his approach is not reduced to something sectorial, because his starting point is the acknowledgement of a holism of values. In this way, he proposes *de facto* "a broad axiology of research according to his conviction that sciences are related to values from diverse angles (cognitive, social, etc.). These values in science are related to other human values and rooted in human needs" (Gonzalez 2013a, p. 1512). Within this framework, the distinctions regarding the different realms or types of values ("structural" versus "dynamic," "internal" as different from "external") do not properly admit a *separation*, but they are interconnected in the system of values.

Although Rescher's axiology of scientific research—which preferentially follows a structural perspective—can be compatible with a dynamic approach, he does not develop in an articulate way this dynamic trait regarding the relation between prediction and values, which emphasizes the *historicity* of scientific activity. Consequently, his approach does not especially take in the external values of science, in general, and of scientific prediction, in particular. For this reason, even though his approach is broader than other axiological proposals, it lacks, in my judgment, the due attention to the dynamic trait (internal and external). Within the dynamic trait, prediction and the connected values have a key role for the evaluation of the aims, processes, and results of scientific prediction. Moreover, this can be seen in three cases: in basic science, in applied science, and in the application of science.

References

Alai, M. 2014. Novel predictions and the no miracle argument. *Erkenntnis* 79 (2): 297–326.
Barnes, E.C. 2008. *The paradox of predictivism*. Cambridge: Cambridge University Press.
Carrier, M. 2014. Prediction in context: On the comparative epistemic merit of predictive success. *Studies in History and Philosophy of Science* 45: 97–102.
Doppelt, G. 2007. The value ladenness of scientific knowledge. In *Value-free science? Ideals and illusions*, ed. H. Kincaid, J. Dupré, and A. Wylie, 188–217. New York: Oxford University Press.
Echeverría, J. 2001. Tecnociencia y sistemas de valores. In *Ciencia, Tecnología, Sociedad y Cultura en el cambio de siglo*, ed. J.A. López Cerezo and J.M. Sánchez Ron, 221–230. Madrid: Biblioteca Nueva.
———. 2002. *Ciencia y valores*. Barcelona: Destino.
———. 2003. *La revolución tecnocientífica*. Madrid: Fondo de Cultura Económica.

Fernández Valbuena, S. 1990. Predicción y Economía. In *Aspectos metodológicos de la investigación científica*, ed. W.J. Gonzalez, 2nd ed., 385–405. Madrid-Murcia: Ediciones Universidad Autónoma de Madrid and Publicaciones Universidad de Murcia.

Friedman, M. 1953. The methodology of positive economics. In *Essays in positive economics*, M. Friedman, 3–43. Chicago: The University of Chicago Press (6th reprint, 1969).

Gómez, A. 2005. Ciencia y valores en los estudios del cerebro. *Arbor* 181 (716): 479–492.

Gonzalez, W.J. 1999a. Racionalidad científica y actividad humana: Ciencia y valores en la Filosofía de Nicholas Rescher. In *Razón y valores en la Era científico-tecnológica*, N. Rescher, 11–44. Barcelona: Paidós.

———. 1999b. Ciencia y valores éticos: De la posibilidad de la Ética de la Ciencia al problema de la valoración ética de la Ciencia Básica. *Arbor* 162 (638): 139–171.

———. 1999c. Valores económicos en la configuración de la Tecnología. *Argumentos de Razón Técnica* 2: 69–96.

———. 2003. Racionalidad y Economía: De la racionalidad de la Economía como Ciencia a la racionalidad de los agentes económicos. In *Racionalidad, historicidad y predicción en Herbert A. Simon*, ed. W.J. Gonzalez, 65–96. A Coruña: Netbiblo.

———. 2004. The many faces of Popper's methodological approach to prediction. In *Karl Popper: Critical appraisals*, ed. Ph. Catton and G. Macdonald, 78–98. London: Routledge.

———. 2005. The philosophical approach to science, technology and society. In *Science, technology and society: A philosophical perspective*, ed. W.J. Gonzalez, 3–49. A Coruña: Netbiblo.

———. 2008a. El enfoque cognitivo en la Ciencia y el problema de la historicidad: Caracterización desde los conceptos. *Letras* 114 (79): 51–80.

———. 2008b. Economic values in the configuration of science. In *Epistemology and the social, Poznan Studies in the Philosophy of the Sciences and the Humanities*, ed. E. Agazzi, J. Echeverría, and A. Gómez, 85–112. Amsterdam: Rodopi.

———. 2008c. Rationality and prediction in the sciences of the artificial. In *Reasoning, rationality and probability*, ed. M.C. Galavotti, R. Scazzieri, and P. Suppes, 165–186. Stanford: CSLI Publications.

———. 2010. *La predicción científica. Concepciones filosófico-metodológicas desde H. Reichenbach a N. Rescher*. Barcelona: Montesinos.

———. 2011. Conceptual changes and scientific diversity: The role of historicity. In *Conceptual revolutions: From cognitive science to medicine*, ed. W.J. Gonzalez, 39–62. A Coruña: Netbiblo.

———. 2012. Las Ciencias de Diseño en cuanto Ciencias de la Complejidad: Análisis de la Economía, Documentación y Comunicación. In *Las Ciencias de la Complejidad: Vertiente dinámica de las Ciencias de Diseño y sobriedad de factores*, ed. W.J. Gonzalez, 7–30. A Coruña: Netbiblo.

———. 2013a. Value ladenness and the value-free ideal in scientific research. In *Handbook of the philosophical foundations of business ethics*, ed. Ch. Lütge, 1503–1521. Dordrecht: Springer.

———. 2013b. The roles of scientific creativity and technological innovation in the context of complexity of science. In *Creativity, innovation, and complexity in science*, ed. W.J. Gonzalez, 11–40. A Coruña: Netbiblo.

———. 2014. The evolution of Lakatos's repercussion on the methodology of economics. *HOPOS: The Journal of the International Society for the History of Philosophy of Science* 4 (1): 1–25.

———. 2015a. *Philosophico-methodological analysis of prediction and its role in economics*. Dordrecht: Springer.

———., ed. 2015b. *New perspectives on technology, values, and ethics: Theoretical and practical. Boston studies in the philosophy of science*. Dordrecht: Springer.

Guillán, A. 2013. Analysis of creativity in the sciences of design. In *Creativity, Innovation, and Complexity in Science*, ed. W. J. Gonzalez, 125–139. A Coruña: Netbiblo.

———. forthcoming. Predicción científica y valores: Análisis de la dimensión estructural y de la componente dinámica. *Contrastes. Revista Internacional de Filosofía*.

Harker, D. 2008. On the predilections for predictions. *British Journal for the Philosophy of Science* 59: 429–453.

Hudson, R.G. 2007. What's really at issue with novel predictions? *Synthese* 155: 1–20.

Kuhn, Th. S. 1970. Postscript—1969. In *The structure of scientific revolutions*, Th. S. Kuhn, 174–210. Chicago: The University of Chicago Press (2nd ed., 1970).

Lakatos, I. 1970. Falsification and the methodology of scientific research programmes. In *Criticism and the growth of knowledge*, ed. I. Lakatos and A. Musgrave, 91–196. London: Cambridge University Press. Reprinted in Lakatos, I. 1978. *Mathematics, science and epistemology. Philosophical papers*, vol. 2. Ed. J. Worrall and G. Currie, 128–200. Cambridge: Cambridge University Press.

Laudan, L. 1984. *Science and values. The aims of science and their role in scientific debate.* Berkeley: University of California Press.

Leibniz, G. W. 1969. *Philosophical Papers and Letters* (Trans. L.E. Loemker). Dordrecht, Reidel.

Martínez Solano, J. F. 2012. La complejidad en la Ciencia de la Economía: De F. A. Hayek a H. A. Simon. In *Las Ciencias de la Complejidad: Vertiente dinámica de las Ciencias de Diseño y sobriedad de factores*, ed. W.J. Gonzalez, 233–266. A Coruña: Netbiblo.

Neira, P. 2003. La racionalidad tecnológica en H. A. Simon y los problemas de predicción. In *Racionalidad, historicidad y predicción en Herbert A. Simon*, ed. W.J. Gonzalez, 147–166. A Coruña: Netbiblo.

———. 2015. Values regarding results of the information and communication technologies: Internal values. In *New perspectives on technology, values, and ethics: Theoretical and practical. Boston studies in the philosophy and history of science*, ed. W.J. Gonzalez, 47–60. Dordrecht: Springer.

Niiniluoto, I. 1993. The aim and structure of applied research. *Erkenntnis* 38 (1): 1–21.

———. 1995. Approximation in applied science. *Poznan Studies in the Philosophy of Sciences and the Humanities* 42: 127–139.

———. 2014. Values in design sciences. *Studies in History and Philosophy of Science* 46: 11–15.

Popper, K.R. 1957. *The poverty of historicism.* London: Routledge and K. Paul (Reprinted by Routledge, 1991.).

Reichenbach, H. 1949. *Experience and prediction. An analysis of the foundations and the structure of knowledge.* Chicago: The University of Chicago Press (1st ed. 1938).

Rescher, N. 1973. *Conceptual idealism.* Oxford: Blackwell.

———. 1977. *Methodological pragmatism. A systems-theoretic approach to the theory of knowledge.* Oxford: Blackwell.

———. 1978. *Scientific progress.* In *A philosophical essay on the economics of the natural science.* Oxford: Blackwell.

———. 1988. *Rationality. A philosophical inquiry into the nature and the rationale of reason.* Oxford: Clarendon Press.

———. 1989. *Cognitive economy. The economic dimension of the theory of knowledge.* Pittsburgh: University of Pittsburgh Press.

———. 1992a. *A system of pragmatic idealism. Vol. I: Human knowledge in idealistic perspective.* Princeton: Princeton University Press.

———. 1992b. Our science as *our* science. In *A system of pragmatic idealism. Vol. I: Human knowledge in idealistic perspective*, N. Rescher, 110–125. Princeton: Princeton University Press.

———. 1993a. *A system of pragmatic idealism. Vol. II: The validity of values: Human values in pragmatic perspective.* Princeton: Princeton University Press.

———. 1993b. How wide is the gap between facts and values. In *A system of pragmatic idealism. Vol II: The validity of values,* N. Rescher, 65–92. Princeton: Princeton University Press.

———. 1993c. Values in the face of natural science. In *A system of pragmatic idealism. Vol II: The validity of values*, N. Rescher, 93–110. Princeton: Princeton University Press.

———. 1994. *A system of pragmatic idealism. Vol. III: Metaphilosophical inquires.* Princeton: Princeton University Press.

———. 1995. *Satisfying reason. Studies in the theory of knowledge.* Dordrecht: Kluwer.

———. 1997. *Objectivity. The obligations of impersonal reason*. Notre Dame: University of Notre Dame Press.

———. 1998a. *Predicting the future. An introduction to the theory of forecasting*. New York: State University of New York Press.

———. 1998b. *Complexity: A philosophical overview*. N. Brunswick: Transaction Publishers.

———. 1998c. Communicative pragmatism. In *Communicative pragmatism and other philosophical essays on language*, N. Rescher, 1–48. Lanham: Rowman and Littlefield.

———. 1999a. *Razón y valores en la Era científico-tecnológica*. Barcelona: Paidós.

———. 1999b. *The limits of science*. revised ed. Pittsburgh: University of Pittsburgh Press.

———. 2003. *Sensible decisions. Issues of rational decision in personal choice and public policy*. Lanham: Rowman and Littlefield.

———. 2004. Pragmatism and practical rationality. *Contemporary Pragmatism* 1 (1): 43–60.

———. 2006. The Berlin School of logical empiricism and its legacy. *Erkenntnis* 64: 281–304.

———. 2012a. The problem of future knowledge. *Mind and Society* 11 (2): 149–163.

———. 2012b. Pragmatism and purpose. In *Pragmatism. The restoration of its scientific roots*, N. Rescher, 21–47. New Brunswick: Transaction Publishers.

———., 2012c. Rationality and moral obligation. In *Studies in philosophical anthropology*, N. Rescher, 79-93. Heusentamm: Ontos Verlag, Heusenstamm.

———. 2014a. Leaping in the dark and other aspects of rational inquiry. In *The pragmatic vision. Themes in philosophical pragmatism*, N. Rescher, 35–45. Lanham: Rowman and Littlefield.

———. 2014b. Fact and value in pragmatic perspective. In *The pragmatic vision. Themes in philosophical pragmatism*, N. Rescher, 67–78. Lanham: Rowman and Littlefield.

Shmueli, G. 2010. To explain or to predict? *Statistical science* 25 (3): 289–310.

Simon, H.A. 1989. The state of economic science. In *The state of economic science. Views of six Nobel laureates*, 97–110. Kalamazoo: W. E. Upjohn Institute for Employment Research.

———. 1990. Prediction and prescription in systems modeling. *Operations Research* 38: 7–14. Compiled Simon, H.A. 1997. *Models of bounded rationality. Vol. 3: Empirically grounded economic reason*, 115–128. Cambridge, MA: The MIT Press.

Simon, H. A. 1996. *The sciences of the artificial*, 3rd ed. Cambridge, MA: The MIT Press (1st ed., 1969; 2nd ed., 1981).

Chapter 9
Analysis of Prediction from Ethics of Scientific Research

Abstract Among the values that modulate scientific prediction there are ethical values. This chapter clarifies the ethical factors that modulate prediction in the context of *scientific activity* and *science as activity*. To do this, two perspectives of analysis are considered: (a) the endogenous ethics, which is oriented towards scientific activity by itself; and (b) the exogenous ethics, which analyses science as an activity connected with other activities (social, cultural, political, economic, ecological, etc.).

The starting point is the study of Rescher's ethics of science, which gives primacy to the internal perspective. This leads to going more deeply into the exogenous perspective, which is also important for the problem of prediction. After that, the research is oriented towards the reflection about the problems posed by the relation between scientific prediction and ethical values. Thus, the repercussions on scientific prediction of the ethical limits of science are considered first. Second, the study of the ethical values of scientific prediction is developed from the dynamic viewpoint, which deals with the evaluation of the aims, processes, and results of the research. In this regard, the differences between basic science, applied science, and the application of science must be taken into account.

Keywords Ethics of science • Scientific prediction • Prescription • Basic science • Applied science • Application of science

In recent decades, the axiological perspective in the analysis of science has emphasized how science, in general, and scientific prediction, in particular, are modulated by values. In effect, a turn has occurred in the contemporary philosophy and methodology of science, from a dominant view of science as value-free (*Wertfrei*) to the opposite conception, according to which science is value-laden (Gonzalez 2013a, pp. 1504–1506). These values can be either *internal* values to scientific activity (cognitive, methodological, etc.) or *external* values, which are those that modulate science as an activity related to other human activities (social, cultural, political, economic, etc.).

© Springer International Publishing AG 2017   285
A. Guillán, *Pragmatic Idealism and Scientific Prediction*, European Studies in
Philosophy of Science 8, DOI 10.1007/978-3-319-63043-4_9

This change of perspective has motivated the development of ethics of science, which studies the more specific realm of ethical values (honesty, credibility, social responsibility, etc.) that are present in scientific activity (see Resnik 1998). Thus, science is susceptible to ethical evaluations "insofar as science is a free human activity, whose values are related to the processes of research (honesty, originality, etc.) or to its nexus with other activities of human life (social, cultural, political, etc.)" (Gonzalez 2015, p. 12). From this perspective, ethics of science can follow two different angles of analysis: the endogenous and the exogenous.

From the endogenous viewpoint, ethics of science deals with the ethical values that have to do with *scientific activity* by itself, so there are values in the aims, processes, and results of the scientific activity. Meanwhile, the exogenous viewpoint involves paying attention to *science as an activity* that is related to others, which means emphasizing that there are ethical values that have to do with the context (social, cultural, political, economic, ecological, etc.) in which the activity is carried out (Gonzalez 2015, p. 19).

Therefore, the ethics of science—endogenous and exogenous—considers the *activity* carried out by free human agents within a socio-historical context. From this perspective, the analysis of the ethical factors that affect scientific prediction must distinguish aspects of prediction: (i) the cognitive content, which is oriented towards the future (real or possible); and (ii) the human activity that leads to obtain or to use this knowledge about the future.

In effect, on the one hand, it does not seem that the *cognitive content* as such can be susceptible to an ethical valuation (that is, scientific knowledge itself is not something morally adequate or inadequate) (Rescher 1999a, pp. 159–162). But, on the other hand, this cognitive content is the result of a human *activity* oriented towards achieving knowledge about the possible future, and there are aims, processes and results, as well as relations with context (social, cultural, political, etc.). Consequently, prediction is linked with the ethics of science "through the presence of prediction in the *research activity*, mainly when the ethical limits of science are discussed" (Gonzalez 2015, p. 20).

Within this framework of the ethics of science, where there is a double dimension of analysis—endogenous and exogenous—this chapter seeks to clarify the ethical factors that modulate prediction in the context of *scientific activity* and *science as activity*. Rescher does not develop this aspect of the philosophical analysis of prediction when he suggests his theory of prediction (mainly Rescher 1998). However, his philosophy of sciences pays much attention to the ethics of scientific research.[1] Thus, the starting point of this study is his proposal about the ethical features of scientific research.

In this regard, several steps are followed. First, the attention is on pragmatic idealism as a framework for ethics, in general, and ethics of science, in particular. Secondly, the analysis seeks to go more deeply into the double dimension of analysis

---

[1] This characteristic can be seen in the large number of publications that Rescher has devoted to the study of the ethics of science. Among then, some can be highlighted here: Rescher (1996, 1999a, 2003a).

regarding the relation between science and ethical values: the endogenous view-point and the exogenous perspective. This involves paying attention to three realms: basic science, applied science, and the application of science, which are different, due to both endogenous considerations (the aims, processes, and results) and to exogenous ones (the relations with the context).

Thirdly, the problem of the ethical limits of scientific research is considered, as well as its incidence on prediction from both the endogenous and the exogenous perspectives. Fourthly, the dynamic trait of the relation between ethical values and scientific prediction is considered. This angle of analysis is difficult to incorporate into Rescher's framework, where the structural dimension has primacy, so he does not grasp the features related to the historicity of the human activity carried through in a social context. This activity involves paying attention to the "internal" dynamics (the aims, processes, and results) [Gonzalez 2013a] and the "external" dynamics (the relations with the context) [Gonzalez 2012, pp. 8–9.] Both the internal and external dynamics of science have incidence on the ethical values that accompany prediction in three different cases: basic science, applied science, and the application of science.

9.1 Pragmatic Idealism as the Framework of the Ethics of Science

The ethics of science proposed by Rescher falls within the framework of his philo-sophical system, which he characterizes as "pragmatic idealism," but which is in fact open to elements of other philosophical orientations, as in the case of ethics.[2] On the one hand, the Kantian influence is present, insofar as he wants to base moral in mind and seeks universal ethical principles and values. And, on the other hand, he thinks that science is modulated and conditioned by ethical values, insofar as it is a free human activity, which is oriented towards ends.

Through the combination of these two background lines—one general and other more oriented towards scientific practice—Rescher shows Kantian and pragmatic elements in his thought about the ethics of scientific research. Thus, he proposes an ethics of science which has two main characteristics: (1) it is an ethics of a norma-tive character, so it is about what *ought to be*, according to values and principles that are universally valid; and (2) it is a teleological ethics, where the ethical adequacy of human actions—actual or potential—always takes into account validity with regard to the aims sought.

---

[2] This feature of Rescher's approach does not mean that he is an eclectic philosopher, but rather that he has his own system of thought. See Chap. 1, Sect. 1.2.

9.1.1   The Grounding of Moral in Mind: The Cognitive Basis as a Support of the Ethical Dimension of Our Science

When Rescher addresses ethics, in general, and the ethics of science, in particular, he commonly uses the terms "ethics" and "moral". However, a distinction can be found in Evandro Agazzi's work. In his judgment, "moral" consists of the set of norms and principles that regulate human behavior; while "ethics" has to do with rational justification of moral, so one of its tasks is lay the foundations of the norms and moral principles (Agazzi 1992; see also Gonzalez 1999b, p. 145). According to this characterization, ethics, in general, and ethics of science, in particular, are a part of philosophy.

In this chapter, the term "ethics" is used in the sense suggested by Agazzi; that is, as a philosophical discipline that makes a critical analysis of the moral norms and principles. In this case, it seems clear that ethics cannot be reduced to a simple description of those norms and principles (even if the description is very detailed). On the contrary, it includes a prescriptive or normative dimension, so it is always open to the realm of the "ought to be." In this regard, there is in Rescher a distinction between "mores" and "moral." He suggests this distinction in order to emphasize the normative character of ethics (Rescher 2008, pp. 397–400).

On the one hand, "mores" have to do with the traditions and conventions that are either implicitly or explicitly accepted by a community. Meanwhile, "moral" or "morality" has a universal support, because, according to Rescher, it is founded on needs that human beings have in common. He highlights, therefore, the pragmatic dimension, which sees the human activity as oriented towards ends. Thus, he thinks that "morality is a functional enterprise whose aim is to channel people's actions toward realizing the best interests of everyone. This turns morality into something quite different from mere mores geared toward communal uniformity and predictability. (After all, morality is not a matter of anthropology; it addresses what people *should* do rather than what they *actually* do)" (Rescher 2008, p. 393).[3]

Therefore, the characterization of moral suggested by Rescher configures ethics of science on the basis of two main characteristics: (a) it is normative, so it deals with the *ought to be*, instead of being about what those things that, in fact, are (see, for example, Rescher 1997b); and (b) it is teleological instead of merely "instrumental", because actually morality is understood by Rescher as a human endeavor oriented towards ends rather than being focused just on means (1997a, pp. 127–130). On the one hand, the normative character of ethics implies that the grounding of morality cannot be about customs or conventions (*mores*), which are intrinsically variable; and, on the other, the configuration of ethics as teleological involves expressly addressing the ends of the human activity, instead of considering just the means (Rescher 1989).

---

[3] "Anthropology" here might refer to social anthropology, so the goal of ethics is not merely the description of social uses and traditions.

On this philosophical basis, the grounding of moral that Rescher suggests combines Kantism with pragmatism, which are the mainstays that support his philosophy, in general, and his philosophy of science, in particular. Kantism can be seen, above all, when he links the conception of "morality" with *human rationality* (Rescher 1997b, pp. 160–163). This feature leads to an ethical intellectualism, because he thinks that there are cognitive roots as a support for ethics, in general, and the ethics of science, in particular (Gonzalez 1999a, p. 13).

Rescher expressly maintains this ethical intellectualism: he claims that "the problem of the *rational grounding* of morality has deep cognitive ramifications. Here knowledge, mind, and even science intervene" (Rescher 1999a, p. 61). In this regard, he thinks that H. A. Prichard was right when he noted that, when we know what is morally adequate (the "correct" thing), we do not need a subsequent reason about why we should do that (Prichard 1912, quoted in Rescher 1999a, p. 62).

But, even when he considers that this line of thought is correct, Rescher seeks another line of argumentation, one not completely internal to morality itself. Thus, he proposes an *ontological* support for morality, which is rooted in the characterization of human beings as *rational* agents. He does this from a perspective that takes into account the individual agents, instead of being oriented towards social agents. There is, then, a philosophical anthropology with a metaphysical background, which has links with the improvement of the agent as a human being.[4]

In effect, Rescher points out that "the mandatory character of morality is ultimately rooted in an *ontological* imperative of making values with regard to the self and the world, and imperative to which free agents as such are subjected. From this ontological perspective, the ultimate basis of the moral duty is rooted in the obligation we have as rational agents (towards ourselves and towards the world in general) of making the most of our opportunities of self-development" (Rescher 1999a, pp. 65–66). This feature leads to the promotion of agents from values.

Rescher links this ontological grounding of morality, which takes into account human agents as rational agents, to an ethical approach of teleological character. He suggests this from a pragmatic perspective, according to which "morality as such consists in the pursuit, through variable and context-relative means, of invariant and objectively implementable ends that are rooted in a commitment to the best interests of people in general" (Rescher 2008, p. 408). In his approach, the adequate ends are those that contribute to universal human needs, so they are oriented towards achieving the best interests of people. In this way, he considers that there are ethical

---

[4]Regarding this issue, the pragmatic aspect of his thought prevails, insofar as it goes from the actions to the subjects of those actions. He thinks of the person as one who unifies the processes, which are developed in the "personhood," which makes the features of that entity possible. In effect, he maintains that "once we conceptualize the core 'self' of a person as a unified manifold of actual and potential process—of action, and of capacities and dispositions to action (both physical and psychic)—then we have a concept of personhood that renders the self or ego experientially accessible. (After all, experiencing itself simply consists of such processes.)," Rescher (1994, p. 178). On the concept of "personhood" in Rescher, see Moutafakis (2007, ch. 2, pp. 63–83; and ch. 6, pp. 139–160). An analysis of the concept of "person" is in Gonzalez (1983).

principles and values that are universal, such as dignity, respect, freedom, or safety.[5]

This acknowledgement of the universal character of certain ethical values involves the acceptance of certain elements of realism with regard to values (Rescher 1993a). Thus, once again, his system of pragmatic idealism is open to realism. This feature can be seen insofar as he accepts the notion of "moral fact," which is about what is morally adequate or correct in an objective way. In that case, there is an objective support to morality, which leads Rescher to reject any form of relativism regarding ethics and to accept a "moral realism" (Rescher 2008, pp. 393–396; see also Rescher 1993b).

But Rescher does not think that "moral facts" are *properties*, either non-natural or supervenient. Instead, he considers that moral acts have a relational character. In his judgment, "the moral status of an act is not the sort of thing that is a property or quality at all, but a relational feature whose determination involves a wide variety of contextual issues: agents, circumstances, motives, alternatives, and the like" (2008, p. 396). Again, this feature links his conception of "morality" with *human rationality* (Rescher 1993c). Thus, in his approach, rationality—which is not independent of the context—involves an evaluative dimension, so it should be oriented towards adequate ends, and the adequacy of the ends should be also considered from an ethical viewpoint (Rescher 1988, pp. 92–106).

Thus, he considers that the direction of human action towards morally adequate ends according to ethical values is a *rational* task. But the ability of human agents to exert rationality is *limited* and is conditional on factors that depend on the context (Rescher 1988, pp. 115–118). In this way, his approach acknowledges that, even when "morality" has a support of a universal character, there is variability according to socio-cultural factors. For this reason, "the universal of morality not only permit but *require* adjustment to local conditions, and at the local level we are concerned not with the validation of morality as such, but with the justification of a particular moral code for a particular group in particular circumstances" (Rescher 2008, pp. 406–407).

Therefore, the universal character of moral principles and values should be combined with the variability—due to socio-historical factors—of the concrete moral codes. To do this, Rescher suggests a "hierarchy of morality." Within this hierarchy, he distinguishes five levels (2008, p. 403). (1) *Defining aims of morality*, which are the aims that identify morality as such. In his judgment, they consist of supporting the best interests of people and avoiding causing them harm. (2) *Fundamental principles or controlling values*, which are the moral virtues (such as fairness, generosity, probity, honesty, etc.). (3) *Governing rules*, which are the rules that guide human deliberations and decisions according to moral principles (for instance, do not kill, do not lie, or do not steal). (4) *Operating directives*, which give us standards and criteria according to the context, so they specify concrete situations that have to do

---

[5] The Kantian influence is highlighted here. As Rescher admits, "this aspect of the present account is thoroughly Kantian. The governing idea of Kant's moral theory is that of *universality*; that when an action is wrong or right it is so always and for everyone," Rescher (1997b, p. 167).

with level 3 (like, for example, 'killing in self-defense is justifiable'). (5) *Concrete rulings*, which are individual resolutions with respect to specific issues in specific cases (for example, 'it is justifiable that X has killed Y in self-defense').

According to this "hierarchy of morality," there are some moral principles that are invariable, because they are grounded in universal features related to *human activity* and they have a rational support. The activity demands the best interest of the agents, who should orient their actions to the achievement of ethically adequate ends. Thus, he thinks that the characteristic aims of morality are in the highest level of the moral principles. In this way, the other levels are justified on the basis of their relation with these principles. However, levels 4) and 5) introduce in the ethics a factor of variability, because, in his approach, the operating directives and the concrete rules take into account the context (cultural, social, political, etc.), which is changeable (Rescher 2008, pp. 404–405).

When this approach is transferred to the area of *scientific activity*, there are clear repercussions for the ethics of science. For Rescher, there are ethical values in science insofar as it is *our* science (Rescher 1990). Science is our science due to two reasons: first, it is the result of an activity carried out by human agents; and, second, science corresponds to our *conceptual* framework, which is different from the conceptual framework that other agents may have. Thus, when he characterizes science as *our* science, the two mainstays of his system of thought are present: pragmatism and Kantism.

On the one hand, he considers that there are ethical values in science like *human activity* oriented to ends. For Rescher, values, in general, and ethical values, in particular, are mainly placed at the ends of the *scientific activity*. In his judgment, "values play a crucial role in science, and (...) this role is not something arbitrary or added, but it is inherent to the goal structure that defines science as a rational search" (Rescher 1999a, p. 95). The goals of scientific activity connect with the evaluative dimension of (human and scientific) rationality, which deals with the ends that should be preferred or values (Rescher 1988, pp. 92–106). Consequently, the ends of scientific activity prevail over the means and over the finally achieved results (and their possible consequences). These features lead Rescher to distance himself from ethical utilitarianism, in general, and ethical consequentialism, in particular.[6]

And, on the other hand, Rescher usually focuses his attention on basic science, so he commonly thinks of the kind of scientific activity that seeks to enlarge or improve *knowledge*, instead of being more concerned about applied science or the application of science. For this reason, in his approach, ethical values in science refer to the cognitive dimension. Thus, he notes that "although the search for knowledge is not our only adequate task, it is however an activity whose normative level is high, because knowledge serves to facilitate the fulfillment of any other legitimate good: each and every one of these goods is cultivated in the most effective way by someone who pursues their fulfillment by means of knowledge" (Rescher 1999a, p. 105).

---

[6]On the characteristics of ethical consequentialism there is an influential paper by Elisabeth Anscombe (1958).

From this perspective, science—as expression of knowledge—connects with the first level of the "hierarchy of morality" proposed by Rescher (2008, p. 403). Scientific knowledge can contribute to the achievement of the *defining aims of morality* (to contribute to the best interests of people and to avoid harm to others). To do this, scientific activity should be modulated by ethical values, which are in the second level of the hierarchy. These are values such as honesty, respect, safety, heath, etc. These ethical values (endogenous and exogenous) should be present in the three successive steps of scientific research: the aims of the research, the means oriented to achieving those aims, and the results achieved (with the due attention to the consequences that those results might have).

On the basis of the ends that, for Rescher, are characteristic of morality—level 1—and the presence of ethical values that modulate the aims, means, and results of the research—level 2—a series of governing rules should be established (level 3). These rules seek to orientate human activity towards morally adequate aims, according to ethical values which are universally valid. In this regard, there are ethical rules that regulate scientific activity (such as, for example, 'do not plagiarize' or 'do not perform nonconsensual experiments on human beings'). The presence and the acceptance of this kind of rules can be seen in the deontological codes of some scientific professions (for example, in medicine).

Finally, there are other two levels: 4 and 5. They are about operating directives and concrete rulings, respectively. Within these two levels there might be variability in the ethics of science, due to concrete situations and contexts. This variability can appear due to *internal* factors to science. Thus, it is possible to think that the operating directives and the concrete rulings about what is ethically adequate can vary according to different contexts; those of basic science, applied science, and the application of science, whose aims, processes, and results have specific characteristics. Concurrently, there can be variations according to external determining factors that are due to the different socio-historical contexts. Thus, the ethics of science must take into account the context (social, cultural, economic, political, etc.) where scientific activity is developed.

This framework about the "hierarchy of morality," where there are different levels, has, in my judgment, two main consequences for the ethics of science. First, Rescher's approach implies that ethical issues are not something "added" to science. This means that ethical values and problems do not affect science "from outside," since the scientific human activity is itself value-laden. Second, it means that the ethics of science cannot be addressed as something isolated, since it is connected with the whole human experience. Thus, there are ethical values in science as such (endogenous values) and there are ethical values that modulate science as an activity connected with others (exogenous values).[7]

In effect, he thinks that scientific knowledge is an important good, but it is a good *among others* (Rescher 1999a, pp. 162–165). Thus, on the one hand, even when Rescher gives priority to the aims of scientific activity, the means and the results of the research (basic or applied) are also subjected to ethical evaluation, insofar as

---

[7] On the theoretical framework of the ethics of science, see Gonzalez (1999b, pp. 147–151).

they can interfere or obstruct the cultivation of other legitimate goods (Rescher 1999b, p. 243). And, on the other hand, besides the attention to the values which modulate *scientific activity*, the ethical issues which affect *science as activity* should also be considered, insofar as science is an activity connected with the whole human experience (social, cultural, political, economic, ecological, etc.). These considerations provide a broad framework for the ethics of science, where there are two dimensions of analysis: (a) the endogenous ethics (which takes into account the scientific activity), and (b) the exogenous ethics (which analyses science as activity).

9.1.2 Science and Ethical Values in Rescher's Approach

Rescher considers that there is an ethical dimension of scientific research insofar as science is a *human creation and activity* (Rescher 2006b, p. 218). As human activity, he sees science as *our science*. Thus, on the one hand, it is a human endeavor with its own characteristics; and, on the other, it is an activity connected with others. In this regard, in his philosophy of science, in general, and his ethics of science, in particular, there is a clear preference for emphasizing the internal factors; those which characterize the scientific activity itself as such.

Usually, Rescher sees the relation between science and ethical values within the framework of *scientific research*. It has three successive steps—of ethical relevance—within its internal dynamics: (i) the choice regarding the possible research goals, which configure the objectives sought, (ii) the selection of the means in order to achieve those goals, which configure the processes chosen, and (iii) the results achieved, which are important within this framework of teleological activity.[8]

But, besides the internal factors, the external factors to science should be acknowledged and taken into account. Thus, scientific research is developed within an institutional framework (public or private) and in a socio-historical context, which is changeable. From this perspective, there are ethical values that modulate science as it is an activity connected with other activities (social, cultural, political, economic, ecological, etc.). Rescher gives priority to the internal factors to scientific activity and he gives less weight to the external factors.

In his approach, he rules out that science can be neutral with regard to values and, therefore, he rejects the possibility that science can be isolated from ethical evaluations. In effect, science is a human activity carried out by *human agents*, so "the scientist does not, and cannot, put aside his common humanity and his evaluative capabilities when he puts on his laboratory coat" (Rescher 2006b, p. 201). From this perspective, Rescher's ethics of science is primarily centered on the *person*; and, from the person, the social and institutional levels are finally reached.

---

[8] These are the three steps that configure the internal dynamics of scientific research. Cf. Gonzalez (2012, pp. 8–9).

But, even when Rescher is above all interested in persons, he acknowledges that the attention to the individual level of the agents is not good enough. There is a social dimension that is ethically important, and it leads one to take into account the institutional modulation. So "phenomena of the collectivization and increasing social diffusion of modern science are the main forces that have resulted in making a good deal of room for ethical consideration within the operational framework of modern science" (Rescher 2006b, p. 203).

Within this framework—which highlights human actions as free: from the individual actions to the social ones—Rescher suggests that scientific research is connected with ethical problems from several viewpoints (2006b, pp. 201–218): (1) The choice of research goals; (2) the staffing of research activities; (3) the selection of research methods; (4) the specification of standards of proof; (5) the dissemination of research finding; (6) the control of scientific misinformation; and (7) the allocation of credit for research accomplishments.

Rescher thinks that these issues are *internal* to science, insofar as they involve moral choices related to persons: either the scientists themselves take part in the decision or they directly affect the work scientists develop. Thus, in his approach, those actions which directly imply the scientists and their investigations are "endogenous" (the *internal* dimension of science from an ethical point of view). Meanwhile, he sees as "exogenous"—the ethical features *external* to science—the issues related to the use of results; that is, the questions that are about "what is *done with* scientific discoveries once they have been achieved" (Rescher 2006b, p. 202).[9]

1. In order to analyze his proposal, the presence of ethical problems in each one of the steps of the dynamics of scientific activity should be acknowledged, where there are ends, means, and consequences in line with the aims, processes, and results of basic science, applied science, and the application of science.[10] In the case of the ethics of science, Rescher is usually focused on the first case and takes into account some aspects of the others; especially, the third one. As his ethics is primarily of a teleological character, the problems related with the choices regarding the aims of the research are especially important.

Regarding the problem of the choice of the aims, Rescher takes into account three possible levels: individual, institutional, and national. The *individual* level has to do with the choices that the scientists make regarding those aims of research that might have an ethical component (for example, when they decide to do applied

---

[9] The external dimension is addressed by Rescher in several publications; of note is Rescher (1996).

[10] It should be emphasized that each one of those steps poses ethical questions. From this perspective, science should not be reduced to "general" ethical problems, since there are specific problems within each one of those realms (basic, applied, and of application). Ethics of science must be able to go in depth into each one, taking into account how these realms are interrelated. In this regard, ethics is not reduced to the structural dimension of scientific research, since it also affects the dynamics over time. Thus, there are ethical factors in the historicity that have to do with the relation of the agent with the context (natural, social, or artificial) and other factors that deal with the historicity of the very relations among the researchers (for example, in the research centers). Gonzalez, *Personal communication*, 2.1.2015.

research oriented toward the development of chemical weapons for military ends) [Rescher 2006b, pp. 205–206].

The *institutional* level affects the laboratories, departments, or institutes of research; that is, it has to do with the organization where science is developed. Rescher thinks that ethical problems can arise at the institutional level when the orientation of the research is considered. Thus, when a topic of research is selected, an ethical problem could be whether it is ethically proper to give priority to funding facilities, instead of taking into account the benefit the investigation would have for society or its contribution to the advancement of science (Rescher 2006b, pp. 204–205).

Then, the *national* level also intervenes. This level has to do with the systems of scientific and technological policy in the different countries. In this regard, Rescher thinks that the decision about the distribution of funding among the different research problems can have ethical implications, because it involves deciding what research will have priority (2006b, pp. 203–204). This framework proposed by Rescher is even more complex if other levels are considered. Thus, in the case of Spain, for example, scientific and technological policy do not have only a national level, but also a regional level (Autonomous Communities) and a supranational level (the European Union).

2. A second group of ethical problems that Rescher takes into account are those related to the staff carrying out the research. He relates this issue to the increasing collectivization of science; that is, to the need for bigger scientific communities in order to carry out the research. Since laboratories, research groups, research centers, etc. consist of a large number of persons, there are a series of administrative issues that have to do with the management of these entities. Such problems did not appear before (or at least they did not have the same importance as now). In this regard, Rescher thinks that, in decision making, there is a risk of giving primacy to *administrative* issues over strictly *scientific* matters (2006b, p. 206).

3. There are also ethical issues related to the research methods (Rescher 1999a, pp. 157–158). He thinks that there are procedures of research that are ethically inadequate or questionable. In this regard, he considers that the most visible and pressing problems are those related to experimentation, above all in sciences such as biology, medicine, or psychology. Thus, among the examples of ethically inadequate research procedures, the nonconsensual experiments on human beings or the consensual experimentation that puts the life or well-being of the subjects at risk can be pointed to.

4. Another kind of ethical problem is that related to the standards of proof. It is an issue that has to do "with the amount of evidence that a scientist accumulates before he deems it appropriate to announce his findings and put forward the claim that such-and-so may be regarded as an established fact" (Rescher 2006b, p. 208). In his judgment, besides the epistemological and methodological components, there is an ethical dimension related to this problem, because "in presenting his own results, a researcher may be under a strong temptation to fail to

do justice to the precise degree of certainty and uncertainty involved" (2006b, p. 208). For this reason, the decisions regarding the standards of proof should be modulated by ethical values such as honesty or credibility.

5. The dissemination of information can also be problematic. In this regard, Rescher thinks that the "ethical problem of favoritism in the sharing of scientific information has come to prominence (...) in our day" (2006b, p. 209). This happens when the scientists or the laboratories share information about the results of their research following ethically questionable reasons, like mere personal benefit. Problems can also arise with regard to information dissemination because of the content of the information (for example, how to proceed once a prediction is achieved about the end of the world in few years' time, due to the impact of a meteorite on Earth) [Rescher 1999a, p. 156].

6. Other problem—related to information dissemination—is the issue of the "control, censorship, and suppression of scientific misinformation" (Rescher 2006b, p. 211). About these matters, he thinks that scientists have the duty to protect their colleges and society in general from the dangers that can derive from the publication of research results that are manifestly false (in this sense, truth can be also seen as an ethical value).[11]

Rescher considers that this way of acting is especially necessary in sciences like medicine, since it has direct repercussions on public health and well-being. This is an especially complex issue, because there are epistemological, methodological, and ethical issues that must be taken into account.[12] Thus, on the one hand, mechanisms of control should be established in order to evaluate the results of the research (for example, through the review procedures established by scientific journals, where values such as reliability and honesty are at stake); and, on the other hand, novelty and innovation in scientific research should also be accepted (Rescher 2006b, pp. 211–212).[13]

7. The last group of ethical problems that Rescher admits is that related to the allocation of credit or authorship of the research results (Rescher 2006a). On this issue, he notes that "the allocation to individuals of credit for the research accomplishments resulting from conjoint, corporate, or combined effort" can be problematic from an ethical viewpoint (2006b, p. 216). This problem is especially pressing nowadays, due to the phenomenon of the "collectivization" of scientific research, which is carried out in laboratories, research groups and institutions with a large number of researchers.[14]

---

[11] The view of truth as an ethical value appears in philosophers like Karl Popper. On this topic, see Martinez Solano (2005, pp. 282–294).

[12] A good example of the combination of these factors is provided by John Worrall when he analyzes clinical trials. See, in this regard, Worrall (2006).

[13] This problem also has implications regarding the limits of science as barriers (*Shranken*), insofar as the distinction between science and non-science (or pseudoscience) is at stake. Cf. Rescher (2006b, p. 215). See, in this regard, Gonzalez (2016).

[14] Ethical problems that have to do with the publication, dissemination, and authorship of the research results are analyzed in Resnik (1998, pp. 96–121).

When Rescher develops these seven aspects of scientific research in which ethical problems can arise, he expressly maintains that they are *internal* issues to science, so the ethical dimension is not something "supervenient" to scientific activity. Rescher acknowledges that he is little interested in the external issues, since he will not analyze the ethical problems that are outside science and that, in his judgment, have to do above all with the use and the applications of research results. Expressly he maintains that "such questions of what is done with the fruits of the tree of science, both bitter and sweet, are not problems that arise *within* science, and are not ethical choices that confront the scientist himself. This fact puts them outside of my limited area of concern. They relate to the exploitation of scientific research, not to its pursuit, and thus they do not arise *within* science in the way that concerns us here" (Rescher 2006b, p. 202).

In order to assess his approach, it should be pointed out, firstly, that Rescher seems to accept a distinction regarding the ethical problems, to the extent that he assumes a difference between the cases of scientific research as such (either basic or applied) and the application of science, which consists of the use that agents make of the results obtained through scientific research.[15] Secondly, he is usually interested in basic science, although he admits that the most visible ethical issues are those related to applied science, because it has a higher social incidence that basic science (Rescher 2006b, p. 202).

Because he usually thinks of a context of basic science, he normally highlights the *endogenous* ethical values, which are those that have to do with scientific activity as such (for example, honesty, reliability, or truthfulness). But, in his approach, he expressly notes that, since science is a human activity, "it must be ultimately subject to the same standards of moral evaluation we use with regard with any other human activity" (Rescher 1999a, p. 163). This involves paying attention to the *exogenous* ethical values, which are those values that modulate science insofar as it is an activity connected with others (social, cultural, political, economic, ecological, etc.).

Consequently, even when Rescher's philosophy of science usually gives primacy to the internal dimension of analysis, his proposal offers an adequate framework to make an ethical analysis of science from two different levels: the endogenous ethics and the exogenous ethics. They have to do with the *scientific activity* as such and *science as an activity* connected with others, respectively (Gonzalez 1999b, pp. 147–151). From this perspective, his ethics of science is open to a dynamic approach, which has an internal dimension—the aims, processes, and results of the research—and an external component, which has to do with the relations with the context (Gonzalez 2012, pp. 8–9).

However, he scarcely pays attention to the *historicity* of human activity, so he does not develop how the historical character of science by itself and with regard to its relations with other activities has incidence on the ethical values, from both the

---

[15] His interest in ethical problems related with the application of science can be seen in several of his publications. Among them, Rescher (1996, 2003a)

endogenous and the exogenous perspectives. And this historicity has to do with ethical issues in the relations with the context (natural, social, or artificial), as well as with the different ethical issues that arise in the relations with other research over time (mainly in decision-making). So the historicity of science should not be reduced to one ethical vector (the "extrinsic" one) since it also takes into account the other vector (the "intrinsic" one), which deals with the interactions among the researcher agents over time.[16]

9.2 Two Levels of Ethical Analysis of Science: The Endogenous Ethics and the Exogenous Ethics

Regarding the ethical study of science, there is, in principle, a double dimension of analysis. In the first place, the endogenous ethics considers scientific activity as such; that is, without taking into account the relation with the context in which research is carried out. From this perspective, it is highlighted how science, as a free and social human activity, is value-laden. Values considered here are endogenous values, since they modulate scientific activity as such. From this endogenous viewpoint, the presence of ethical values such as honesty, reliability, personal dignity, the acknowledgment of other's contributions, etc. is highlighted (Gonzalez 1999b, p. 147).

Meanwhile, from the second perspective of analysis—the exogenous ethics—science is seen as a human activity *among others*. This orientation contemplates "scientific activity 'towards the outside,' so it is interested in the set of problems posed by the ethical limits of scientific research, according to the incidence for the people, human society in general (and the environment, insofar as it has repercussions on people or on the social life)" (Gonzalez 1999b, p. 149). From this perspective, there are exogenous ethical values in science, insofar as science is related to the whole human experience (for example, solidarity, legality, or social responsibility).

---

[16]This means that, in this regard, agents' historicity is twofold: the relations *between the researches* have features of variability over time that make those relations different from the relations between the agents and the context (natural, social, or artificial). Frequently, the relations of the former case are intense and revisable, possibly to a larger extent than in the latter case. This type of variability can be seen in the changes in the groups of research and in the internal dynamic of the research centers. History of science selects the representative actions of both types of historicity, insofar as they are important for the development of the scientific activity. Gonzalez, *Personal communication*, 2.1.2015.

9.2.1   Endogenous Ethical Values in Scientific Activity

The endogenous ethics of science considers scientific activity by itself, so it addresses the values that have to do with the very process of research (either in basic science or in applied research). It is a perspective that contemplates "scientific activity 'towards the inside' with the consequent interest in issues such as the honesty of scientific endeavor (reliability in the publication of the data really obtained; the originality of the work, insofar as it has to do with the credibility of the researcher, etc.) or the appropriateness of addressing certain ends or means, according to the moral rules of behavior" (Gonzalez 1999b, p. 149).

Within the framework of a teleological conception of ethics like that proposed by Rescher, the role of the ends of scientific research is emphasized. In principle, they are more relevant than the means and, in addition, they prevail over the results and their possible consequences. In this way, the ethical evaluation of the ends of scientific research is a critical issue. In this regard, his philosophical approach clearly differentiates two realms with regard to their ethical implications: the aims of basic science and the goals of the applied scientific research.

In accordance with his ethical intellectualism, scientific knowledge is, for Rescher, a good (as are, for example, justice, health, or well-being) [Rescher 1999a, pp. 105–106]. For this reason, it is possible to think that basic science seeks a goal that is itself ethically adequate, to the extent that it is a type of research oriented towards increasing or improving the available knowledge. Meanwhile, "the applications of science and technology present a different setting, since there are more questions involved in them than the simple increase of knowledge" (1999a, p. 165).

When Rescher addresses how science is affected by ethical issues, his analysis usually gives primacy to the endogenous ethics; above all, insofar as he thinks of scientific knowledge as a *good*. Thus, he sees knowledge as a good in itself, which also has an instrumental component, since it contributes to the achievement of other human goods. In his judgment, "whatever other projects we may have in view—justice, health, environmental attractiveness, the cultivation of human relations, and so on—it is pretty much inevitable that their realization will be facilitated by the knowledge of relevant facts" (Rescher 1999b, p. 243).

This feature emphasizes that his interest is usually focused on basic science, where scientific research seeks either to increase our knowledge (hence the importance of prediction as an aim) or improve the knowledge available now (which emphasizes prediction as scientific test). Thus, in the case of Rescher, the preference for the endogenous viewpoint of analysis goes hand in hand with the higher interest in basic scientific research, where exogenous ethical values have less presence than in the case of applied science. In this way, "he assumes that ethics of science is directly connected with epistemology (that is, with the cognitive component of science) when he wonders if knowledge as such can be ethically inadequate" (Gonzalez 1999a, p. 13).

In this regard, it is possible to consider that Rescher gives truth an ethical value, besides its cognitive value, since the *goodness* of scientific knowledge is linked with its truth value. Karl Popper expressly did this. In his approach, "ethical principles form the basis of science. The idea of truth as the fundamental regulative principle—the principle that guides our search—can be regarded as an ethical principle. The search for truth and the idea of approximation to the truth are also ethical principles; as are the ideas of intellectual integrity and of fallibility, which lead us to a self-critical attitude and to toleration" (Popper 1984, p. 226, 1992, p. 199. See Gonzalez 1999b, pp. 155–156).

In Rescher's approach, the value of true knowledge is emphasized. So, in his judgment, "we cannot work in this world without an adequate cognitive accommodation. Our mind needs each piece of information as much as our body needs food" (Rescher 1999a, p. 166). In this regard, scientific prediction, as *cognitive content*, can also have an ethical role, within the context of the human activity.[17] This can be seen when he notes that the anticipation of the future is a *human need*: "to act, to plan, to survive, we must anticipate the future" (1998, p. 64). Therefore, scientific prediction carries a cognitive content that directly serves the best interest of people.

But, although Rescher's approach has characteristic features of ethical intellectualism—where there is a cognitive support for ethics, in general, and for the ethics of science, in particular—he expressly acknowledges that "inquiry is a human activity. It must, in the final analysis, be subject to the same general standards of moral appraisal that we use in relation to any other human activity" (Rescher 1996, p. 150). Therefore, the activity displayed by scientists when developing research must be regulated by the fundamental principles or controlling values that Rescher places in the second level of his "hierarchy of morality." These ethical principles and values are (Rescher 2008, p. 403):

(i) Do not cause people needless pain (*gentleness*); (ii) do not endanger people's lives or their well-being unnecessarily (*care for safety*); (iii) honor your genuine commitments to people; in dealing with people, give them their just due (*probity*); (iv) help others when you reasonably can (*generosity*); (v) do not take improper advantage of other (*fairness*); (vi) treat others with respect; (vii) do not violate the duly established rights and claims of others; (viii) do not unjustly deprive others of life, liberty, or the opportunity for self-development; and (ix) do not deliberately aid and abet others in wrongdoing.

Scientists must assume ethical principles and values when they are carrying out their research activity. Hence they regulate the human processes of making science, also in the case of basic scientific research. Thus, there are a series of endogenous ethical values in science (honesty, gentleness, care for safety, probity, fairness, etc.). These values modulate the aims and the means of research, as well as the behavior of the researchers, insofar as science is a human activity. From this perspective, an

---

[17] In Rescher's approach, the cognitive content of basic science as such (that is, considered by itself) is not susceptible of ethical evaluation. Ethical evaluation appears when basic science is seen as related with human activity (where ends, means, and results with their consequences are at stake).

ethical problem is whether certain aims and means are ethically adequate in accordance with endogenous criteria.

Consequently, Rescher's ethical intellectualism—which can be seen, above all, when he addresses ethical problems related to basic science—does not involve a view of scientific activity as ethically well-oriented (or as an activity that is neutral with regard to ethical issues). So, although he thinks of the possession of scientific knowledge as something neutral with regard to ethical values (that is, having knowledge is not something morally adequate or inadequate in itself) [Rescher 1999a, pp. 159–162], *scientific activity* (basic or applied) is certainly value-laden, and it is conditional on ethical values (both endogenous and exogenous).

From the endogenous perspective, there might be ethically inadequate behaviors in scientific research (such as, for example, in the cases of plagiarism or when fraud is committed in the publication of the achieved results). In order to evaluate those behaviors, Rescher usually gives primacy to the personal level—the agent's responsibility—over the institutional component. For this reason, he focuses on the individual actions and choices that can be subject to an ethical analysis (for example, when a scientist has to decide if he will use information that was obtained by ethically reprehensible means, such as experimentation on prisoners of war) [Rescher 1999a, p. 155].

But Rescher's approach is not reduced to the endogenous dimension of analysis. He also contemplates science as an activity connected with others, so there is interdependence with exogenous factors. This relation with the context is more important in applied science (where the relation with technology is also more relevant) and, above all, in the application of science, insofar as the *use* of scientific knowledge can be oriented towards issues that are questionable from an ethical viewpoint. In a clearer way than in the case of the endogenous ethics, the exogenous ethics of science involves paying attention to the institutional level (the laboratories, companies, research centers, etc.), where problems arise such as the credit for the results, allocation of responsibilities, or corporate interests.

9.2.2 Science and Exogenous Ethical Values: Science as Activity

Besides the endogenous ethics—which is oriented towards *scientific activity* as such—there is the exogenous dimension of analysis. From the exogenous dimension, ethics of sciences addresses *science as activity*, so the relations with the whole human experience (social, cultural, political, economic, ecological, etc.) are emphasized. Thus, there are more elements at stake in the ethics of science than those circumscribed to the specific realm of scientific activity. In this way, when science is considered as an activity among others, "there is, in effect, a linkage with ethical concerns of the general realm (social responsibility, solidarity, promotion of personal dignity, equal rights, ...)" (Gonzalez 1999b, p. 150).

This perspective is expressly assumed by Rescher, since he insists on the search for scientific knowledge as a human project *among others* (Rescher 1999a, pp. 99–121). For this reason, he thinks that science "must, in the final analysis, be subject to the same general standards of moral appraisal that we use in relation to any other human activity" (Rescher 1996, p. 150). However, he does not develop in an exhaustive way the problems that arise in the exogenous dimension of the ethics of science, because he is usually focused on the realm of basic science, instead of paying attention to questions that affect more specifically the applied research.

In effect, "it is in the *exogenous realm* where many of the reflections about the ethical valuations of applied science should be situated, because they have a direct social incidence (not only in the direct field of the human life, but also in what has to do with the environment—for example, contaminant chemicals—or what might have technological projection, above all when it is clearly inadvisable—for example, bacteriological weapons)" (Gonzalez 1999b, p. 151).

Thus, the exogenous ethics goes further than the simple attention to scientific activity as such, since it sees science as an activity that is developed in a wider context (social, cultural, political, economic, ecological, etc.). In this regard, an especially important problem is that related to social responsibility, both of the individual agents and of the groups of agents. In effect, scientific activity can affect the social life, above all when the activity is applied (medicine, applied physics, applied economics, etc.).

This is even clearer when applied science has technological projection,[18] because the social incidence of the technological devices is even more direct for citizens (for instance, in the areas of transport, communications, domestic devices, etc.). In addition, this social repercussion can be seen in the case of the application of science (for example, in issues of policy, like the Intergovernmental Panel of Climate Change).[19] Thus, applied science, application of science, and technology, can affect people, society, and environment from an ethical viewpoint, either in a positive or in a negative way.

On this issue of ethical responsibility, although Rescher is usually focused on questions related to basic science, there is in his approach a concern about the problems that have to do with collective responsibility (that is, the responsibility of the groups of agents). This problem is crucial for the ethics of science in its exogenous dimension, since scientific activity (basic, applied, or of application), as well as technological doing, are usually developed in institutions (companies, laboratories, research centers, etc.), so there is a social dimension that is characteristic of science as human activity.

---

[18] This projection has been clear during the last decades in the realm of communications (for example, between the signal theory and mobile telephony devices). But this feature does not involve reducing technology to applied science or considering that technology is a pure application of science. The relation between science and technology was emphasized in chapter one of this book, and it is described by Rescher as two legs of the same body (1999a, p. 100).

[19] This can be seen in the successive international meetings on climate change, where the meeting in Río de Janeiro (1992), Kyoto (1997), Copenhagen (2009), and Paris (2015) can be highlighted.

In order to address this problem, Rescher thinks that "the natural step is to begin by basing our understanding of group responsibility upon that of individual responsibility" (Rescher 2003d, p. 125). Since his ethical proposal is focused on the person, he considers the problem of social responsibility as related to scientists as individuals, so the set is reached from the individual agents. Thus, the ethical evaluation of entities is possible (in the case of science, laboratories, research centers, etc.) to the extent that ethical characteristics can be attributed to the individuals who form those entities.

Regarding this issue, Rescher considers that, besides the element of "causality," the notion of "intentionality" is crucial in order to have a genuine *collective responsibility* (2003d, p. 126). Thus, the fact that the actions of a group of individuals are the cause of a certain result is a *necessary* condition to consider the group as ethically responsible for this result, but it is not a sufficient condition to attribute responsibility. Moreover, in his judgment, "without a normative responsibility that transcends mere causality, there can be no actual guilt" (2003d, p. 130).

Seen from an ethical perspective, science has intentionality.[20] As an intentional human activity it is oriented towards the search of aims (theoretical or practical), which can be ethically adequate (the research for the development of a new vaccine) or clearly inadequate (the research oriented to the subsequent development of bacteriological weapons) [Gonzalez 1999b, p. 143]. From this viewpoint, it seems clear to me that laboratories, research centers, etc., can be "moral subjects," since the individual members who form the group actively and intentionally seek a previously established goal.

But, different from the intentionally of individuals, the intentionality of a group requires some kind of coordination among the individual members of that group. Thus, "when the products of group activity are concerned, it only makes sense to speak of group intentions in the case of coordinated productions. Without the synthesis or unification of actions there is no meaningful collective intention" (Rescher 2003d, p. 130). In the case of science (either basic or applied), coordination and cooperation among scientists is usually present, since scientific activity is commonly developed by communities within an institutional framework (universities, laboratories, research centers, companies, etc.).

In this regard, Rescher notes that "there are principally two kinds of teams: those that are leader directed and/or hierarchical and those that are purely cooperative and unstructured" (Rescher 2003d, p. 137). The kind of coordination that appears among the members who form a group has consequences for the allocation of ethical responsibilities. Because, in the framework that Rescher expressly suggests, a group of agents can be ethically responsible to the extent that its individual members are responsible. So, in order to attribute social responsibility to a group, from an ethical viewpoint, the individual agents that form the group must be responsible to some extent for the activity carried out and for the result finally achieved.

---

[20] The concept of "intentionality" is understood as a feature of a human activity, a feature which is developed in the action of an agent or a group of agents, so it is not equivalent to "intention" as a simple mental act (and, therefore, unobservable). On this issue, see Elskamp (1986).

Consequently, Rescher thinks that, when the group has a hierarchical structure, the allocation of responsibility is, in principle, easier than when the group works in a cooperative way, without a well-delimited hierarchy or without a leader who guides the activity (Rescher 2003d, pp. 137–138). The same happens when the issue is to allocate credit or authorship of a result obtained through research (Rescher 2006b, p. 216). Thus, the institution where research is developed can be considered either responsible for an ethically reprehensive result (for example, the pollution of the environment) or can receive credit for the positive results achieved by its researchers (for example, a new vaccine).

But, in order to specify to what extent there is an ethical responsibility, we have to determine the specific individuals who have cooperated in the activity carried out and their contribution to the final result. In principle, this task is easier when there is a leader or a hierarchical structure within the group, while responsibility and credit tend to fade when the group works through the cooperation of its members, without a clear internal structure (Rescher 2003d, pp. 137–138).

Regarding this problem, the distinction between "legal responsibility" and "moral responsibility" is important. For Rescher, "groups can be *legal* persons and thus bear legal responsibility. Legal responsibility is alienable and capable of transfer and delegation" (2003d, pp. 132–133). Meanwhile, moral responsibility is, in his judgment, inalienable. Thus, "groups can only bear [moral] responsibility derivatively—either by way of aggregation (consensus) or by way of delegation (via representation)" (Rescher 2003d, p. 132). Moreover, the existence of moral responsibility—either individual or of a group—does not necessarily involve legal responsibility, and vice versa.

However, the legal features have repercussions on the ethics of science; because science is developed within legal frameworks (autonomic or regional, national, and international) that have consequences for the scientific research. The legal framework affects, for example, issues such as the funding of science, insofar as it can provide the conditions for allowing or prohibiting ethically questionable researches (for instance, those oriented toward the development of chemical weapons) [Balmer 2015]. In this regard, Rescher seems to assume that the legal realm is more variable; whereas ethics has a more stable basis, since it is supported by universal ethical principles and values.

Due to Rescher's insistence on the ethical features that are universal, his approach to ethics, in general, and to the ethics of science, in particular, does not grasp adequately the variability in the field of human values (among them, ethical values). In my judgment, when the ethics of science is analyzed from the exogenous perspective, we can clearly see how there might be variability in the ethical values, insofar as science is a *human activity* with a dynamic component. The dynamic component of science has an internal dimension (the articulation in aims, processes, and results) and an exogenous component, which has to do with the relations with the sociohistorical context where scientific activity is developed (Gonzalez 2012, pp. 8–9).

From this perspective, the exogenous values (solidarity, legality, social responsibility, etc.), which regulate or should regulate science according to its social incidence, are, in principle, more variable than the endogenous ethical values (honesty, coherence, etc.). In effect, the exogenous ethical values modulate science as an activity developed in a context—social, cultural, political, economic, ecological, etc.—that is changeable. This feature implies that the ethics of science is open to the historicity of human activity, which is compatible with the acknowledgement of the objectivity of values.[21]

However, Rescher's approach is more static. Thus, he first gives primacy to the idea of a system. The idea of a system is important both in his conception of human values, in general, and ethical values, in particular, which are structured into a hierarchy of values. Hence, the universal values, which are absolute and inalienable, are at the first level of the hierarchy; whereas those issues that are subject to variation are at the lowest levels of the hierarchy (Rescher 2008, p. 403). And, secondly, he is interested in rejecting ethical relativism, so he usually insists on the universality of values (see, for example, Rescher 1997a).

Consequently, although he acknowledges that there are variations in values, Rescher sees this problem in contextual terms. So, in his judgment, there are contextual matters that affect concrete ethical codes. But this is not an acceptance of the historicity (of science, the agents, and the reality itself that is researched). In my judgment, the acknowledgement of a genuine *historicity* is required in order to address the ethical analysis of science, both from the endogenous viewpoint and— above all—from the exogenous perspective. Historicity is crucial in order to consider the problem of the limits of science, since it allows us to address the factors of variability, both of scientific activity as such (the endogenous viewpoint) and of science as activity related to other activities (the exogenous perspective).

9.3   The Ethical Limits of Scientific Research: Their Incidence for Scientific Prediction

When the ethical limits of scientific research are considered, a double perspective of analysis should be taken into account. On the one hand, the ethical limits can be *internal* to scientific activity, where the aims, processes, and results are addressed. But, on the other hand, the ethical limits can be *external*. They appear insofar as science is an activity related to other activities (social, cultural, political, economic, ecological, etc.). These limits—internal and external—have incidence for scientific prediction, insofar as prediction is the result of an *activity* that seeks to achieve justified answers to meaningful questions regarding future events (Rescher 1998, pp. 37–39).

As activity, scientific prediction can be subjected to an ethical evaluation of its aims, processes, and results. Here there can be limits according to ethical considerations. In this regard, the differences between the basic scientific activity, applied scientific research, and the application of science should be taken into account,

---

[21] The problem of how historicity can be compatible with objectivity is analyzed in Gonzalez (2008b).

insofar as they affect the articulation of the aims, processes, and results (and their possible consequences. These differences have repercussions on the ethical values linked with prediction in each one of the three contexts—basic science, applied science, and application of science—both from the internal and the external perspective.

9.3.1 Approaches to the Ethical Limits of Scientific Research: Importance for Ends and Means

Regarding the problem of the ethical limits of scientific research, Rescher thinks that, in principle, three different approaches are available (1999a, pp. 152–154): (a) panregulation, which considers that scientific knowledge should be completely under control and regulated; (b) *laissez-faire*, which considers that controlling or limiting knowledge is never adequate; and c) centrism or moderate option, whose starting point is a view of scientific knowledge as a good among others.

Panregulation—the first possible option—consists of a totalitarian view of scientific knowledge. According to this approach, "knowledge is power, and the use of power in a community *always* should be controlled and regulated. Moreover, knowledge belongs to that especial category of things that require a particular treatment, and should be regulated with exceptional rigor" (Rescher 1999a, p. 152). Thus, the advocates of this approach maintain that every aspect of scientific research must be subject to an external control, so science loses its autonomy.[22]

Meanwhile, the second approach noted by Rescher—*laissez-faire*—maintains that scientific knowledge should never be limited, since it "belongs to the especial ethical category of things that—such as life or freedom—are subject to a fundamental right that is inalienable" (Rescher 1999a, p. 152). On this basis, limits cannot be set to scientific research, since scientific knowledge is considered as a supreme good. Therefore, science would be completely autonomous and any other attempt to set limits to scientific activity would be considered as something ethically inadequate.

These approaches—for Rescher—are unacceptable. Thus, he opts for the third approach (centrism or moderate option) according to which scientific knowledge should be considered as a good among others. On this basis, it is possible to preserve the autonomy of science and to acknowledge, at the same time, that there are ethical limits that should be respected.[23] From this perspective, the search for

---

[22] Niiniluoto is a philosopher who has insisted especially on the autonomy of science as a characteristic feature of what science is or should be. In his judgment, "the community of investigators ceases to be a scientific community if it gives up—or is forced to give up—this principle of autonomy," Niiniluoto (1984, p. 6). On the concept of "autonomy," see Gonzalez (1990, pp. 100–104 and 108–109, 2015, pp. 25–28) and O'Neill (2003).

[23] Rescher considers that autonomy is a defining feature of science. In his judgment, "the acceptability of the scientific proposals is an issue that should be completely resolved at the level of

knowledge should be subjected to ethically grounded considerations of general interest. Thus, "exactly in the same way as it happens with other activities, public interest is a potential source of adequate restrictions [to scientific research]" (Rescher 1999a, p. 152).

Rescher considers the three options regarding the problem of the ethical limits of science as conceptions of general character about scientific knowledge, which has also incidence for the problem of the ethical limits of scientific research. In this regard, it seems clear that his ethics of science is supported by epistemology. In this way, when he thinks of the ethical limits of scientific research (either in the realm of basic science or in the context of applied science), he usually gives primacy to the cognitive component of science.

Certainly, his approach to the limits of science is in accordance with his ethical intellectualism, insofar as there is a cognitive support for the ethics of science. But, in Rescher's approach, scientific knowledge is a good among others. This involves taking into account the *activity* developed by science, which should not hinder other human goods (freedom, safety, well-being, etc.).

Therefore, the centrist approach to the problem of the limits of science involves, on the one hand, considering knowledge as a good; because obtaining knowledge is a human need, due to the kind of creatures we are. This feature is especially clear when scientific knowledge about the possible future is considered, since the ability to anticipate the future is a crucial aspect of human survival (Rescher 1998, p. 64). Due to the high value of knowledge, he thinks that we should adopt a wary attitude regarding the issue of the ethical limits. In this judgment, "the idea of 'ethical limits' should be always implemented with care. Free research is a delicate and vulnerable plant, which should be fed and protected as much as we can" (Rescher 1999a, p. 166).

But, on the other hand, Rescher insists that knowledge is not a supreme good, so he sees the search for scientific knowledge as a valuable human goal *among others*. This means that "*there are* ethical limits to research that legitimate *can* be implemented in some cases" (Rescher 1999a, p. 167). Therefore, we should find a balance between the value of free research—and the due care and respect for the autonomy of science—and the acknowledgement of other values, which can lead to establishing limits to scientific research on the basis of ethical considerations (Gonzalez 1999b, pp. 154–155).

This involves considering the activity developed by science, both by itself and according to its relations with other human activities. From this perspective, it seems clear that there is a double dimension of analysis when the ethical limits of science are considered: (i) *scientific activity*, where there is an articulation of aims, processes, and results; and (ii) *science as activity*, which is oriented towards the

---

internal considerations to scientific endeavor. A 'science' subject to external criteria of correctness is simply not worthy of that name," (1999a, p. 115). The search for a balance between the autonomy of science and the limits due to ethical issues is a main concern in the recent literature on the ethics of science. See, for example, Douglas (2007, pp. 126–131).

relations between science and the context (social, cultural, political, economic, ecological, etc.).

Too often Rescher gives primacy to the internal dimension of analysis. On the one hand, his preference for the internal dimension is due to his structural conception about science, where the idea of a system prevails. And, on the other, he is more interested in basic research than in applied science, which has more repercussions in the social milieu. But, to the extent that he defends a holism of values, the external dimension cannot be omitted, because the scientific values and the other human values are interrelated (also in the case of the ethical values).

Within this framework, Rescher thinks that the most important ethical limits that can be put on scientific research are those related to three aspects of knowledge (1999a, pp. 154–159): (I) empirical information, because the use of information can be restricted when it is known information was obtained or fund through illegitimate means; (II) topics of research, insofar as there are some topics that can lead to damaging applications to people, society or the environment; and (III) research methods or procedures, because some methods of research can be ethically inadequate (for example, nonconsensual experimentation on human beings).

Undoubtedly, ethical limits can be put on scientific activity regarding these aspects of knowledge. But, in my judgment, the ethical limits of science can be better seen when the internal dynamics of *scientific activity* is considered. This perspective involves that the ethical limits—when they are contemplated from the internal dimension to science—have to do with the aims, processes, and results of scientific research, which are different in the context of basic science from the realm of applied science (see Gonzalez 2013b, pp. 17–18). From this perspective, which is oriented towards the three successive steps in the internal dynamics of scientific activity, the ethical problems related to prediction and prescription also become clearer.

Regarding the aims of research, a question is whether there are scientific goals that should be limited according to ethical values. Rescher's concern about the ethically inadequate topics of research can be placed at this level. But, in this regard, a distinction should be made between the goals of basic science and the aims of applied science. So, to the extent that basic science is directed toward increasing or improving the available knowledge, the goals of this kind of research do not seem to be ethically problematic by themselves.

From this perspective, basic research cannot be limited on the basis of its aims, which are usually the explanation and prediction of phenomena and events. Furthermore, insofar as it is oriented towards these aims, basic research provides knowledge, which is valuable in itself and in addition can contribute to cultivating other human goods (Rescher 1999a, pp. 105–106). Meanwhile, in the case of applied research, the aims sought have a more practical character. Therefore, it can be seen how applied research, according to the sought aims, can have direct repercussions (either beneficial or damaging) on people, society, and the environment.

On the one hand, according to the *aims* sought (for example, the development of a new vaccine), applied research can contribute to human well-being to a greater extent than basic science (for instance, astronomy).[24] But, on the other, there are aims of applied research that are ethically questionable, due to their negative repercussions on people, society, or the environment (for example, research about certain chemicals). In some cases it is clear whether the aim sought is ethically inadequate (an example could be human cloning), but there are cases where the search for a goal leads to an unexpected unfavorable result.

In this regard, prediction has a role for the anticipation of the possible unfavorable effects of research also in the case of basic science. So, although the aims of basic research (mainly, explanation and prediction) are not problematic from an ethical viewpoint, they can lead to undesired or undesirable applications. Thus, prediction can contribute to reducing the uncertainty about future phenomena, so it may be crucial in order to minimize the problem of risk (above all, in applied research) [Shrader-Frechette 1994, ch. 6, pp. 101–117; especially, pp. 111–117]. One must note, however, the difficulty in making predictions in the scientific realm, so it might be difficult to predict that a certain research will lead to damaging applications in the future (Rescher 1999a, pp. 158–159).[25] For this reason, he thinks that we should be especially wary if we limit scientific research on account of the concrete aim sought (Rescher 1999a, p. 163).

Once the aim of the research is established, *processes* must be selected to achieve that aim. In this regard, Rescher maintains that there should be limits to scientific research according to the processes or methods, since there may be ethically inadequate research methods. He provides several examples of ethical limits regarding the processes, such as the nonconsensual experimentation on human beings, or the experimentation that inflicts excessive or avoidable harm on animals (Rescher 1999a, p. 157). Ethical limitations regarding the processes also include the use of information obtained by ethically questionable procedures (for example, the experimentation on prisoners of war) [Rescher 1999a, p. 155].

After establishing the aims and selecting the processes, scientific research leads to a *result*, which can also be evaluated from an ethical perspective (mainly according to its consequences). This evaluation of the results can be either internal or either external to scientific activity (Neira 2015, p. 47). From the internal perspective, the adjustment of the result achieved to the previously established aim is assessed. Here, endogenous ethical values such as honesty or reliability are taken into account. According to these values, it can be claimed that there are questionable practices, like fraud in the publication of results, which are outside the limits of scientific research.

In addition, an external evaluation of the results can be made. This leads to the exogenous dimension regarding the limits of scientific research, which takes into

---

[24] This is expressly accepted by Rescher. In his judgment, "one very pervasive problem at this institutional level is the classical issue of pure, *or* basic, versus applied, *or* practical, research. This problem is always with us and is always difficult, for the more 'applied' the research contribution, the more it can yield immediate benefits to man; the more 'fundamental', the deeper is its scientific significance and the more it can contribute to the development of science itself," (2006b, p. 204).

[25] On the problems related to the prediction of future science, see Rescher (2012a).

account that the results of scientific research (and their possible consequences) can have negative repercussions on people, society, or the environment, so they interfere with other human values (social, cultural, ecological, etc.). This feature can be clearer seen in the case of applied science—where the relations with technology are more important—than in the realm of basic science, where the endogenous factors often have primacy.

But, besides basic science and applied science, the application of science must also be considered (the use of scientific knowledge by agents; for example, in professional contexts such as medical practice) [See Niiniluoto 1993]. The perspective of the application of science highlights how basic research, as well as applied science, may lead to undesired or undesirable applications. In this regard, there are ethical elements that modulate scientific prediction in the application of science, since prediction can serve as a guide for agents' decision-making when they apply scientific knowledge (for example, in issues related to climate change).

9.3.2 From the Internal Viewpoint to the External Perspective

Undoubtedly, over the last decades the external perspective of analysis has motivated the development of the ethics of science, in general, and the concern about the ethical limits of scientific research, in particular. In effect, "one of the reasons for the increasing interest in the ethics of science is the existence of a negative social perception regarding some recent scientific developments, above all in their technological projection (especially in the realm of the genetic engineering and in the cases of environmental damage due to industrial products)" (Gonzalez 1999b, p. 152).

Therefore, the need to establish external ethical limits to scientific activity seems to be clear when science is contemplated as an activity connected with other human activities. From this perspective, the relations with the context (social, political, cultural, economic, ecological, etc.) must be considered. Thus, it can be seen how the results of the scientific activity (and the possible consequences of those results) can have repercussions on people, society, and the environment, either in a positive or in a negative way.

Certainly, science can have a *positive* effect in other realms of human experience, above all when it is applied science (for example, medical research has decisively contributed to health, which is a human good). But scientific research can have also a *negative* effect for people, society, and the environment, to the extent that it can interfere with other human values, such as safety, well-being, or sustainability. Thus, it seems clear that, besides the endogenous limits to scientific activity, there are also exogenous limits, which consider the repercussions of science in the whole human experience. These are clearer in applied science—which usually has technological projection—than in the context of basic research, where the internal factors have usually primacy.

Insofar as the external ethical obstacles limit *science as activity* (that is, due to its relations with the context), they are, in principle, more variable than the limits due to internal factors (such as honesty, reliability, or truthfulness). In this regard, Agazzi's proposal is interesting. He highlights the variability that has to do with contextual factors and that has incidence for the problem of the limits of science. In his judgment, besides the aims, processes, and results, the analysis of the ethical limits of science must take into account the *conditions* (those factors that make an action possible) and the *circumstances* (the context where the action is developed) [Agazzi 1999, p. 255. See also Gonzalez 1999b, pp. 160–161].

According to Agazzi, "in order to evaluate an action from a moral viewpoint it is almost always needed to take into account its *conditions* and *circumstances*. Sometimes, the concrete conditions that would make an action possible (fair by itself) will not be morally acceptable: in this case, the mentioned action should not be performed" (1999, p. 255). Among conditions, it is possible to highlight, in my judgment, the means and resources required to develop the research, which can be ethically questionable. An example in this regard is that of illicitly funded research. Thus, although the aims, processes, and results of a specific piece of research are adequate from an ethical viewpoint, the research should not be developed if, for example, the funding has been achieved in an ethically inadmissible way.

Agazzi also notices that "sometimes circumstances surrounding action must be considered: it is possible that a course of action morally allowed under 'normal' circumstances becomes morally reprehensible under 'exceptional' circumstances, or vice versa" (1999, p. 255). This might happen, for example, when priorities must be established in the systems of scientific and technological policy (at the regional, national, and international levels) in order to support research problems. In this regard, circumstances can be crucial and have incidence for the consideration of the economic distribution as adequate or inadequate from an ethical viewpoint (for example, when a country is at war, assigning many resources to research oriented to military ends might be considered as correct).

Rescher also pays attention to the context in which human activity is developed, and he is aware that context introduces elements of variability in ethics, in general, and in ethics of science, in particular. This appears explicitly in his "hierarchy of morality," where the values of the lowest levels vary according to contextual factors (Rescher 2008, pp. 403–405). Thus, his ethics of science combines the existence of universal ethical values with issues open to changes due to context-dependent factors. Moreover, regarding the ethical limits of scientific research, he notes that issues such as the means or resources that make scientific research possible should also be taken into account (Rescher 1999a, pp. 155–156), so his approach is in some sense open to considering the "conditions" and "circumstances" of actions.

However, although Rescher's approach is open to variable elements regarding ethical values, which have repercussions on the variability of the limits of science (at least, when they are considered from the external dimension), he is usually more concerned with ethical values and principles that are stable, insofar as they have features of universality. This is related to his rejection of a moral relativism of either social or historical character. In effect, in his proposal, the objectivity of values, in general, and of ethical values, in particular, is crucial, to the extent that it allows the rational deliberation about problems that have to do with value matters (Rescher 2008).

From this perspective, ethical limits to scientific research—internal and external—have a *rational* basis, which is directed to what is objectively preferable, instead of being merely focused on what is preferred. In this way, the ethical limits of science are not something constructed, neither are they the result of consensus (explicit or implicit), but they are established through a rational deliberation that seeks optimal solutions taking into account the concrete circumstances and the context (Rescher 1988, pp. 169–175, 1999a, pp. 90–96). Thus, in his approach, Rescher combines the objectivity of values—grounded in universal human needs—with the contextual character of the human activity performed in a social milieu.

But, although Rescher takes into account the context (social, cultural, political, economic, ecological, etc.), he does not pay attention to the concept of *historicity* that, in my judgment, is required for adequately developing the dynamical component of the ethics of science. Instead of being a conception open to historicity, Rescher's approach is more static. Thus, on the one hand, his analysis is mainly structural, so ethical values are part of the wider system of human values. And, on the other hand, he searches for invariable ethical values and principles, so there can be objective basis for the rational deliberation about ethical issues.

In this regard, it should be noted that historicity is compatible with the objective character of the problems of evaluation, so the acknowledgement of historicity (of science, agents, and the reality researched) does not necessarily lead to ethical relativistic conceptions.[26] Furthermore, the objectivity of the issues of ethical evaluation that have to do with science require, in my judgment, taking into account the internal dynamics of scientific activity—the changes regarding the aims, processes, and results of the research—and the external dynamics (the relations with the context, which is also changeable).

Concurrently, the relation of scientific prediction with ethical problems can be best seen—in my judgment—when historicity is acknowledged. This aspect of the philosophico-methodological analysis of scientific prediction is not developed by Rescher. In effect, in his main book about prediction, there is no specific attention to the study of scientific prediction from the viewpoint of the ethics of science (Rescher 1998). In my judgment, in order to develop an ethical analysis of scientific prediction, the dynamical component (both internal and external) of the scientific activity must be considered, paying attention to the notion of historicity. Moreover, this should be done regarding three cases: basic science, applied science, and the applications of science.[27]

---

[26] On how historicity is compatible with objectivity, see Gonzalez (2008b).

[27] These three cases should be taken into account in the development of the axiology of research as well. See Chap. 8, Sect. 8.5 of this book.

9.4 Prediction and Ethical Values from the Dynamic Perspective: Evaluation of the Aims, Processes, and Results

Ethics of science can be addressed in relation to three different realms: basic science, applied science, and the application of science. The relation between science and ethical values has different characteristics in each one of these realms, since they are different activities (see Chap. 5, Sect. 5.2, and Chap. 8, Sect. 8.5). Their differences have repercussions on the relation between prediction and ethical values in the three realms. This feature can be seen from the dynamic perspective, which has an internal dimension (the articulation in aims, processes, and results) and an external one, which has to do with the relations with the context. The different internal configuration of the three types of activities (basic, applied, and of application) as well as their own features regarding the external dynamics involve specific ethical problems. However, they have also some aspects in common, since they are free human activities of a social character.

9.4.1 Ethical Values and Prediction in the Realm of Basic Science

Commonly, basic science is considered as the type of scientific activity directed at increasing or improving the available knowledge. For this reason, within the ethical concern with basic science, one problem which has received especial attention is that related to whether the possession of knowledge can be ethically inadequate as such. In this regard, as Wenceslao J. Gonzalez notes, the aspect of the cognitive content should be distinguished from the dimension of the free human activity (1999b, p. 161).

In effect, "from the point of view of the very *content* of knowledge (as its truth value, for example, is considered), ethics does not have a realm where it can work on: ethics does not evaluate the cognitive content as such. Meanwhile, to the extent that knowledge is connected with a *free human activity*, which has some goals or aims, some means, and some results—and can have also links with other free human activities—then the situation is different. From this perspective, the ethical evaluation is possible, which is not oriented towards the possession of information *per se*, but towards the human activity that leads to information search, its acquisition, and its use" (Gonzalez 1999b, p. 161).

When Rescher takes into account the problem of the ethical evaluation of basic science, his approach follows this very direction. Thus, on the one hand, he sees scientific knowledge as neutral with regard to ethical values. He insists that the possession of knowledge cannot be valued as something morally adequate or inadequate in itself (Rescher 1999a, pp. 159–162). But, insofar as science is a free human activity, he considers that basic scientific research is value-laden and conditional on ethical values, which are mainly endogenous (1999a, pp. 162–165).

Thus, although scientific knowledge as such can be considered as neutral regarding ethical values, the ethical evaluation of basic science is possible when the free and intentional human activity is considered. This involves taking into account the three successive steps of the internal dynamics of scientific activity, where ethical values can intervene: the concrete aims of the research, the processes oriented towards achieving those aims and the result finally achieved (and its consequences) [Gonzalez 1999b].

Regarding the aims of the basic research, scientific prediction has an important role. On the one hand, basic research is oriented toward increasing our knowledge, so prediction is a relevant aim; and, on the other, it seeks to improve the available knowledge, so it emphasizes prediction as scientific test. Insofar as it provides knowledge about the possible future, scientific prediction as an aim of science does not pose any ethical problems.

Moreover, it is possible to think—as Rescher does—that knowledge about the future can raise a positive ethical evaluation. Thus, on the one hand, it allows us to know the reality better and, on the other, it favors agents' adaptability and survival (Rescher 1998, p. 64). However, basic scientific activity (also when it is oriented towards knowing the possible future) requires some means, which may be reprehensible form an ethical viewpoint. This is expressly maintained by Agazzi. In his judgment, the problem of the moral judgment regarding the *means* not only refers to applied science, but also to basic science (Agazzi 1992).

Thereafter, the means generate some results that can also have inadvisable consequences. In this regard, Rescher's reflection on the ethical problems related to the standards of proof can have repercussions for the problem of the ethical values of prediction and the realm of basic science. According to Rescher, "in presenting particular scientific results, and especially in presenting his own results, a researcher may be under a strong temptation to fail to do justice to the precise degree of certainty and uncertainty involved" (2006b, p. 208).

Uncertainty is one of the main obstacles to scientific prediction (Rescher 1998, p. 135). Furthermore, to the extent that prediction is about the future (something that is not yet), it seems that there is always certain degree of uncertainty that accompanies a prediction. Thus, the temptation to improperly reduce the degree of uncertainty that accompanies the results can appear when those results are about the possible future. Regarding this issue, there are endogenous ethical values that should be present in the presentation and communication of the scientific results, such as reliability or credibility.

The results of basic science can have consequences that might be evaluated from an ethical perspective. In order to address this problem, the distinction between "non-oriented basic research" and "oriented basic research" is certainly important (Gonzalez 2007, p. 191). The former is a type of research that is not connected with applied research, even in the long run; while the latter does lead to an applied dimension, at least potentially. This distinction can be useful to evaluate prediction, because "it seems obvious that some predictions belong to 'non-oriented basic research' (for example, predictions about 'black holes' in distant galaxies) and others fall within the 'oriented basic research,' which is the most frequent case in economic predictions" (Gonzalez 2007, p. 191).

According to this distinction, it seems clear to me that scientific prediction in basic science, from the point of view of the consequences, can be subjected to an ethical evaluation, above all when the basic research is connected with applied science. In this regard, anticipating the possible objectionable consequences of a research is especially important, since they can lead to limiting a research or to making changes regarding the aims of the means in order to minimize those possible negative consequences. However, as Rescher notes, anticipating the development of future science is something especially difficult (2012a).

On this issue, Agazzi also notes that the possible negative consequences of the results of basic science are not themselves foreseeable or necessary, since they depend on the free and conscious choices of the agents (Agazzi 1992. See Gonzalez 1999b, pp. 161–162). Consequently, when ethical evaluation has to do with the consequences of scientific research, a wary attitude seems the best option. As Rescher maintains, "the aspirations of unrestrained research are essential. They have weight enough and are valid enough so they never should be put aside because of something that lacks clear and actual dangers" (1999a, p. 166).

9.4.2   Ethical Elements of Prediction in Applied Science and Repercussions on Scientific Prescription

Certainly, ethical problems that have to do with science, in general, and with scientific prediction, in particular, are more visible in applied science than in the realm of basic research. This higher visibility of the ethical matters in applied science is because, compared with basic research, the social repercussions of applied research are greater. Thus, there are values and ethical problems related to the *exogenous* dimension of analysis (the relations with the context), since applied scientific research can directly affect people, society, and the environment.

Furthermore, within this scientific realm, prediction has a direct relation with prescription, since the applied scientific activity is oriented towards solving concrete problems, so it needs guides for action in order to achieve its aims (Gonzalez 2015, p. 317). Thus, ethical values that affect scientific prediction in applied science can be seen, above all, when its relation to *prescription* is considered. Prescription involves a normative aspect—it suggest the guides for action that *should* be followed in order to solve some concrete problem—which is connected with the realm of values, were ethical values can have an important role.

In this regard, as A. Sen has emphasized in the case of applied economics, the prescriptive dimension always leads to an ethical component. In his judgment, "any prescriptive activity must, of course, go well beyond pure prediction, because no prescription can be made without evaluation and an assessment of the good and the bad" (Sen 1986, p. 3). In turn, the evaluative aspect, which encompasses an ethical component—endogenous and exogenous—implies that "the 'normative' domain cannot be filled completely by prescriptions, insofar as prescriptions are made after an evaluation of the goals to be achieved by them" (Gonzalez 2015, p. 327). Thus, the prescriptive activity of applied science (which need knowledge about the possible future) requires an evaluation of ends in which ethical criteria (endogenous and exogenous) intervene.

This approach involves assuming a rationality of ends, so scientific rationality is not a simple instrumental rationality. In the case of applied economics, Gonzalez notices that "rationality of ends would take charge of drawing the aims of economic prescriptions. This includes several levels: first, the selection of possible *aims*; second, the elaboration of *priorities*, either in terms of a hierarchy of ends within a delimited field or through a consideration of the realms which are or should be priorities; and third, evaluation of the *consequences* which derive from those aims (in economic terms or in other terms: social, cultural, political, etc.)" (Gonzalez 2015, p. 333). This approach is valid for any applied science, so it is not confined to the realm of economics.

In effect, within the sphere of applied science, prediction is related methodologically to the future. Thus, the prediction about the possible future is needed to establish what should be done (Gonzalez 2008a, p. 181). Thus, prescription seeks to offer solutions in order to resolve the concrete problems posed; so it has a clear teleological character. In this regard, first, the concrete aims are selected; secondly, priorities are established regarding the aims; and, thirdly, the possible consequences are evaluated. Certainly, ethical values and problems intervene at each of these levels.

Regarding the aims, it seems clear that applied science is more connected with the context than basic science, so its social repercussions are also higher. Social incidence of applied research can be either positive (prescription can lead to solutions that provide a social or environmental benefit, for example, the research about renewable energy), or negative or questionable from an ethical viewpoint (for example, the research with military ends). This feature is acknowledged by Rescher, although he is above all interested in the specific problems of basic research.

As Rescher notes, the individual level—the decisions of the scientists—should be distinguished from the institutional level (the research centers where science is developed). Since his ethics of science focuses on people, he thinks that, at the individual level, "the ethical question of research goals and the allocation of effort—namely that of the individual himself—can arise and present difficulties of the most painful kind" (Rescher 2006b, p. 205). Thus, when the scientists choose the aims of the research, the decision-making can take into account ethical criteria (for example, "in the choice of a military over against a nonmilitary problem context—A-bombs versus X-rays, poison gas versus pain killers" [Rescher 2006b, p. 205]) .

Meanwhile, at the institutional level, the research centers (public or private) should select the research lines and the concrete ends. In this regard, Rescher notes that frequently the choice "is resolved in favor of the applied end of the spectrum by the mundane, but inescapable, fact that this is the easier to finance" (Rescher 2006b,

p. 205), instead of taking into account the social benefit that the research can produce.

Once the realm of the investigation and the concrete aims have been decided, it is possible to order them according to a hierarchy of priorities. Again, decision-making of scientists themselves and of the institutions where science is developed, as well as the systems of scientific and technological policy (regional, national, and international), intervene. These decisions can be guided by ethically adequate criteria (for example, giving priority to research that can contribute to the common good or can motivate the progress of the discipline) or other criteria, which might be questionable from an ethical viewpoint (for instance, the search for personal or institutional prestige, economic benefit, etc.).

Thereafter, prescription requires an evaluation of the possible consequences of the results eventually achieved. In this regard, prediction is crucial; because, as Simon states "we construct and run models because we want to understand the consequences of taking one decision or another" (Simon 1990, p. 10). Thus, in applied science, prediction allows us to anticipate the possible problems, so it precedes and serves as a basis for the prescriptive task of science (for example, research about climate change). But prediction also has a role for the anticipation of the possible consequences of the solutions offered. In this way, it also contributes to avoid or, at least minimize, the negative consequences of prescription.

Where the ethical problems posed by applied science are more pressing is, in my judgment, at the level of the consequences; above all, due to exogenous considerations, which have to do with the relation between science and the wider context. Thus, the consequences of the applied research can produce harm to people, society, and the environment. This feature has links with the problem of social responsibility, both of the individual agents (the scientists) and of the groups of agents (laboratories, research groups, etc.).

The problem of social responsibility also affects the means of the research, above all, when experimentation on human subjects is required. A good example in this regard is provided by clinical research in medicine and pharmacology, where "the fundamental ethical challenge posed by clinical research is whether it is acceptable to expose some to research risks for the benefit of others" (Wendler 2014, sec. 9). Prediction also has a role here since, in order to weigh adequately the risks and the potential benefits, we need a reliable anticipation of the risks that will be assumed by the experimental subjects and the benefits that will be obtained once the investigation is finished.

Again, the difficulties of making prediction regarding the development of science must be noticed. For this reason, it seems to me that Agazzi is right when he proposes that we should "study the 'reasonably foreseeable' consequences of our choices and select those that do not involve seriously negative consequences, even in the long run" (1999, p. 258). Regarding this issue, Agazzi's approach has links with Rescher's proposal, which insists that scientific research should be only limited on the basis of serious and actual dangers (Rescher 1999a, p. 163).

However, when the research has social incidence (as it is usually the case in applied science), an especially important issue is that related to who should participate in decision-making about the risks that can be considered assumable. Thus, on the one hand, science itself can inform us about the risks and their scope. This feature means that the standards of proof in applied research should be established taking into account ethical principles and values. In effect, as Carl F. Cranor notes, "otherwise we risk imposing the consequences of poor information, uncertainty, small samples, and scientific skepticism on an innocent public" (1994, p. 185).

But, on the other hand, when science is considered as an activity connected with other human activities (social, cultural, political, economic, ecological, etc.), it does not seem adequate that only scientists or experts intervene in the decision-making, because their decisions can affect the wider human experience. For this reason, Rescher is right, in my judgment, when he maintains that "the competing interests should be weighed here: the value of knowledge against people's well-being. (Those who develop a research are not necessarily the most qualified and the best situated to do the indispensable analysis.)" (Rescher 1999a, p. 163).

This feature means that it is indispensable that we find a balance between the autonomy of science and the respect for the exogenous ethical limits, which require science to take into account external considerations to scientific activity itself. Thus, on the one hand, "the acceptability of the scientific proposals is an issue that should be completely resolved at the level of internal considerations to scientific endeavor" (Rescher 1999a, p. 115), so science is autonomous regarding the criteria of correctness of knowledge. And, on the other hand, it seems clear that society as a whole—not only the scientists—should participate in decision-making regarding the ethical limits of scientific research (above all, if it is applied research).

9.4.3   The Problem of the Application of Scientific Prediction: The Variability of Contexts and the Adaptability of the Agents

Besides basic science and applied science, the problem of the application of science must also be considered. Application of science "is concerned with the use of scientific knowledge and methods for the solving of practical problems of action (e.g., in engineering or business)" (Niiniluoto 1993, p. 9). In this context of the application of science by agents (either by an individual or collectively), prediction again has a role, because it can serve as a support in decision-making (see Gonzalez 2013b, pp. 17–18). This is what usually happens in professional contexts (for example, in medical practice) and in the realms of policy (in economic matters, environmental policy, public health, etc.).

Certainly, the applications of science have clear ethical applications, insofar as the decision-making can affect people, society, and the environment. This feature has to do with three successive levels of the application of science: aims, processes, and results (and their possible consequences). Thus, there are ethical values and criteria at the three levels, so the selection of aims and processes, as well as the evaluation of the results (and the possible consequences) must take into account ethical values and principles (Gonzalez 2005, p. 26). Because the application of science is directly connected with science as activity (that is, it deals with the relations between science and the context), the values related to social responsibility (which can even lead to a legal responsibility) are especially important.

When the application of science is developed in professional contexts, the existence of deontological codes is especially relevant since they regulate the professional practices according to ethically adequate criteria at the three levels mentioned (aims, processes, and results). However, "like all rules (...) those in professional codes of ethics will never specify how to apply them in all situations. Thus, professional codes of ethics often provide no clear way to distinguish priorities among conflicting obligations (e.g., to employers versus the public), even though they proscribe certain behavior" (Shrader-Frechette 1994, p. 41).

Thus, the criterion of the professionals prevails, which transcends the structural dimension of the research and its applications, because it involves a dynamic component, which is oriented towards the free human activity of a social character that is developed in a changing context. In this way, deontological codes might not grasp all the situations, since the decisions with an ethical component have to do, on the one hand, with the relations between the agent or the groups of agents with the *context* (natural, social, or artificial) and are open to changes; and, on the other, there are variations because of the historicity of the very relations of the *researchers*, which have interactions among them when developing their professional activity (in laboratories, hospitals, universities, etc.).[28]

These features are even clearer when the application of science deals with problems related to public health, since in this case there are more agents involved and the relations with the context are also broader, so there are a large number of variables that are not easily predictable. The interest of Rescher in the application of science is at this level, because he mainly focuses on the use of the knowledge about the future in matters related to policy, instead of being focused on professional practices (Rescher 2012b).

Rescher admits that there might be complexity regarding knowledge when decision-making is at stake: "the decision problems that we face in contemporary public affairs are often too complex to allow a resolution by way of rational calculation and what might be called the application of 'scientific principles'" (2012b, p. 205). From this perspective, the application of science, in general, and of scientific predictions, in particular, must deal with the problem of complexity, which has incidence on phenomena in the short, middle, and long run.

However, Rescher does not relate the complexity of the application of scientific knowledge with the *historicity of the human activity*. In effect, he does not address how ethical factors are modulated by the relations with the context (natural, social, or artificial) and by the interactions among the agents themselves who apply scientific knowledge. This is, in my judgment, an indispensable element in order to

---

[28] Gonzalez, *Personal communication*, 2.1.2015.

understand the problems that have to do with the application of science; especially, when those problems are of an ethical character.

In effect, when the applications of science have to do with complex systems (as is the usual case in the context of policy), the obstacles that make prediction difficult are often related with the historical character of those systems. Thus, prediction and the subsequent prescription—with its expression in terms of planning in order to solve practical problems—should deal with many variables that are open to changes because of historicity.

This feature poses problems for the application of science, which can also have an ethical component (for example, in issues related to public health, as happens when there is risk of an epidemic, or in questions related to climate change).[29] In effect, knowledge about the possible future—prediction—is required in order to anticipate the possible problem and the consequences of the solutions that might be adopted. However, it is not an easy task, because, as Rescher notes, "the eventual effects of the measures we take to address the challenges become lost in a fog of unpredictability" (2003b, p. 89).

The lack of reliable predictions that serve as a guide in order to search solutions to the practical problems poses clear difficulties for the process of decision-making. Thus, in this way, the agents' action prevails. They may, on the basis of the same applied knowledge, apply knowledge in different ways taking into account different contexts. So, in the first place, faced with the same problem (for example, risk of epidemic), the different experts or entities which manage an issue might suggest different solutions. And, in the second place, it is also possible that the experts have opposite opinions regarding the consequences of the solutions suggested.[30]

On this problem, Rescher thinks that very often a direct transfer of the solutions suggested by scientists to solve practical problems is not possible. For this reason, he notes that "we very much need them [the experts] to indicate alternatives, clarify issues, assess consequences, evaluate assets and liabilities, and generally work to inform the public debate on the issues. But we emphatically do *not* need them to decide matters. By all means let them [the experts] do their work and have their say about it. But when this is said and done, then by all means let the people decide" (Rescher 2003c, p. 121). Here, the main problem is the allocation of responsibility, because, when decisions are collectively made, responsibility (both moral and legal) tends to fade (Rescher 2003d, pp. 137–138).

In my judgment, Rescher suggests a rejection of a naïve view regarding the ability of science to offer solutions to practical problems of action. Thus, he thinks that the rational procedure in decision-making regarding issues of public interest "is to feel our way cautiously step by step—to experiment, to try plausible measures on a small scale and see what happens, and to let experience be our guide" (Rescher 2012b, p. 212). Therefore, scientific knowledge, in general, and knowledge about

---

[29] See, in this regard, Mearns (2010) and Parker (2011, 2014).

[30] This kind of problems is also frequent in professional practices (for example, when several physicians disagree about the best way of treating a patient).

the future, in particular, can serve as a guide for decision-making, but commonly they do not offer univocal solutions to the problems.

All these considerations emphasize the importance of the ethical values and principles in addressing the problems related to the application of science. But this task requires taking into account the historicity of the human activity developed within a context (social, cultural, political, economic, ecological, etc.), which is open to changes. Thus, on the one hand, historicity makes prediction difficult, so it hinders its role as a guide for the solution of practical problems; and, on the other, historicity modulates the ethical factors that intervene in the application of science according to the aims, processes, and results (and their possible consequences). Due to his lack of attention to historicity, Rescher's approach does not encompass—in my judgment—all the elements at stake when the ethical problems posed by prediction in the realm of the application of science are analyzed.

References

Agazzi, E. 1992. *Il bene, il male e la scienza. Le dimensioni etiche dell'impresa scientifico-tecnologica*. Milán: Rusconi. Spanish version of Ramón Queraltó: *El bien, el mal y la Ciencia. Las dimensiones éticas de la empresa científico-tecnológica* 1996. Madrid: Tecnos.

———. 1999. Límites éticos del quehacer científico y tecnológico. In *Ciencia y valores éticos*, ed. W.J. Gonzalez. Monographic issue of Arbor 162 (638): 241–262.

Anscombe, G.E.M. 1958. Modern moral philosophy. *Philosophy* 33: 1–19. Reprinted in Wallace, G. and Walker, A. D. M. eds. 1970. *The Definition of Morality*, 211–234. London: Mathuen.

Balmer, B. 2015. The social dimension of technology: The control of chemical and biological weapons. In *New perspectives on technology, values, and ethics: Theoretical and practical. Boston studies in the philosophy of science*, ed. W.J. Gonzalez, 167–182. Dordrecht: Springer.

Cranor, C.F. 1994. Public health rsearch and uncertainty. In *Ethics of scientific research*, ed. K.S. Shrader-Frechette, ch. 10, 169–185. Lanham: Rowman and Littlefield.

Douglas, H. 2007. Rejecting the ideal of value-free science. In *Value-free science? Ideals and illusions*, ed. H. Kincaid, J. Dupré, A. Wylie, 120–139. New York: Oxford University Press.

Elskamp, R.G. 1986. Intención e intencionalidad: Estudio comparativo. *Anales de Filosofía* 4: 147–156.

Gonzalez, W.J. 1983. La primitividad lógica del concepto de persona. *Anales de Filosofía* 1: 79–118.

———. 1990. Progreso científico, autonomía de la Ciencia y realismo. *Arbor* 135(532): 91–109.

———. 1999a. Racionalidad científica y actividad humana: Ciencia y valores en la Filosofía de Nicholas Rescher. In *Razón y valores en la Era científico-tecnológica*, N. Rescher, 11–44. Barcelona: Paidós.

———. 1999b. Ciencia y valores éticos: De la posibilidad de la Ética de la Ciencia al problema de la valoración ética de la Ciencia Básica. In *Ciencia y valores éticos*, ed. W.J. Gonzalez. Monographic issue of Arbor 162 (638): 139–161.

———. 2005. The philosophical approach to science, technology and society. In *Science, technology and society: A philosophical perspective*, ed. W.J. Gonzalez, 3–49. A Coruña: Netbiblo.

———. 2007. La contribución de la predicción al diseño en las Ciencias de lo Artificial. In *Las Ciencias de Diseño: Racionalidad limitada, predicción y prescripción*, ed. W.J. Gonzalez, 183–202. A Coruña: Netbiblo.

———. 2008a. Rationality and prediction in the sciences of the artificial. In *Reasoning, rationality and probability*, ed. M.C. Galavotti, R. Scazzieri, and P. Suppes, 165–186. Stanford: CSLI Publications.

———. 2008b. El enfoque cognitivo en la Ciencia y el problema de la historicidad: Caracterización desde los conceptos. *Letras* 114 (79): 51–80.

———. 2012. Las Ciencias de Diseño en cuanto Ciencias de la Complejidad: Análisis de la Economía, Documentación y Comunicación. In *Las Ciencias de la Complejidad: Vertiente dinámica de las Ciencias de Diseño y sobriedad de factores*, ed. W.J. Gonzalez, 7–30. A Coruña: Netbiblo.

———. 2013a. Value ladenness and the value-free ideal in scientific research. In *Handbook of the philosophical foundations of business ethics*, ed. Ch. Lütge, 1503–1521. Dordrecht: Springer.

———. 2013b. The roles of scientific creativity and technological innovation in the context of complexity of science. In *Creativity, innovation, and complexity in science*, ed. W.J Gonzalez, 11–40. A Coruña; Netbiblo.

———. 2015. *Philosophico-Methodological Analysis of Prediction and its Role in Economics*, Springer, Dordrecht, 2015.

———. 2016. Rethinking the limits of science: From the difficulties for the frontiers to the concern on the confines. In *The limits of science: An analysis from "barriers" to "confines"*, Poznan studies in the philosophy of the sciences and the humanities, ed. W.J. Gonzalez, 3–30. Leiden: Brill.

Martinez Solano, J.F. 2005. *El problema de la verdad en K. R. Popper: Reconstrucción histórico-sistemática*. A Coruña: Netbiblo.

Mearns, L.O. 2010. Quantification of uncertainties of future climate change: Challenges and applications. *Philosophy of Science* 77 (5): 998–1011.

Moutafakis, N.J. 2007. *Rescher on rationality, values, and social responsibility. A philosophical portrait*. Heusenstamm: Ontos Verlag.

Neira, P. 2015. Values regarding results of the information and communication technologies: Internal values. In *New perspectives on technology, values, and ethics: Theoretical and practical.Boston studies in the philosophy and history of science*, ed. W.J. Gonzalez, 47–60. Dordrecht: Springer.

Niiniluoto, I. 1984. *Is science progressive?* Dordrecht: Reidel.

———. 1993. The aim and structure of applied research. *Erkenntnis* 38 (1): 1–21.

O'Neill, O. 2003. Autonomy: The Emperor's new clothes. *Proceedings of the Aristotelian Society* 77: 1–21.

Parker, W.S. 2011. When climate models agree: The significance of robust model predictions. *Philosophy of Science* 78 (4): 579–600.

———. 2014. Values and uncertainties in climate prediction, revisited. *Studies in History and Philosophy of Science, Part A* 46: 24–30.

Popper, K.R. 1984. *Auf der Suche nach einer besseren Welt*. Munich: Piper. English translation by Laura J. Bennett and reviewed by K. Popper and M. Mew. 1992. *In Search of a Better World*. London: Routledge.

Prichard, H.A. 1912. Does Moral Philosophy Rest on a Mistake? *Mind* 21: 21–37.

Rescher, N. 1988. *Rationality. A philosophical inquiry into the nature and the rationale of reason*. Oxford: Clarendon Press.

———. 1989. *Moral absolutes. An essay on the nature and rationale of morality*. New York: Peter Lang.

———. 1990. Our science as O-U-R science. In *A useful inheritance. Evolutionary aspects of the theory of knowledge*, N. Rescher, 77–104. Savage: Rowman and Littlefield.

———. 1993a. Moral objectivity (Are there moral facts?). In *A system of pragmatic idealism. Vol. II: The validity of values: Human values in pragmatic perspective*, N. Rescher, 173–186. Princeton: Princeton University Press.

———. 1993b. Moral values as immune to relativism. In *A system of pragmatic idealism. Vol. II: The validity of values: Human values in pragmatic perspective*, N. Rescher, 187–205. Princeton: Princeton University Press.

———. 1993c. Moral rationality: Why be moral? In *A system of pragmatic idealism. Vol. II: The validity of values: Human values in pragmatic perspective*, N. Rescher, 206–232. Princeton: Princeton University Press.

———. 1994. Philosophy in process. In *A system of pragmatic idealism. Vol. III: Metaphilosophical Inquires*, N. Rescher, 168–181. Princeton: Princeton University Press.

———. 1996. *Public concerns. Philosophical studies of social issues*. Lanham: Rowman and Littlefield.

———. 1997a. Against moral relativism. In *Objectivity. The obligations of impersonal reason*, N. Rescher, 124–150. Notre Dame: University of Notre Dame Press.

———. 1997b. Moral objectivity: The rationality and universality of moral principles. In *Objectivity. The obligations of impersonal reason*, N. Rescher, 151–171. Notre Dame: University of Notre Dame Press.

———. 1998. *Predicting the future. An introduction to the theory of forecasting*. New York: State University of New York Press.

———. 1999a. *Razón y valores en la Era científico-tecnológica*. Barcelona: Paidós.

———. 1999b. *The limits of science*. Revised ed. Pittsburgh: University of Pittsburgh Press. 1st ed. 1984.

———. 2003a. *Sensible decisions. Issues of rational decision in personal choice and public policy*. Lanham: Rowman and Littlefield.

———. 2003b. Technology, complexity, and social decision. In *Sensible decisions, issues of rational decision in personal choice and public policy*, N. Rescher, 81–98. Lanham: Rowman and Littlefield.

———. 2003c. Risking democracy (Some reflections on contemporary problems of political decision). In *Sensible decisions. Issues of rational decision in personal choice and public policy*, N. Rescher, 113–124. Lanham: Rowman and Littlefield.

———. 2003d. Collective responsibility. In *Sensible decisions. Issues of rational decision in personal choice and public policy*, N. Rescher, 125–138. Lanham: Rowman and Littlefield.

———. 2006a. Credit for making a discovery. In *Studies in the philosophy of science*, N. Rescher, 181–200. Heusenstamm: Ontos Verlag.

———. 2006b. The ethical dimension of scientific research. In *Studies in the philosophy of science*, N. Rescher, 201–218. Heusenstamm: Ontos Verlag.

———. 2008. Moral objectivity. *Social Philosophy and Policy Foundation* 25 (1): 393–409.

———. 2012a. The problem of future knowledge. *Mind and Society* 11 (2): 149–163.

———. 2012b. Political pragmatism. In *Pragmatism. The restoration of its scientific roots*, N. Rescher, 205–215. N. Brunswick: Transaction Publishers.

Resnik, D.B. 1998. *The ethics of science. An introduction*. London: Routledge.

Sen, A. 1986. Prediction and economic theory. In *Predictability in Science and Society*, ed. J. Mason, P. Mathias, and J.H. Westcott, 3–23. London: The Royal Society and The British Academy.

Shrader-Frechette, K.S. 1994. *Ethics of scientific research*. Lanham: Rowman and Littlefield.

Simon, H.A. 1990. Prediction and prescription in systems modeling, *Operations Research* 38: 7–14. Compiled in Simon, H.A., *Models of bounded rationality. Vol. 3: Empirically grounded economic reason*, The MIT Press, Cambridge, MA, 1997, pp. 115–128.

Wendler, D. 2014. The ethics of clinical research. In *The Stanford Encyclopedia of philosophy*, ed. E.N. Zalta (fall edition 2014), http://plato.stanford.edu/archives/fall2012/entries/clinical-research/. Accessed on 20 Jan 2015.

Worrall, J. 2006. Why randomize? Evidence and ethics in clinical trials. In *Contemporary perspectives in philosophy and methodology of science*, ed. W.J. Gonzalez and J. Alcolea, 65–82. A Coruña: Netbiblo.

Author Index

© Springer International Publishing AG 2017
A. Guillán, *Pragmatic Idealism and Scientific Prediction*, European Studies in Philosophy of Science 8, DOI 10.1007/978-3-319-63043-4

Subject Index

© Springer International Publishing AG 2017
A. Guillán, *Pragmatic Idealism and Scientific Prediction*, European Studies in
Philosophy of Science 8, DOI 10.1007/978-3-319-63043-4

Printed in the United States
By Bookmasters